CORN Publication Series 16

All authors publishing in the CORN Series have been invited by the editors. All articles are intensively discussed at preparatory meetings, reviewed by the book editors and double blind peer reviewed by external reviewers.

D/2018/0095/105

DOI 10.1484/M.CORN-EB.5.113376

ISBN 978-2-503-57600-8

eISBN 978-2-503-57601-5

Printed in the EU on acid-free paper.

Peasants and their fields

The rationale of open-field agriculture, *c*. 700–1800

Edited by

Christopher Dyer, Erik Thoen and Tom Williamson

BREPOLS

CONTENTS

LIST OF CONTRIBUTORS

HANS ANTONSON	Lund University and KMF forum AB, Sweden
HANNE COTTYN	Ghent University, Belgium
CHRISTOPHER DYER	University of Leicester, UK
CARL-JOHAN GADD	University of Gothenburg, Sweden
JUNICHI KANZAKA	Soka University in Tokyo, Japan
HANS RENES	Utrecht University / Vrije Universiteit Amsterdam, The Netherlands
NICOLAS SCHROEDER	FRS-FNRS, Université Libre de Bruxelles, Belgium
PETRI TALVITIE	University of Helsinki, Finland
ERIK THOEN	Ghent University, Belgium
TOM WILLIAMSON	University of East Anglia, UK

LIST OF FIGURES

LIST OF TABLES

The rationale of open fields. A collection of essays

Christopher Dyer, Erik Thoen
and Tom Williamson

In the Middle Ages and the early modern period open fields could be found in many countries of Europe. They took a wide variety of forms, but can in essence be defined as areas of cultivated land in which the intermingled plots of different cultivators, without upstanding physical boundaries, were subject to some degree of communal management in terms of cropping and grazing. Sometimes such fields occupied a high proportion of the land in a district, but often they formed a relatively minor element in landscapes which also contained enclosed fields, woodland or expanses of pasture. In some areas, open-field agriculture had already been abandoned before the end of the Middle Ages, but in others it continued to flourish into the nineteenth or even twentieth centuries.

Open fields have been studied since the nineteenth century by geographers, historians and archaeologists, and differences of approach and attitudes mean that the subject has always been debated. Much about the origins, development and rationale of open fields remains contentious. This is partly due to the fact that the topic is rather difficult to study because the sources are relatively scarce. One of the reasons for this is that, as is the case with many other practices in the Middle Ages and the early modern period, open-field farming was based on customary bylaws. These were often written down in a late stage of their existence, or sometimes only when changes to customary practices had become necessary, so they are not good evidence for the rules observed in earlier times. Some features of open-field farming are apparently described in manorial accounts, surveys and rentals which often make oblique and rather brief references to cultivation practices and have to be used with great caution. With some exceptions, maps only become available from the sixteenth century, and maps which depict field systems in detail, such as those made to record boundaries on estates or to levy tithes, are scarce or in many areas non-existent. Of course other sources can give us glimpses of open-field practices, both written documents and the remains of the fields themselves, visible in the landscape or recoverable through archaeological work, but they are often difficult to understand. Moreover, the available evidence gives rather limited information about the contemporary rationale behind open-field management, which lies at the core of this book. Why, across wide areas of Europe, did such fields sometimes become central to the experience of so many of our ancestors, shaping not only farming practices but also the basic structures of their everyday lives? And why, in contrast, did they fail to develop, or had a less significant role, elsewhere? What part did open-field agriculture play in agricultural production, enabling our ancestors to survive, and yield a surplus from which rents and tithes could support the higher ranks of society? In other words what was the rationale behind open-field farming?

This book gives attention to a number of specific questions, the answers to which can contribute to the overarching issue of the reasons for the emergence, survival and development of open fields. These questions relate, in particular, to identifying the period when open fields began and were operated, as other problems might be more easily solved if we know about the circumstances of the time in which decisions were being taken. Different elements of the fields may belong to different periods, however: for example, the regular pattern of the distribution of strips, and the extension of common grazing across all of the arable in a township, which are a feature of many of the more complex systems may have been introduced at a relatively late stage. The coincidences between developments in fields and other trends need to be established, such as the development of the nucleated villages, and the growth in their size. The connection between villages and the more complex and extensive forms of open field is well known, but which came first, and when did they both emerge? More general changes in society and the economy may have made the adoption of open fields desirable and even necessary. Was there a connection, for example, between the growth in the power of lords, visible in a number of centuries including the ninth, the thirteenth, and in some countries the sixteenth, and the more sophisticated management of estates? Were estate managers encouraged to bring into use the rotations of crops originally practised on the lords' demesnes more widely, on the lands cultivated by the peasants who were dependent on the lords? At all periods between the ninth and the seventeenth centuries we are aware of fluctuations in the functions and cohesion of the village community, sometimes in reaction to pressure from lords or the state, sometimes self-generated – the obvious enquiry is whether the organization of open fields represents an initiative from these villagers. Historians are divided between those who favour a top-down approach, expecting that initiative will tend to come from those with power and education, and those who expect that the people who worked the fields would have the incentive and the knowledge to innovate, maintain and adapt their methods of cultivation. The problem is connected with early notions of property, as the open fields were based on a balance of rights between lords and peasants. Within the community a delicate equilibrium needed to be established between individual possession of land and collective control over its use. One approach to the history of open fields has been to emphasize the social benefits that they brought, as they involved a sacrifice of individual ambitions in favour of the common good. Should we see them, not as a means to produce crops and create wealth, but as a form of social security? Were features of the system, such as the scattering of strips, designed to avoid risk and protect the cultivators from harvest failure?

Whether fields were created and run by lords or by villagers, we must ask what influenced their decisions in relation to the creation or remodelling of open fields. The environment was clearly a factor, as is apparent from the strongly regional character of variations in their character and layout. How did such influences as the heaviness of the soil, or the balance between arable, pasture and woodland, shape attitudes which led to the development of open fields? Densities of population also vary from region to region, and it is obviously necessary to ask whether open fields were a feature of more densely settled countryside, and stimulated by pressure of rising population in particular places and times. Levels of population were not the only stimulus to the increase in production from the land. Fluctuations in the market for agricultural produce could have been another spur to the adoption of open-field farming. Open fields are often assumed to have been designed to provide a subsistence for the cultivators, but there are many indications that in times of rising market demand, from an urban population for example, cultivators sought ways to increase their surplus as well as provide for their own needs of food and raw materials. Can we find evidence for rising productivity from land in open fields, or of specialization in crops or livestock that would generate an income from sales? The market affected the management of land in a number of other ways, in the supply of labour for example, and we ought to consider the provision of labour to cultivate the fields in terms of hired workers as well as family members. Did the famous 'inefficiency' involved in travelling from one dispersed strip to another not only lead to an expenditure of time by the peasant and his family, but also incur costs in terms of the money needed to reward workers? Did the peasants who earned money from sales of crops or wool or cheese invest some of the proceeds on capital investment, in livestock, implements and buildings? Or were they, like most peasants, chronically short of money to spend in such productive ways? Open fields represent a dimension of medieval agrarian technology, and we should ask if methods of ploughing and the development of rotations prepared the way for the laying out of the fields, as is often said. Perhaps also changes in technique had a bearing on the long-term development of open fields, bearing in mind their well-known capacity to adapt and modify their form and management. New technologies, such as the transformation in the later Middle Ages and early modern periods in the balance between arable and pasture, may have rung the death knell of the open fields. Other factors contributing to their demise included changes in patterns of land holding and in attitudes to property, and the general development in the economy.

Over recent decades open fields have been investigated in new, interdisciplinary ways, and as a Europe-wide phenomenon. In this book, more than ever before, their development and operation are explained in terms of economic, social, agrarian and environmental developments which were common, to varying degrees, to all parts of Europe. It contains ten new studies from a wide range of regions, together with important comparative research from south America and Japan. It contributes to our understanding of the local and regional variants of open fields within their environmental context. We believe that by adopting this specific but fundamental approach, this collection of essays could represent a milestone in the study of open-field agriculture, contributing to our understanding of general structures and tendencies in pre-industrial society. Such is the importance of open fields as one of the building blocks for the social and economic fabric of the common European past.

The book is the result of an international collaboration within the CORN network which, for more than twenty years, has sought to promote collaboration between rural historians in Europe (Comparative Rural History Network). The book is the result of two workshops, one in Bern (S) in 2013 during the Rural History Conference (19–21 Aug) and one in Ghent (B) in mid-2014 (13–14 June). The initial papers could be improved thanks to the remarks made by the audience and by invited specialists during these workshops. To increase the range and quality of the book we invited a few other authors, who had not been able to take part in the preparatory workshops, to make a contribution. Anonymous peer reviewers guaranteed the quality of all the final articles, recommending very interesting improvements. We sincerely want to thank all people mentioned for their support and not a least the authors for their patience and willingness to collaborate.

This volume brings together eleven original chapters, written by ten authors from six countries. Although occasionally references are made to other European countries, it deals in particular with open fields in the Low Countries (Belgium and the Netherlands), the Scandinavian countries (Norway, Denmark, Sweden, Finland) as well as England, where in the past it has probably been studied in the most intensive way. It is probably the first time that studies of open fields in a variety of countries have been brought together. The book ends with some concluding remarks of the editors about the central themes. An original feature is the inclusion of two non- European examples (part of Bolivia and part of Japan) to show that it is wrong to regard open fields as a purely European phenomenon. These two contributions provide a valuable comparative dimension, because they show that features such as intermingled strips and communal management of cropping developed independently in circumstances very different from those prevailing in Flanders, Sweden and England.

The book is fully illustrated often with original figures and maps, which was necessary because many of the arguments depend on visual and topographical evidence. We are very grateful to lic. Hans Blomme (Ghent University) who kindly (re-)drew many of the figures and to Sander Berghmans who helped to correct the proofs. Of course we finally are grateful to Brepols Publishers and especially to its editor Chris VandenBorre who were prepared to publish this book in the (renewed) CORN publication series where it rightly belongs.

1 Open fields in England: an overview

Tom Williamson, University of East Anglia, UK

I. The diversity of open fields

Open fields – areas in which the arable land of different proprietors lay intermingled, as unhedged strips – could be found in most, if not all, regions of England by the thirteenth century. As in Continental Europe, they took a wide variety of forms. In some districts they were extensive and continuous, accounting for most or all of the arable land in a township. In such circumstances farming was usually carried out on highly cooperative lines. The strips or 'lands' were grouped into blocks called furlongs (in some areas, 'shots' or 'wangs') which were in turn aggregated into larger units called 'fields', usually two or three in number, one of which lay fallow each year and was grazed in common by the livestock of the community. This not only provided additional pasturage in areas in which grassland was sometimes in relatively short supply; it also, more importantly, ensured that the land was dunged, thus helping to restore the nitrogen and other nutrients which were constantly being lost through cropping. Where holdings lay in a multiplicity of small intermingled strips, and were subject to rights of communal grazing at certain times, regular meetings of farmers were required to decide the order in which land should be ploughed, harrowed and seeded, and the courses of cropping to be followed within each furlong (see Dyer, this volume). Manorial demesnes were drawn into such communal regimes to varying extents. They sometimes took the form of dispersed strips but they could also comprise compact blocks near to manor houses.

In Midland districts, where the land was often heavy and poorly-draining, extensive open fields took a distinctive form. The individual strips were usually ploughed in ridges, a practice principally intended to improve drainage: and where (as was often the case) open fields were laid to pasture following enclosure these have been preserved as the earthworks known to modern scholars as 'ridge and furrow'. Perhaps more importantly, the holdings or 'yardlands' in a township – each of which owed the same rents and services to the local lord, and formed the basis for allocating local offices – were in this region usually laid out within the fields in a particularly structured fashion, according to a regular sequence. The strips of the tenants fell in a fixed, recurrent order, sometimes – as in parts of Continental Europe – described in terms of 'sun-division' or *solskifte* (i.e., in a notional sequence following the progress of the sun from south and east, to north and west (Hall, 1995: 82–86; Roberts and Wrathmell, 2000: 125)).

Much attention has been focused on fields like these. But extensive open fields displayed much variation in medieval England. In many areas, as in western East Anglia, holdings were evenly scattered through the fields but without the same degree of rigid regularity. In districts where soils were light and freely-draining, moreover, the individual lands were seldom ploughed in ridges, and extensive tracts of unploughed ground – downs and heaths – often existed beyond the arable. This was largely because in such areas the land was leached of nutrients with particular rapidity, so that it could only be kept in heart by

Peasants and their fields. The rationale of open-field agriculture, ed. by Christopher Dyer, Erik Thoen and Tom Williamson, Turnhout, 2017 (*CORN* Publication Series, 16), pp. 5-28

BREPOLS PUBLISHERS DOI: 10.1484/M.CORN-EB.5.114266

substantial and regular applications of dung. These were the classic areas of 'sheep-corn' husbandry, in which flocks were grazed on the pastures by day and close-folded on the arable land by night, the treading of the sheep serving to incorporate the dung and urine effectively into the ploughsoil (Kerridge, 1992). Close-folding was a less common practice in Midland areas, in part because it was less necessary here but also because it tended to impact the soils, where these were particularly heavy and poorly-draining (Kerridge, 1992: 77–79; Fox, 1984: 130–33).

Over large areas of England open fields did not take forms in which holdings were widely scattered, through arable fields which were continuous and extensive. Sometimes open fields accounted for most of the arable land, but the holdings of particular farms were concentrated in limited areas within them, rather than being extensively intermingled, and there were many 'fields' rather than just two or three – arrangements described by some historians as 'irregular' or 'patchwork' field systems (Postgate, 1973; Roberts, 1973; Hunter, 2003; Roden, 1973; Martin and Satchell, 2008: 147–73; Thoen, this volume). Fields of this type could be accompanied by varying numbers of enclosed fields, in what might be called 'intermediate' landscapes; and landscapes like this merged gradually, without a break, into those of 'bocage' or 'woodland' type, in which all the arable land lay in walled or hedged parcels. They also tended to evolve into them, over time. 'Irregular' open fields were particularly susceptible to early, piecemeal consolidation and enclosure because properties were clustered in limited areas. Each proprietor thus had few neighbours, and few transactions would be required to ensure amalgamation of lands, and subsequent enclosure. Moreover, while communal arrangements for cropping and fallowing usually existed in open fields of this type, the number of individuals involved was often small, reducing opposition to enclosure. Conversely, where open fields were extensive, and holdings were spread more evenly if not regularly, each area of the fields contained the strips of many farmers, making piecemeal consolidation more difficult. The complex and highly communal character of agrarian organization also militated again individual enclosure. Most townships with extensive open fields, in which the holdings of each farm were widely scattered and extensively intermingled, thus usually retained them well into the post-medieval period, when their final removal was brought about by some form of planned reorganization of the landscape, such as parliamentary enclosure. These differences in enclosure history, contingent upon the extent to which holdings lay intermingled, served to remove patchwork and intermediate forms of open field, and to create a simple distinction between what came to be known to early-modern topographers as 'woodland' landscapes – the bocage – and open-field, champagne or 'champion' districts, so that by 1597 William Harrison could make a simple, bipartite distinction:

> It is so, that our soile being divided into champaine ground and woodland, the houses of the first lie uniformelie builded in everie town together, with streets and lanes; whereas in the woodland countries (except here and there, in great market towns) they stand scattered abroad, each one dwelling in the midst of his owne occupieng (Edelen ed., 1994: 217).

II. Open fields and settlement

As the quotation from Harrison implies, there was a close relationship between the character of field systems, and patterns of rural settlement. At one extreme, extensive areas of open field were usually associated with nucleated villages. Numerous farms were grouped

Figure 1.1 Landscape regions in England

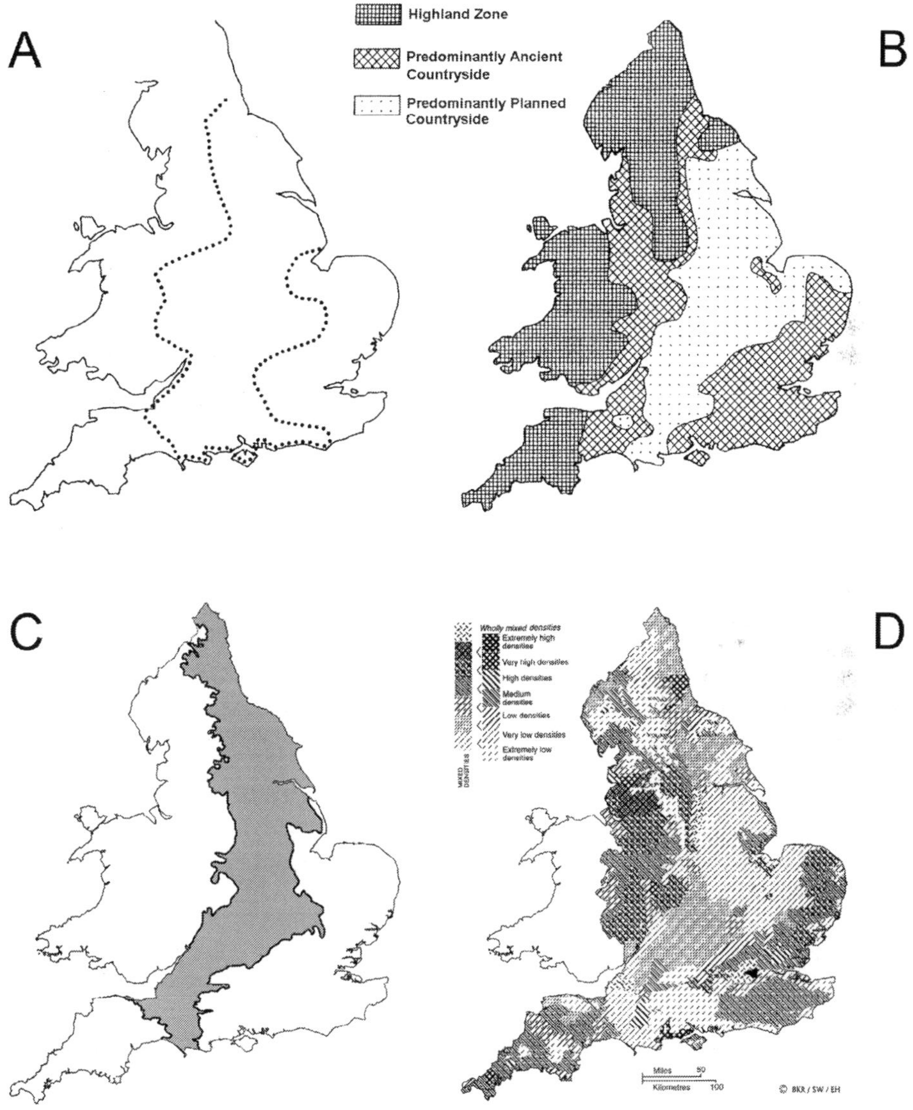

(**A**) the boundaries of Howard Gray's 'Midland System'; (**B**) Oliver Rackham's (1986) distinction between the 'planned' countryside – landscapes of late enclosure which replaced extensive open fields, and the 'ancient' countryside of early enclosure; (**C**) the 'Central Province' of nucleated villages and extensive open fields, as defined by Roberts and Wrathmell (2000); (**D**) densities of settlement dispersion, as mapped by Roberts and Wrathmell (2000)

together as a single settlement, with their lands lying intermingled, far and wide, across the surrounding landscape. Patchwork open fields, in contrast – like bocage landscapes – were associated with more dispersed patterns of settlement, and the holdings in the fields were clustered, near to the farmhouse.

Nucleated villages farming extensive open fields dominated the landscape of a broad swathe of England running from Yorkshire and Durham, through the Midland counties, to the south coast, and with an outlying extension along the South Downs in south-east England (Figure 1.1). The south and east of England, including most of East Anglia, and the western areas of the country generally had more dispersed patterns of settlement, and alternative forms of field system ('patchwork', intermediate, or bocage). Yet it is important to note that the boundaries of the central 'champion' belt – also referred to as the 'Central Province' by some writers – have been mapped in slightly different ways by different people, uncertain whether to include, or exclude, regions with intermediate characteristics. Simple regional maps obscure a further complication. Extensive areas of countryside displaying dispersed settlement and complex, multiple field systems could be found deep within the 'champion', as for example in north Bedfordshire (Brown and Taylor, 1989). Conversely, albeit to a lesser extent, nucleated villages with extensive open fields occurred well outside the mapped boundaries of the 'champion', in major river valleys like that of the Lea in Hertfordshire and Essex especially. Generalized maps reflect broad trends in settlement patterns and field systems, not the existence of hermetically sealed and distinct landscape regions, as is sometimes implied (Homans, 1941: 12–28; Rackham, 1986: 4–5; 178).

Even a broad distinction between 'nucleated' and 'dispersed' settlement is problematic. Hamlets and 'nucleated villages' lay on a continuum; most areas of dispersed settlement, as already noted, also featured some larger nucleations; while in champion areas – especially those characterised by light soils or loams – some villages were only loosely nucleated, many for example comprising a number of semi-detached and individually named 'ends'. More importantly, similar plan elements could be found in both dispersed and nucleated settlements. In particular, many villages were clustered around a large central area of common land – a 'village green' – or around a number of small greens (Williamson et al., 2013: 81–83; Oosthuizen, 2002a and 2002b). Such arrangements were mirrored, in areas of dispersed settlement, by the loose scattering of farms around the margins of extensive commons (a particularly frequent form in East Anglia) on the one hand; and by hamlets clustered around diminutive 'greens', a form of settlement characteristic of areas like western Essex or eastern Hertfordshire, on the other (Rowe and Williamson, 2013: 71–78; Warner, 1987). Settlement in medieval England took a wide variety of forms, but all can be read as expressions of the same essential underlying grammar.

III. The historiography of open-field origins

Debate has continued in England for over a century about the origins of medieval field systems, a debate fuelled to a large extent by an absence of hard evidence. Open fields only appear clearly in the documentary record in the thirteenth and fourteenth centuries, by which time they seem, in their various local and regional manifestations, to be fully formed. Before this we have to rely on archaeological evidence that is mainly indirect or ambiguous in character; or on stray and uncertain references in documents. In particular, a law of Ine, seventh-century king of Kent, implies the existence of intermixed arable, laying down penalties for *ceorls* who failed to adequately fence the outer boundaries of meadow 'or other land divided into shares' (Whitelock, 1955: 368); while the boundary clauses of tenth- and early eleventh-century charters suggest the existence of extensive areas of open field, especially on the chalklands of southern England, where intermingled arable strips sometimes appear to have extended to the boundaries of the estates being defined (Hooke, 1998: 121; Rackham, 1986: 172–75). But even references of the latter kind are insufficient to inform us about the operation and character of open fields, or even arguably about their incidence or general extent.

Until the 1950s and 1960s the origins of open fields in England, and of regional variations in medieval fields and settlement more generally, were sought in the remote past. Seebohm in the 1890s believed that open fields had Romano-British origins, and had developed mainly because of the need for peasants to share the use of ploughteams and ploughs (Seebohm, 1890: 120–22; 409–11). Vinogradoff, writing around the same time, argued that the complex intermixture of properties arose from a primitive concept of tribal shareholding (Vinogradoff, 1892: 236; Vinogradoff, 1905: 150). These writers concentrated most of their attention on the well-developed fields of the 'champion' but Howard Gray expanded his interests to other districts, where 'bocage', 'patchwork' and intermediate landscapes had existed, arguing that variations in medieval landscapes had mainly *ethnic* explanations: they represented the areas settled by particular tribal groups (or combinations of tribes) in the post-Roman period. The 'Midland System' – Gray's term for the 'champion' landscapes of nucleated villages and extensive open fields – was thus a direct importation from the Anglo-Saxon homelands in northern Germany and southern Scandinavia, and its concentration in the central regions of England, he believed, reflected their 'thorough Germanization' in the fifth and sixth centuries (Gray, 1915: 415). In Kent and East Anglia, he argued, where patchwork and irregular field systems predominated, existing Romano-British patterns of land holding had initially been taken over wholesale by Saxon settlers, but properties had subsequently fragmented through the effects of partible inheritance. In East Anglia, a subsequent wave of Danish settlement had produced a high density of free tenures, and more complex systems of manorial organization, leading to further changes in agrarian arrangements. In Kent, in contrast, fields and settlement continued to develop largely on indigenous lines until the tenurial upheavals of the Norman Conquest (Gray, 1915: 415–16). In the middle decades of the twentieth century many scholars similarly emphasized the importance of race and tribal custom in the genesis of regional variation, although there was potential for disagreement over precisely *which* racial groups were responsible for which landscapes, with Homans for example arguing that the distinctive social institutions, field systems and settlement patterns of East Anglia were the consequence of its settlement by Dark Age immigrants from *Frisia* (Homans, 1969).

Other explanations for medieval field systems advanced in the first half of the twentieth century placed less emphasis on culture and ethnicity, but likewise tended to accept their early origins. The Orwins, for example, argued like Seebohm that communal field systems arose from the sharing of ploughs, but believed that this had only become important during the post-Roman period, when the heavy mouldboard plough came into widespread use. They also argued, more generally, that villages and open fields reflected the need for cooperation by pioneer farmers in a hostile environment (Orwin and Orwin, 1938). It would be surprising, in fact, if open fields and the new forms of heavier plough coming into widespread use in the tenth and eleventh centuries had not been connected in some way, given that open fields were arable fields: and the Orwins – like others after them – emphasized the intimate connection between the division of land *as strips* and the use of a large wheeled plough, difficult to manoeuvre in small enclosed parcels.

Such ideas about the origins of open fields, and especially about the early genesis of extensive open fields, were robustly challenged in the 1960s by Joan Thirsk (Thirsk, 1964). She noted that there was no real evidence that open fields on the medieval 'champion' pattern had existed in early Saxon times and, influenced in part by the work of German scholars like Meitzen, argued that villages had developed gradually, as a result of population growth. Crucially, it was this expansion of individual settlements which led to the development of splintered and intermixed holdings. This was in part because of the practice of partible inheritance, and the need to ensure a fair allocation of land of different quality between co-heirs; and in part because, as the frontier of cultivation expanded, newly won land was allocated, in a fair and reasonable manner, to those who had formerly exercised common rights over the area in question, and who were involved in reclaiming it. Thirsk acknowledged that, by the thirteenth century, the peasantry across much of England practised primogeniture but she suggested – as had others before her, and others since (Faith, 1966 and Faith, 1997) -that this was a development of the eleventh and twelfth centuries, associated with the growth of local lordship and the imposition of a network of standard tenements, virgates and the like, on local communities.

Thirsk explained, succinctly but convincingly, how the intermixture of properties in narrow unhedged parcels produced numerous problems for farmers, especially as reserves of pasture dwindled, so that there was an increasing need to utilize more efficiently the grazing afforded by the stubbles left after harvest, and by the fallows. Eventually, as fields multiplied whenever new land was taken into cultivation from the waste, and as the parcels of each cultivator became more and more scattered, regulations had to be introduced to ensure that all had access to their own land and to water, and that meadows and ploughland were protected from damage by stock. The community was drawn together by sheer necessity to cooperate in the control of farming practices. All the fields were brought together into two or three large units. A regular crop rotation was agreed by all and it became possible to organize more efficiently the grazing of stubble and aftermath. Thereafter, the scattering of strips, which had at one time been a handicap, became a highly desirable arrangement, since it gave each individual a proportion of land under each crop in the rotation (Thirsk, 1966:14).

Some reorganization of landholding would have been necessary to bring such a system into existence, to ensure that each farm had an equal proportion of land in each of the fields, or cropping sectors. While this might have been effected through piecemeal exchange, in some circumstances it could have been brought about by planned reorganization. Perhaps more importantly, and in marked contrast to earlier scholars, Thirsk argued that complex open fields on the Midland pattern had emerged over an extended period of time, and that some of the earlier stages in their development were preserved in the variant field systems found outside the Midland belt. The final stages in the development of Midland fields had only occurred at a relatively late date: 'We can point to the twelfth and first half of the thirteenth century as possibly the crucial ones in the development of the first common-field systems'.

Thirsk's powerful and elegant model was immensely influential, and most of the contributors to Baker and Butlin's monumental volume of 1973, *Studies in British Field Systems*, appear to have accepted it in whole or part (Baker and Butlin ed., 1973). But one of its most important implications has received little attention. Thirsk's explanations for the progressive intermixture of holdings only really makes sense if we assume that, for some reason, farmers were unable or unwilling to migrate away from existing settlements as the frontiers of cultivation expanded, so that ever larger villages gradually developed. If farms had been able to spread away, across the landscape, then holdings would, when divided by inheritance, probably have been broken up into larger parcels – divided by field, that is, rather than *within* each field. The expansion of the cultivated area and the division of existing holdings was certainly accompanied in other parts of England by the migration of farms away from old sites, often to the edges of 'greens' or commons, or to the sides of the roads accessing them: a process which began as early as the tenth century in some districts, not until the eleventh or even twelfth in others, but which was generally occurring parallel with the development of extensive open fields in areas of nucleated settlement (Rogerson, 1995: 105, 108, 111; Warner, 1987: 17–18; Dyer, 1990: 111). It is easy to see how the slow expansion of single farms, or small groups of farms, into sizeable villages would have led to the emergence of a pattern of extensively intermingled holdings. Whether new farms were being created through the division of holdings between members of a family, or through the allocation of land to dependant tenants, they could hardly have taken the form of ring-fence properties lying at a distance from the settlement, as these would have been remote from the farmstead itself and thus inconvenient to manure, cultivate and (in particular) reach at harvest time. Saying this, of course, is to do no more than reassert, in a slightly different form, the close association already noted between nucleated villages on the one hand, and extensive forms of open field on the other.

The chronology which Thirsk suggested for the development of open-field landscapes was decisively challenged in the late 1970s and 1980s by two researchers working in the midland county of Northamptonshire, Glenn Foard and David Hall. They suggested, on the basis of extensive field surveys, that the initial Anglo-Saxon settlement pattern of the area had consisted of numerous small sites, represented by scatters of pottery found in the ploughsoil away from the modern settlements (Foard, 1976). These hamlets or farmsteads were abandoned some time between the seventh and the ninth century, and settlement concentrated in villages (Brown and Foard, 1998: 73–82). This nucleation 'event' – the 'great re-planning', to use Foard's term – was initiated by local lords in an attempt to improve agricultural efficiency, and extensive open fields had originated at the same time: they were the consequence of sudden planning rather than of organic, gradual development (Hall, 1981: 36–37; Hall, 1995: 129–39). Supporting evidence for this interpretation was provided by the regular arrangements of virgate holdings already noted which, Hall argued, were clearly in place by the late Saxon period, for they often bore a close mathematical relationship to the number of hides (units of taxation) at which each township was assessed in 1086, according to Domesday Book. In some places there appear to have been eight, in some ten, and in others twelve or sixteen yardlands to the hide (Hall, 1995: 82–86). These arrangements had been created to facilitate the extraction of rents and taxes, and perhaps the allocation of village offices.

Such evidence for large-scale lordly planning accorded well with the suggestions being made around the same time by Bruce Campbell, who expressed doubts over how far peasant communities would by themselves have been able to effect the exchange of lands necessary for the transition to a 'regular' open field system through piecemeal exchange, in the way that Thirsk suggested, arguing instead that the intervention of local lords would have been necessary to bring about such a transition (Campbell, 1981). Harold Fox, in a chapter which appeared in the same volume as Campbell's, also emphasized the importance of the emergence of local manors in the later Saxon period, in place of more diffuse forms of territorial lordship, in the formation of developed open field systems (Fox, 1981). The hand of lordship was also detected by a number of writers in the layout of nucleated villages, both in Northamptonshire and elsewhere in England, which often displayed remarkably regular arrangements of similarly-sized tofts, laid out neatly at right angles to the street, apparently the consequence of deliberate planning (Brown and Foard 1998: 75–77, 91–92; Everson, Taylor and Dunn, 1991: 16–17; Taylor, 1984).

IV. Challenging the 'nucleation' hypothesis

The idea that extensive open fields in England originated in the period between the eighth and tenth centuries, through a 'replanning' of the landscape involving the relocation of outlying farms into nucleated villages, is now well entrenched in the archaeological and historical literature (Rippon, 2007 and 2008). Indeed, settlement 'nucleation' is accepted as an explanatory model even where archaeological evidence for abandoned outlying sites is lacking (see, for example, Gerrard and Aston, 2007). Yet somewhat surprisingly, Thirsk's explanation for the development of extensive and regular open fields, based on a crisis of resources and a need to improve agricultural efficiency, is still widely accepted as an explanation for this development (e.g., Lewis et al., 1997: 199): surprisingly, because it relies on population growth as a principal driver, and in the eighth and ninth centuries population levels must have been considerably lower than they were to be by the twelfth or thirteenth centuries.

Recent research suggests, however, that the 'Northamptonshire model' may require a measure of qualification (Williamson et al., 2013). Re-examination of the archaeological evidence suggests that this cannot at present sustain the idea that large numbers of scattered farms were routinely concentrated – voluntarily or otherwise – into 'villages', at least as this term is usually understood. In spite of the intensity of expert fieldwork which has been carried out in the county, over half the townships (51 per cent) have produced no evidence for abandoned, outlying early Saxon sites at all. Where such sites have been recovered, moreover, they are usually present in only small numbers, with only 5 per cent of townships being able to boast more than three (Williamson et al., 2013, 57). Even allowing for the problems in locating, through surface surveys, sites of this period, these are low numbers on which to build a general hypothesis of abandonment and nucleation. In addition, excavations in Northamptonshire and elsewhere suggest that most early Saxon settlements were comparatively mobile in character, changing location ever few generations (Brown and Foard, 2004; Hamerow, 1991; Taylor, 1983). The evidence from field surveys in Northamptonshire is thus better interpreted as indicating a pattern in which a relatively small number of mobile sites *stabilized* in the seventh or eighth centuries, rather than a dispersed pattern of numerous contemporary settlements

Figure 1.2 Village plans in Northamptonshire

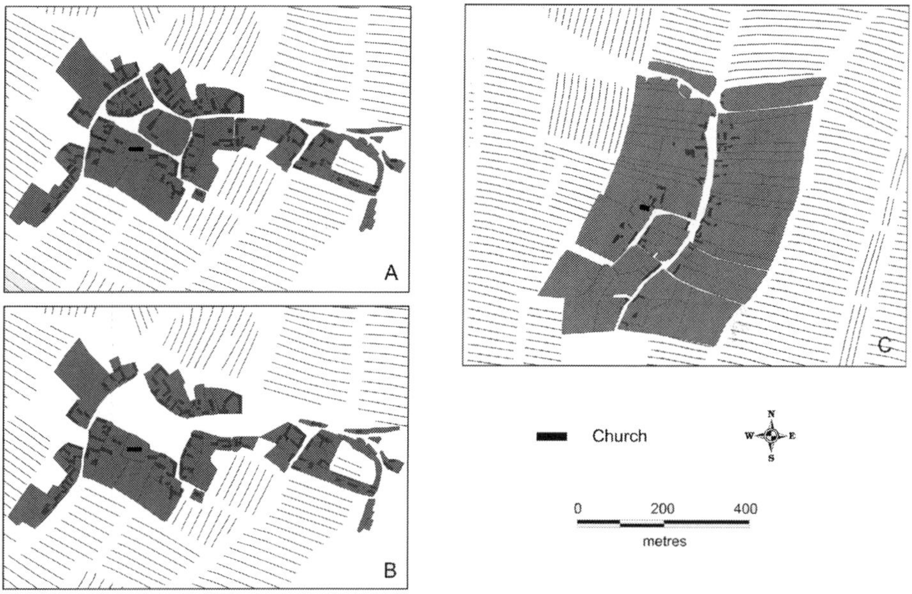

The village of Wappenham, as shown on nineteenth-century maps (A), is the result of the infilling – by two large 'ovoids' of houses – of a central green, and encroachment on narrower strips of common land.

In contrast, villages like Grafton Underwood (C), made up largely of parallel, 'strippy' tofts, are usually interpreted as a consequence of 'planning'. When considered in the context of the surrounding open-fields, however, it becomes clear that a small settlement nucleus – represented by the more irregular enclosures near the church – has simply expanded over its own fields (one strip on the map is equivalent to four on the ground).

which was suddenly 'nucleated'. Villages developed gradually, through the expansion of these relatively small nuclei. Where early Saxon sites were relatively numerous, on the lighter soils, two or occasionally three settlements, lying a few hundred metres apart, expanded and fused to form a 'polyfocal' village (*sensu* Taylor, 1977). More usually, and especially on the more difficult soils, which were only now being brought into cultivation, growth from a single nucleus is indicated. Northamptonshire village plans often seem to have developed through the expansion of houses around, and then into, areas of pasture, corresponding to patches of damp or otherwise difficult soils which were better used for grazing than for arable. Many 'champion' villages thus came to be clustered around large central 'greens' but in most cases these were subsequently filled with houses as the population continued to grow, leaving only small residual patches of pasture, linked by roads (Williamson, 2013: 82–84; see also Oosthuizen, 2002a). In some cases, however,

Figure 1.3 The configuration of common land around Hardwick in south Norfolk and of medieval pastures in western Northamptonshire

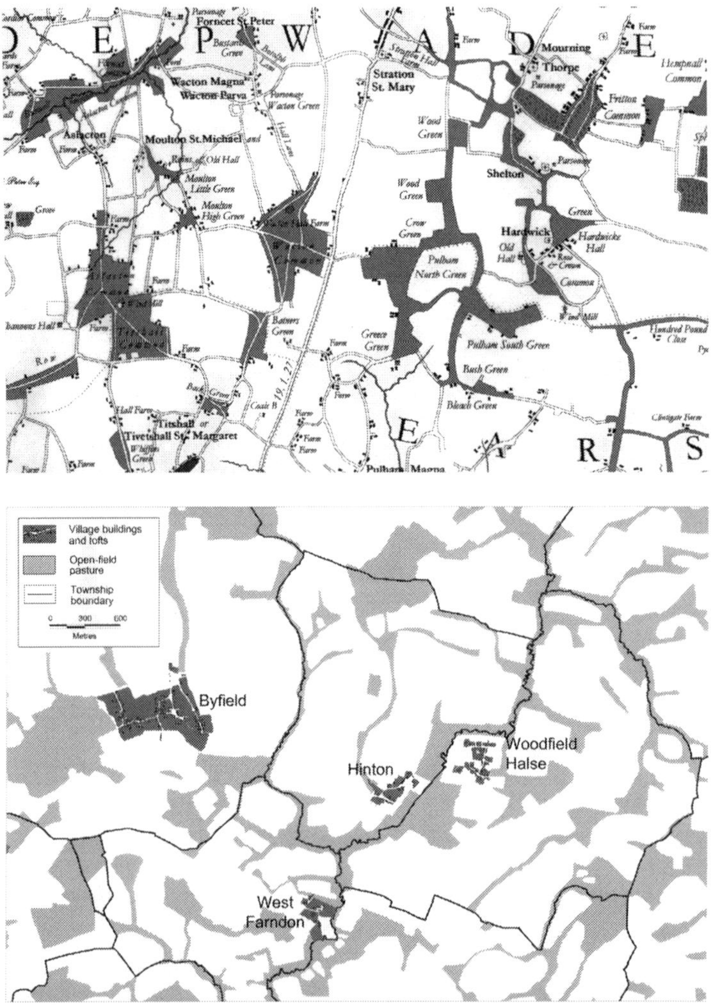

Even in the high Middle Ages the landscapes of 'champion' districts, and those of dispersed settlement, often had more in common than historians sometimes assume. The configuration of common land around Hardwick in south Norfolk (as shown on William Faden's county map of 1797) (above) and of medieval pastures in western Northamptonshire, around Byfield and Hinton, reconstructed from archaeological evidence (below), is remarkably similar. The principal difference was that in the former area medieval settlement dispersed around the remaining fragments of common pasture, whereas in the latter it had remained clustered.

the character of local soils ensured that all the land lying adjacent to the initial Saxon 'cores' was quickly ploughed up, so that, as population grew, settlement was obliged to expand across arable strips – thus creating the pattern of neat, parallel tofts widely interpreted as a consequence of, and evidence for, lordly 'planning' Williamson et al., 2013: 84–87) (Figure 1.2). Some genuinely 'planned villages' may well exist but the majority of observed regularity in settlement plans appears to be a consequence of organic processes.

Villages thus developed over time, as population expanded through the later Saxon periods and into the twelfth century, as Thirsk's model had originally suggested. This in turn implies that her suggested chronology for the development of extensive open fields must also, in broad terms at least, be correct. The 'regular' arrangements of holdings, with repeated sequences of strips in each furlong, which late medieval documents indicate in many Midland fields must likewise have a relatively late origin, rather than dating to the time of the supposed 'great planning' in the middle-Saxon period: for they can only have been imposed once the cultivated area had reached its maximum extent, given that they generally extended throughout the whole of a township's fields. Indeed, the documentary evidence from Northamptonshire leaves little doubt that regular cycles of holdings were mainly the consequence of post-Conquest, probably twelfth-century reorganization. The numbers of 'virgates' or yardlands scattered regularly through the furlongs described in late medieval documents are invariably larger than the number of holdings recorded in the townships in question by Domesday Book in 1086, usually by a large margin (Williamson et al., 2013: 120–25). The few pieces of documentary evidence from England more generally which seem to record the recasting of existing open-fields in more 'regular' forms, moreover – from Segenhoe in Bedfordshire and Dry Drayton in Cambridgeshire – also date from the twelfth century (Fox, 1981: 95–98). Thirsk may have placed the final stages in the development of Midland open fields a little late, but her emphasis on the importance of the post- rather than pre-Conquest period seems broadly correct.

Landscapes characterized by extensive open fields, in Northamptonshire at least, did not therefore appear, fully formed, at a stroke. They developed over time, and they diverged only gradually from other kinds of landscape – 'patchwork', bocage and the rest. In their early phases of development much settlement in champion areas, like much of that in areas of dispersed settlement, comprised loose scatters of farms around areas of common land. But whereas in the latter regions farms continued to disperse across the land surface, often hugging the margins of the dwindling patches of 'waste', in champion districts they remained clustered, with the central areas of common land becoming filled with dwellings, thus leading to the development of strongly nucleated villages surrounded by intermingled properties. In areas of dispersed settlement, in contrast, although farms likewise became increasingly intermingled through partible inheritance, assarting and the like, they did so to a lesser extent (Figure 1.3).

V. Open fields and agricultural specialization

It has been suggested by some researchers that living in villages and farming extensive open fields was a fashion, a 'choice' made by farmers or their masters which spread by emulation; and that the boundaries of the central zone, of large villages and extensive open fields, simply mark where the 'village moment' passed, and this form of agrarian organization became unfashionable (Taylor, 2002: see also, for a different version of this argument, Jones, 2011). But if villages developed gradually over time, through slow expansion, rather than being created suddenly, through a process of 'nucleation', then such arguments make much less sense. Other historians have argued, more cogently, that landscapes of extensive open fields emerged as – in the course of the Middle Ages – urban markets expanded and more complex economic forms of organization developed, allowing particular areas to specialize in cereal growing. Open fields, that is, were a manifestation of the 'cerealization' of particular regional economies (Renes, 2010). Such a suggestion is demonstrably true in the sense that districts which remained thinly settled, and never developed large areas of arable fields – such as the uplands of the north and west of England – by definition never possessed extensive and continuous areas of open field. But its efficacy as an overarching explanatory model is limited by the fact that there is no evidence that extensive and highly communal open fields, and nucleated village, invariably developed in the most intensively arable areas. The latter, indeed, could just as easily be characterized by landscapes of less regular open fields or even enclosed fields.

If, as seems likely, different varieties of landscape and settlement were already becoming distinct by the end of the eleventh century then we might reasonably expect that Domesday Book, for all its faults, would provide some indication that the 'champion' regions of England were the most intensively arable and, perhaps, the most populous. In fact, there is no clear correlation between different types of landscape and the densities of either plough-teams, or of population, which it records (Darby, 1977). If anything, the most intensively ploughed regions in the late eleventh or twelfth centuries were characterized by dispersed forms of settlement, and by fields system which deviated in other ways from the champion 'norm'. It is important in this context to note that in many districts dominated by extensive and highly communal field systems, farmed from nucleated villages, large tracts of unploughed ground, grazed by livestock, survived right through the Middle Ages. This was especially true in areas of light land – on chalk or sand – where significant reserves of pasture formed an integral part of the arable farming system, as well as allowing the production of livestock for the market. They provided the grazing for the large folding flocks without which, as already noted, these thin, easily leached arable soils of such districts could not have been kept in permanent cultivation. In areas like the East Anglian Breckland, nucleated villages and extensive open fields were thus separated by great tracts of heathland, and as a consequence population densities were amongst the lowest in England.

Even in the Midlands, where close-folding was seldom practised, pasture might be more extensive than is often assumed. Recent mapping of open fields in Northamptonshire has shown that while in some parts of the county (such as the valley of the Nene) landscapes of almost unrelieved arable had indeed developed by the twelfth or thirteenth centuries, elsewhere numerous bands of unploughed ground, managed as pasture or meadow, ran through the furlongs. Many townships in the west of the county had little more than 50 per cent of their land area under cultivation. Yet these were unquestionably areas of nucleated settlement, in which all the arable land lay in large open fields, divided into two cropping sectors and containing holdings arranged in highly regular patterns (Figure 1.4). Given that such townships had only around a quarter of their land area under cultivation at any one time, they might be better described as pastoral rather than as specialized arable settlements (Williamson et al., 2013: 109–10). Even where less pasture remained into

Figure 1.4 The landscape of eastern Northamptonshire and the landscape of western Northamptonshire compared

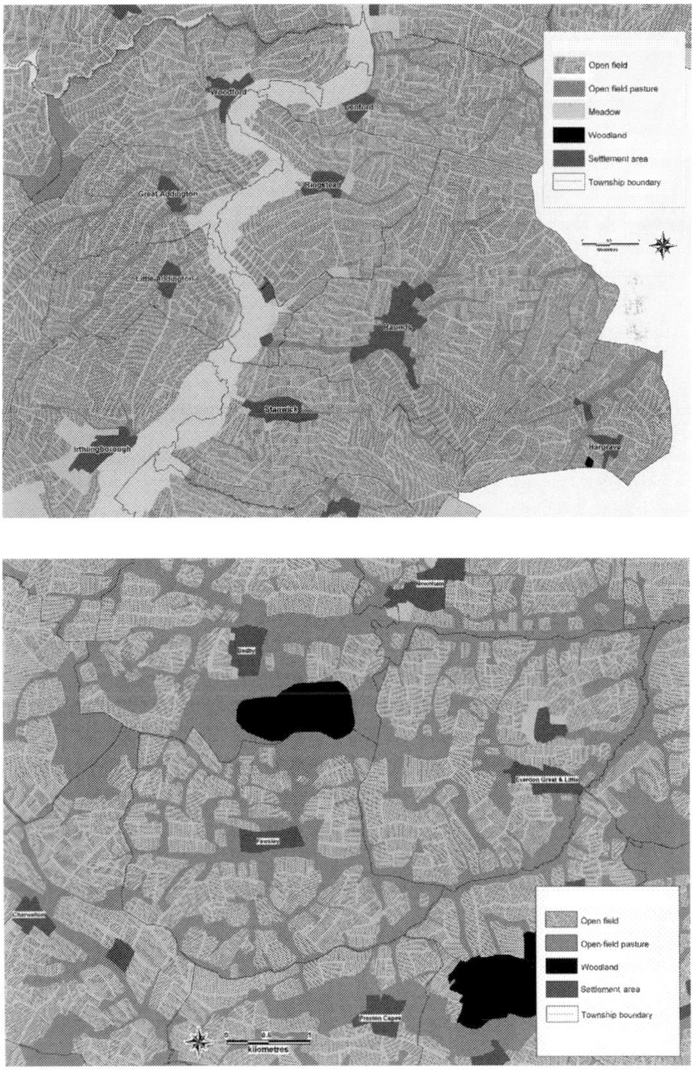

Above: the landscape of eastern Northamptonshire, around the river Nene, in the high middle ages. Planar topography, relatively low rainfall and soils formed in boulder clay ensured a near-continuous furlong pattern, interrupted by only a few, narrow ribbons of pasture. **Below:** the landscape of western Northamptonshire. Broken topography and areas of damp, intractable soils formed in Jurassic mud-stones ensured numerous areas of unploughed ground and a complex, indented furlong pattern (one strip on the map is equivalent to four on the ground)

the thirteenth century, many townships in the Midlands, practising a two-course rotation, had around half their cropped area under fallow at any one time. Such fields would have provided grazing of better quality than that which can be found in similar situations – of 'set-aside' – today, because of the abundance of weeds which would have developed in the previous year's cereal crop in the absence of herbicides. In the course of the Middle Ages many of these places changed from a two- to a three-field system, but we should not necessarily assume that this represents a greater concentration on the production of cereals for human consumption, for such a change may have been made in order to increase the production of peas, beans, oats or other fodder.

Even extensive open-fields thus represented mixed-farming systems, sometimes with a significant pastoral component. Conversely, 'patchwork' landscapes, or even landscapes consisting entirely of enclosures, do not necessarily signify the dominance of pastoral production. The great city of London was surrounded, to north and south, by countryside characterized by dispersed settlements and complex mixtures of small open fields and enclosures. Yet it was here that much of the grain required by the capital was cultivated. Even on the poor, heavy soils of the London clays 'the pre-eminence of enclosed land did not mean that the economy was based primarily on pastoral or woodland activities [...] it was not until the second half of the fourteenth centuries that the larger farmers on the London Clay began to adopt the pastoral bias for which they were later famous' (Roden, 1973: 341). In Suffolk, as Mark Bailey has shown, dispersed settlement, enclosures and 'irregular' open fields were associated with a densely populated landscape, in which most of the land area was under cultivation (Bailey, 2007). We might also note, in passing, that there is no necessary reason why differences in farming economy should have determined the extent to which medieval settlement took a nucleated or a dispersed form: and it was this, after all, which appears to have been the main factor determining whether open fields were extensive, or otherwise, in the landscape. Large and complex open fields were associated with large settlement clusters and it is the presence or absence of these, as much as that of the fields themselves, which needs to be explained.

VI. Villages, farming and the environment

Agrarian historians have discussed a number of factors which may help to explain why, in some districts, it made more sense for farms to remain closely clustered as the population grew during the early middle ages. As early as the 1930s the Orwins argued for the importance of 'co-aration' or plough sharing in the shaping of medieval landscapes (Orwin and Orwin, 1938). They noted that the sharing of ploughs amongst farmers was common practice in early medieval England because the plough and the animals that pulled together constituted – like a modern combine harvester – an expensive piece of equipment which was only employed during limited periods of the year (Orwin and Orwin, 1938: 12–14, 51–52). As larger and heavier ploughs came into widespread use in the course of the Saxon period farmers might have been encouraged to remain living in close proximity in order to facilitate sharing.

The Orwin's arguments have been rejected by many scholars, largely on the grounds that co-aration was practiced in all districts of England, those characterized by nucleated and dispersed settlement patterns alike (Homans, 1941: 81; Dodgshon, 1980: 31–33). Yet it is arguable that *in certain circumstances* the sharing of ploughs may have encouraged people to live close to their neighbours to a greater extent than in others. In this context we should note that large areas of England have a geology of pre-Cretaceous clays and mudstones which give rise to particularly difficult soils which not only, like all soils formed in clay, suffer from seasonal waterlogging, but which are in addition silty or clayey to the upper horizons – they are *pelostagnogleys* or non-calcareous *pelosols* – and thus particularly susceptible to structural damage when ploughed wet (Williamson, 2013: 196–201; Hodge et al., 1984: 155–58; 186–92; 285–88; 293–96; 351–54). Compaction serves to exacerbate the problems of seasonal waterlogging; but more importantly, impacted soils can dry to a hard, brick-like mass, from which it is difficult or impossible for germinating cereals to emerge. These problems are exacerbated by high levels of precipitation and where such soils occur in areas with a particularly damp climate, in the north and west of the country, population densities were usually low and arable land use was restricted to small pockets of better-drained soils: medieval settlement was scattered, and large areas were given over to pastoral pursuits. But where soils of these broad types existed in drier areas, were reasonably fertile, and thus carried moderate or even high population densities, it is easy to see why farms might have congregated in villages. Where there was only a short period of time during which the land could be safely cultivated, especially in the spring, the clustering of farms would facilitate the rapid mobilization of shared ploughs and teams. In addition, co-aration and a particularly short 'window' for cultivation may have had a *direct* impact on the development of the fields around the village, encouraging in particular their eventual reorganization in highly regular forms, in which holdings were scattered in an equitable manner across land of varying aspect and drainage potential, and lying at varying distances from the village. Ploughed in sequence, the lands of all those who contributed to the common ploughs had an equal likelihood of being ready for seeding in reasonable time and reasonable condition. Figure 1.5 shows those areas of England which are dominated by the relevant varieties of clay soils, and also receive less than 700 mm precipitation *per annum*. The association with landscapes of villages, and extensive open fields, is striking.

Other bottlenecks in the farming year, where tools and labour needed to be mobilised rapidly, may have similarly encouraged a greater degree of settlement clustering in some areas of England than in others. It is noticeable, for example, that meadow land was generally present in larger and more concentrated blocks across the 'champion' belt of central England than it was in the districts to either side. Bruce Campbell, mapping the evidence for land use provided by fourteenth-century *Inquisitions Post Mortems,* has noted how:

> From Somerset and east Devon in the south-west to the Vale of Pickering in Yorkshire's North Riding in the north-east, it was in the clay vales of this broad diagonal band of country that meadowland was most consistently represented. Except on the wolds, few demesnes were without at least some meadow (Campbell, 2000: 75–76).

Hay needed to be cut, repeatedly turned, carted and stacked with great speed: poor weather could ruin the harvest. Where meadows were distributed in large, continuous blocks and provided the majority of winter feed for village livestock, efficiencies in the organization of labour, and arising from the sharing of carts and other equipment, may have encouraged farms to congregate in nucleations.

A much more significant factor encouraging village living has, however, been highlighted by Eric Kerridge (Kerridge, 1992). Most areas of light, freely draining soils in England – especially the chalklands of southern and eastern England, but to a significant extent districts such as the East Anglian Breckland, where chalk is overlain by deposits of sandy drift – were characterized by nucleated villages and extensive open fields. As the area under cultivation expanded, and holdings increasingly took the form of numerous intermixed strips, farmers in such places had difficulty in ensuring that their land was effectively manured – a key issue in these environments, where nutrients were rapidly lost through leaching. The most efficient way of manuring in such circumstances was, as already noted, to pen sheep, grazed on the neighbouring heaths and downs by day, tightly together each night on the fallows and aftermath, in moveable folds (Kerridge, 1992). This would have been a difficult and labour-intensive procedure if each cultivator had carried it out himself, on an individual basis, moving his own small flock every day from the open pastures to a fold erected on one of his scattered strips. In addition, as Kerridge has commented, the farmer would also have had 'all the lambing and shearing to attend to. All this would have preoccupied him to such an extent as to leave him little time for growing cereals' (Kerridge, 1992: 26). Common flocks under the control of communal shepherds were an obvious solution. The movement of flocks, the erection of folds, and the decisions required, often on a day-to-day basis, regarding the timing of farming operations relating to sowing and harvesting – all this would have necessitated regular contact between members of an emerging farming community, and would have been most effectively organized where people lived in villages. But here, once again, agrarian imperatives may also have had a direct impact on the layout of the holdings in the fields, for it was important to ensure that all who contributed to a common flock received their fair share of manure, and be seen so to do, as the folds ere moved steadily across the arable land of the township. This would have encouraged – either through time, by piecemeal exchange, or through wholesale reallocation – even spreads of property, across which various forms of communal crop rotation were imposed, in order to create continuous blocks of land under the same 'season' which could be conveniently folded in turn (Figure 1.5).

In these areas of light land, on permeable geologies, the clustering of farms may also have been the consequence of a much simpler factor, strangely overlooked by researchers. Over much of England water can be obtained almost anywhere, from shallow wells. But in areas of very permeable geology, such as chalk and limestone, abundant and reliable water sources are more restricted. Farms were obliged to cluster on spring and seepage lines, often at the foot of major escarpments. In addition, and somewhat counter-intuitively, the same is true where pre-Cretaceous clays form deep impermeable masses, without significant aquifers; good supplies of water tend only to be found at their junction with permeable formations (Williamson, 2013: 184–93). Figure 1.6 – based on early twentieth-century literature on hydrology (especially Woodward, 1910) – shows in very general terms those areas of England in which water supplies are most restricted in this way. Of course, in chalk country as much as in Midland clay vales, medieval settlement *could* have been more dispersed than it actually was. There were, for example, usually a number of viable springs issuing at a particular contact line, some providing supplies of water which had been utilised in prehistoric, Roman or early Anglo-Saxon times, but which lacked attendant settlements by the Middle Ages. Hydrological circumstances seem, however, to have contributed to the clustering of settlement, in combination with the kinds of agrarian

Figure 1.5 The distribution of soil types which may have encouraged the development of clustered settlement

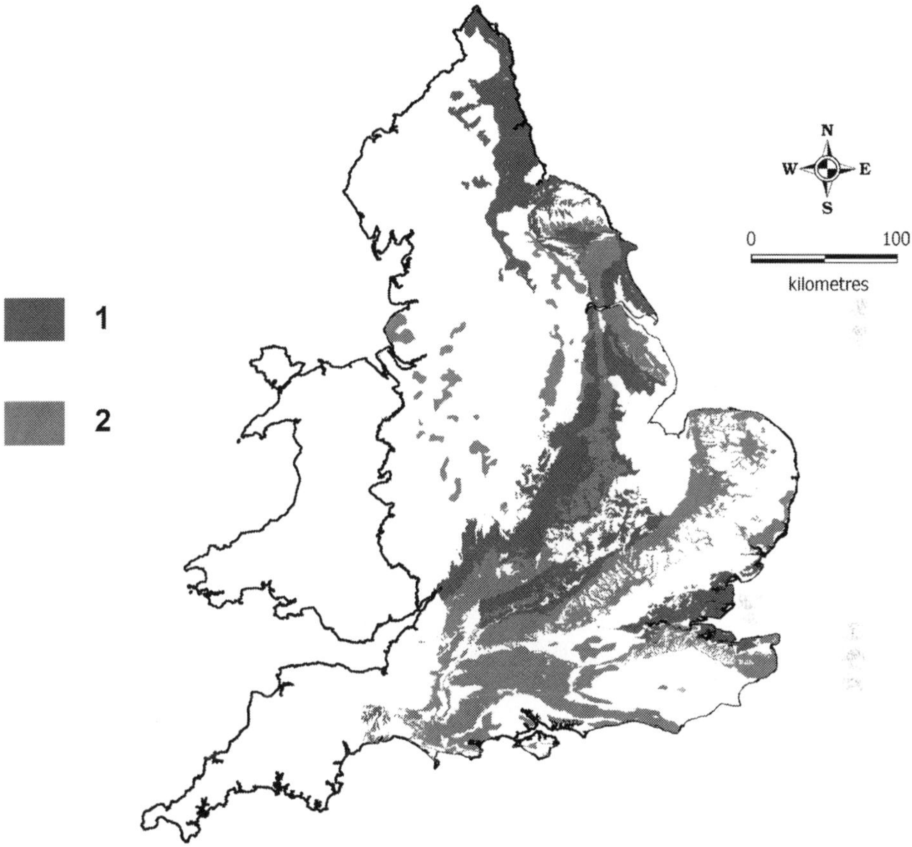

1. – Areas dominated by pelostagnogleys or non-calcareous pelosols and with less than 700mm average rainfall *per annum*. **2.** – Light, freely-draining soils formed in chalk, limestone or sands in which close folding of sheep was practised in order to maintain fertility (compare with Figure 1.1)

imperative already described: that is, a *range* of factors in any district influenced the extent to which settlement took a clustered or a dispersed form. The various encouragements to village living were, for geological reasons, particularly strong in the scarp-and-dip landscapes of the Midland counties, and on the chalklands of England: and it is in these areas that the most extensive and regular forms of open field thus eventually emerged.

Where both agrarian factors and issues of water supply ensured the development of the most tightly-nucleated villages, housing the most agriculturally inter-dependent groups of farming families, field systems were most likely to be remodelled in highly

Figure 1.6 Areas of England in which supplies of water are limited or concentrated

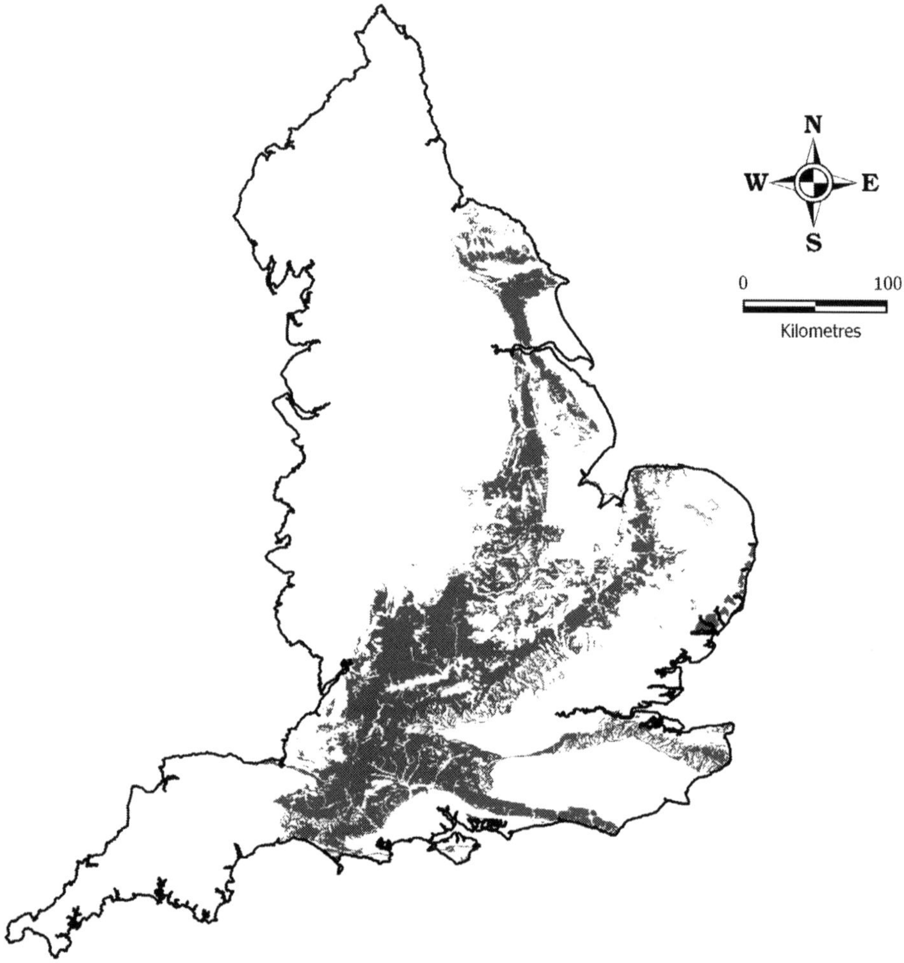

Note: Compare with the landscape regions depicted in Figure 1.1

regular forms: that is, in parts of the Midlands, with their difficult clay soils and extensive meadows. While, as noted above, the need to ensure fair access to communal ploughs may have been an important reason for the regular distributions of strips found in such villages, it perhaps mainly reflected a particular concern for fairness on the part of such communities, and their desire to match the value of holdings, to rents and obligations, in a careful and equitable manner. In other regions, with less communal forms of farming, this issue seems to have been less important to farmers. Even in 'champion' areas on lighter soils such highly regular sequences were seldom found.

VII. Discussion

The arguments advanced above suggest that medieval variations in fields, and in settlement patterns, were intimately connected; and that rather than developing in the ninth century or before, they largely emerged in the period between the tenth and the twelfth centuries, as the population grew rapidly. While markets, social structure, and tenure doubtless played their part in generating diverse forms, the supremacy of environmental factors in generating regional landscapes is evident from even a cursory perusal of soil or geology maps: major changes in field systems and settlement patterns normally correlate with boundaries of geological formations or soil types. Such arguments are often castigated, by British archaeologists especially, as being 'environmentally determinist'. But the suggestion that early medieval farmers needed to live where water was freely available, and adapted their farming practices in order to maximize the productive potential of the land, is hardly controversial.

This said, there are important social implications which remain to be teased from this essentially agrarian narrative. One relates, not simply to the role of 'lordship' in the formation of settlements and field systems, but rather to what the development of these can tell us about the origins of local lordship itself. Historians and archaeologists in England have often emphasized the decline of 'extensive lordship' in later Saxon England, the growth of 'territorial lordship' and the proliferation of local lords, many as we have noted seeing the emergence of villages and communal field systems as one manifestation of this process. The development of local lordship is usually described in essentially 'top down' terms: aristocratic families gained proprietorial rights over territories once subject only to royal control, territories which were then prone to the disintegrative effects of sale and inheritance. As local lordships proliferated, formerly free, or semi-free, peasants were downgraded to dependent cultivators. The small, scattered early and middle-Saxon sites which preceded the development of villages presumably represent the settlements of free individuals of the former kind, and their dependents.

It is striking, therefore, that where medieval villages have been excavated and the pre-village nuclei from which they developed examined, these often seem to have emerged as the principal manorial centres within the settlements in question: the tofts and crofts added to these initial 'cores', as they expanded into true 'villages', were occupied by men of lesser status, usually customary or bond tenants, a pattern evident at both Raunds in Northamptonshire and Wharram Percy in Yorkshire (Auduoy and Chapman, 2009: 39–45; Beresford and Hurst, 1990: 69–78). This observation at least raises the possibility that the development of lordship in the later Saxon period may have been a more complex process than is sometimes supposed. Manors may often have evolved through social differentiation amongst the peasantry, rather than through the fragmentation of large 'multiple estates' which were the property of a permanent elite (Jones, 1976; Faith, 1997). In this context, we should perhaps note the suggestions made by Runciman in 1984 that the period between the eighth and the eleventh century was one of increasing social mobility in England, in which some members of the peasant class acquired large amounts of land and achieved enhanced status, while others declined into bondage (Runciman, 1984); and also the comments of Kerridge, that intermixed parcels of land in the first open fields may have arisen not simply from partible inheritance, but also from other forms of equitable division, associated with the dissolution of what had formerly been jointly held family farms, or the allotment of land to under-tenants or freed slaves (Kerridge, 1992: 48–49). Indeed, as Kerridge also stressed, population growth could not in itself have led to subdivision and intermixture of holdings. This would only have happened if properties had been held by individual peasants, rather than jointly by extended families. Open fields in their various forms were, perhaps paradoxically, expressions of highly individualized forms of tenure, and of societies in which land was farmed in a plethora of small units.

There are likewise social implications in the later development of field systems and settlement patterns. In areas where farming was most cooperative in character, social relations within the community – and between communities and their feudal superiors – came to be mapped out on the landscape in patterns of holdings scattered in regular sequence through the fields. And where settlement developed in dispersed forms, outlying farms might often take the form of free tenements or sub-manors, outside both the economic and, to an extent, the tenurial frameworks of the initial lordships. Such arguments should not be taken too far, however. Extensive open fields and nucleated villages were often found in areas characterized, in the eleventh and twelfth centuries, by an abundance of free tenures, such as Lincolnshire. Conversely, Essex – with a landscape characterised by 'patchwork' open fields and bocage – was one of the most highly manorialized counties in England, with particularly heavy labour services (Williamson, 2013: 242).

The decline of open fields from the late Middle Ages resulted not simply from the development of a market economy *per se*, because by the thirteenth century local societies were already deeply penetrated by the market. It began, rather, with changes in the character of the market in the later middle ages, which encouraged a shift to specialized pasture farming on heavier soils; was accelerated from the seventeenth century by the introduction of new crops and rotations, especially on lighter land, innovations which were difficult to introduce into established two- or three-course field systems; but it was above all driven by a gradual but inexorable reduction in the numbers of farms and a concomitant increase in their average size, something which continued throughout the late medieval and post medieval centuries and which rendered cooperative farming systems, of all kinds, increasingly irrelevant (Williamson, 2000). This said, open fields continued to operate into the twentieth century in a surprising number of places: those at Clothall and Bygrave in Hertfordshire, for example, within 50 km of the centre of London, were only removed at the start of the First World War (Rowe and Williamson, 2013: 51–53). Accustomed as we are to systems of agriculture based on fields cultivated in severalty, we perhaps find open-field agriculture stranger than we should.

VIII. Conclusion

How far any of these observations, suggestions and speculations are relevant to areas outside England is uncertain. One of the great obstacles to understanding the character of landscapes across Europe as a whole is that of language; but this is compounded by the particularly insular character of landscape studies – literally so in the case of England – a problem which, of course, this volume is designed to address. There are, however, many clear and obvious parallels in aspects of settlement, as well as of fields, between England and mainland Europe. The 'infilled greens' which seem to underlie many nucleated villages in England also occur in parts of Germany (Rennes, 2010: 56); in Flanders, the pattern of manorial halls, often located beside churches, lying separate from tenements clustered on the margins of greens is strongly reminiscent of the arrangements found in many areas of dispersed settlement in England, especially in East Anglia; while in France, as in England, landscapes of nucleation appear to be associated with certain kinds of geology more than others: in the Parisian Basin the traditional distinction between the *Champagne humide* and *Champagne sèche* mirrors the differences I have outlined in England between extensive open fields formed in clay, and in chalk, respectively. All this said, it is perfectly possible that constraints on settlement dispersion – and encouragements for farms to congregate in large nucleations – may have taken other forms in different parts of Europe, often perhaps more social and political, than agrarian and environmental, in character. Either way, future research into the genesis of field systems must, of necessity, pay as much attention to the development of the settlements from which these were farmed, in all their rich complexity.

In terms of the 'rationale' of open fields in England, it seems clear that the highly regular systems found in many Midland, 'champion' villages were created through the deliberate re-modelling of earlier, more irregular forms, and were connected with the extraction of surpluses by elites. Re-planning of this kind was, perhaps, largely intended to share burdens equitably between farming families within what were developing as close-knit agrarian communities. Elsewhere, the various forms of open field were not usually designed, at a stroke, to deal with any particular problem, or to achieve any single aim, such as the minimalization of risk; although everywhere fairness in the allocation of land, when farms were being divided by inheritance for example, seems to have been an important structuring influence. Fields developed in a gradual, piecemeal fashion. Differing degrees of settlement dispersion, mainly the consequence of environmental factors but in part of variations in population pressure and, perhaps, inheritance practices served gradually to intermix peasant holdings to varying extents. The resultant difficulties of farming in intermixed parcels, too small to enclose with hedges, were themselves regionally varied, due to environmental factors. In some, for example, the need to organize folding was of critical importance, in others it was a minor issue. The organizational structures created to solve these problems – the 'field systems' themselves – were largely if not entirely the work of farmers, rather than being imposed by a social elite.

Bibliography

Audouy, M. and Chapman, A. (2009), *Raunds: the origins and growth of a medieval village AD 450–1500*, Oxford.

Beresford, M. and Hurst, J. (1990), *Wharram Percy: deserted medieval village*, London.

Bailey, M. (2007), *Medieval Suffolk: an economic and social history 1200–1500,* Woodbridge.

Baker, A. R. H. Baker and Butlin, R. A. ed. (1973), *Studies of field Systems in the British Isles*, Cambridge.

Brown, A. E. and Taylor, C. C. (1989), 'The origins of dispersed settlement: some results from Bedfordshire', *Landscape History,* 11, pp. 61–82.

Brown, A. E. and Foard, G. (2004), 'The Anglo-Saxon period' in: M. E.Tingle ed., *The archaeology of Northamptonshire*, Northampton, pp. 78–101.

Brown, T. and Foard, G. (1998), 'The Saxon landscape: a regional perspective' in: P. Everson and T. Williamson ed., *The archaeology of landscape*, Manchester, pp. 67–94.

Campbell, B. M. S. (1981), 'Commonfield origins – the regional dimension' in: T.Rowley ed., *The origins of open-field agriculture*, London, pp. 112–29.

Campbell, B. M. S. (2000), *English seigniorial agriculture*, Cambridge.

Darby, H. C. (1977), *Domesday England*, Cambridge.

Dodgshon, R. (1980), *The origins of British field systems: an interpretation*, London.

Dyer, C. (1990), 'Dispersed settlements in medieval England. A case study of Pendock, Worcestershire', *Medieval Archaeology,* 34, pp. 97–121.

Edelen, G. ed. (1994), *William Harrison's Description of England*, London.

Everson, P., Taylor, C. C. and Dunn, C. J. (1991), *Change and continuity: rural settlement in north-west Lincolnshire*, London.

Faith, R. (1966), 'Peasant families and inheritance customs in medieval England', *Agricultural History Review*, 14, pp. 77–95.

Faith, R. (1997), *The English peasantry and the growth of lordship*, Leicester.

Foard, G. (1978), 'Systematic fieldwalking and the investigation of Saxon settlement in Northamptonshire', *World Archaeology*, 9, pp. 357–74.

Fox, H. S. A.(1981), 'Approaches to the adoption of the Midland system' in: T.Rowley ed., *The origins of open-field agriculture*, London, pp. 64–111.

Fox, H. S. A. (1984), 'Some ecological dimensions of English medieval field systems' in: K. Biddick ed., *Archaeological approaches to medieval Europe*, Kalamazoo, pp. 129–58.

Gerrard, C. and Aston, M. (2007), *The Shapwick project*, London.

Gray, H. L. (1915), *English field systems*, Cambridge (Mass.).

Hall, D. (1982), *Medieval fields*, Aylesbury.

Hall, D. (1995), *The open fields of Northamptonshire*, Northampton.

Hamerow, H. (1991), 'Settlement mobility and the 'middle Saxon shift': rural settlements and settlement patterns in Anglo-Saxon England', *Anglo-Saxon England*, 20, pp. 1–17.

Hodge, C. A. H., Burton, R. G. O., Corbett, W. M., Evans, R. and Searle, R. S. (1984), *Soils and their use in eastern England*, Harpenden.

Homans, G. C. (1941), *English villagers of the thirteenth century*, Cambridge (Mass.).

Homans, G. C. (1969), 'The explanation of English regional differences', *Past and Present*, 42, pp. 18–34.

Hooke, D. (1998), *The landscape of Anglo-Saxon England*, Leicester.

Hunter, J. (2003), *Field systems in Essex*, Colchester.

Jones, G. R. J. (1976), 'Multiple estates and early settlement' in: P. H. Sawyer ed., *Medieval settlement: continuity and change*, London, pp. 15–40.

Jones, R. (2011), 'The village and the butterfly: nucleation out of chaos and complexity', *Landscapes*, 11, pp. 25–46.

Kerridge, E. (1992), *The common fields of England*, Manchester.

Lewis, C., Mitchell-Fox, P. and Dyer, C. (2001), *Village, hamlet and field: changing settlements in central England*, Macclesfield.

Martin. E. and Satchell, M. (2008), *Wheare most inclosures be. East Anglian fields, history, morphology and management*, published as *East Anglian Archaeology,* 124, Ipswich.

Oosthuizen, S. (2002a), 'Ancient greens in Midland landscapes: Barrington, Cambridgeshire', *Medieval Archaeology*, 46, pp. 110–15.

Oosthuizen, S. (2002b), 'Medieval greens and moats in the Central Province: evidence from the Bourne valley, Cambridgeshire', *Landscape History*, 24, pp. 73–87.

Orwin, C. S. and Orwin, C. S. (1938), *The open fields*, Oxford.

Postgate, M. R. (1973), 'Field systems of East Anglia' in: A. R. H. Baker and R. A.Butlin, ed., *Studies of field systems in the British Isles*, Cambridge, pp. 281–324.

Rackham, O. (1986), *The history of the countryside*, London.

Renes, H. (2010), 'Grainlands. The landscape of open fields in a European perspective', *Landscape History*, 31, pp. 37–70.

Rippon, S. (2007), 'Emerging regional variation in historic landscape character: the possible significance of the "long eighth century"' in: M. Gardiner and S. Rippon ed., *Landscape history after Hoskins, Vol. 2: Medieval landscapes*, Macclesfield, pp. 105–21.

Rippon, S. (2008), *Beyond the medieval village: the diversification of landscape character in southern Britain*, Oxford.

Roberts, B. K. (1973), 'Field systems of the west Midlands' in: A. R. H. Baker and R. A. Butlin, R. A. ed., *Studies of field systems in the British Isles*, Cambridge, pp. 188–231.

Roberts, B. K. and Wrathmell, S. (2000), *An atlas of rural settlement in England*, London.

Roberts, B. K. and Wrathmell, S. (2002), *Region and place; a study of English rural settlement*, London.

Roden, D. (1973), 'Field systems of the Chilterns and their environs' in: A. R. H. Baker and R. A.Butlin, ed., *Studies of field systems in the British Isles*, Cambridge, pp. 325–76.

Rogerson, A. (1995), *Fransham: an archaeological and historical study of a parish on the Norfolk boulder clay*, unpublished PhD thesis, University of East Anglia, Norwich.

Rowe, A. and Williamson, T. (2013), *Hertfordshire: a landscape history*, Hatfield.

Runciman, W. G. (1984), 'Accelerating social mobility: the case of Anglo-Saxon England', *Past and Present*, 104, pp. 3–30.

Seebohm, F. (1883), *The English village community*, London.

Taylor, C. (1977), 'Polyfocal settlement and the English village', *Medieval Archaeology*, 21, pp. 189–93.

Taylor, C. (1983), *Village and farmstead*, London.

Taylor, C. (1994), 'The regular village plan: Dorset revisited and revised' in M. Aston and C. Lewis ed., *The medieval landscape of Wessex*, Oxford, pp. 213–18.

Taylor, C. (2002), 'Nucleated settlement: a view from the frontier', *Landscape History*, 24, pp. 53–71.

Thirsk, J. (1964), 'The common fields', *Past and Present*, 29, pp. 3–29.

Vinogradoff, P. (1892), *Villainage in England*, Oxford.

Vinogradoff, P. (1905), *The growth of the manor*, London.

Warner, P. (1987), *Greens, commons and clayland colonization*, Leicester.

Whitelock, D. ed. (1955), *English historical documents*, Vol. I, 500–1042, London.

Williamson, T. (2000), 'Understanding enclosure', *Landscapes*, 1, pp. 56–79.

Williamson, T. (2013), *Environment, society and landscape in early medieval England*, Woodbridge.

Williamson, T. Liddiard, R. and Partida, T. (2013), *Champion. The making and unmaking of the English Midland landscape*, Liverpool.

Woodward, H. B. (1910), *The geology of water supply*, London.

2 Open fields in their social and economic context: the west Midlands of England

Christopher Dyer, University of Leicester, UK

This article is concerned with the rationale of the open fields as a contribution to resolving the dilemma facing all interpretations of pre-modern rural society: to what extent was the purpose of agriculture to achieve self-sufficiency, or to gain profits in the market? Firstly, the scene will be set in the west Midland region of England, defining an outline of the character, origins and development of its open fields, and in the second part the social and economic functions of the open fields will be explored, making special use of the by-laws which played an important part in the management of the fields.

I. The region and its fields

The English counties of Gloucestershire, Warwickshire and Worcestershire in combination are given some degree of unity because their boundaries coincide approximately with the valley of the river Severn and its tributaries, including the river Avon. The region is well suited to be a test bed for ideas about medieval fields as it is is divided almost equally between open fields associated with nucleated villages, and areas of dispersed settlement with irregular open fields mixed with old enclosures, allowing comparison between contrasting fields and landscapes in close proximity. Another reason for valuing the contribution that the region can make to the study of fields is the large body of evidence that is available, and its high quality, not just from the records of the large church estates, but also from those of lesser lay landlords. Numerous early estate maps show the fields before modern enclosure, the pattern of ridge and furrow survived long enough to be photographed from the air, and settlements are available for archaeological investigation. The estates, manors, villages and landscapes of the region have been much studied and historical research can be built on high-quality interpretations (e.g. Harley, 1958; Hilton, 1966; Hooke, 1985; Roberts, 1973).

Open fields predominated in the south and east of the region, in the Avon valley and to the south of that river (see Figure 2.1): the south and east of Warwickshire was called the Feldon (described as 'a plain champion country' in the sixteenth century) (Gover, Mawer and Stenton, 1936: 15). The same type of agricultural landscape stretched into the adjoining Vale of Evesham ('the granary of Worcestershire'), and beyond into the centre of the county and towards the Severn valley. In Gloucestershire much of the farming of the Cotswold hills was organized in open fields, and open fields also dominated in the clay lowlands to the north and west of the hills. The general topographical terms 'champion' and 'wold' can be applied to these open-field areas which lie on the western fringes of the 'central province' or the area of 'the Midland system'. Similarly 'woodland' can be used for the north and west of the region, which included the royal forests and private chases of Corse, Dean, Feckenham, Horwood, and Malvern, and the north-western part of Warwickshire which was known as the Forest of Arden though it was never a royal forest.

Peasants and their fields. The rationale of open-field agriculture, ed. by Christopher Dyer, Erik Thoen and Tom Williamson, Turnhout, 2017 (*CORN* Publication Series, 16), pp. 29-47

BREPOLS PUBLISHERS DOI: 10.1484/M.CORN-EB.5.114267

Figure 2.1 The west Midland region (counties of Gloucestershire, Warwickshire and Worcestershire) indicating the main types of historic landscapes

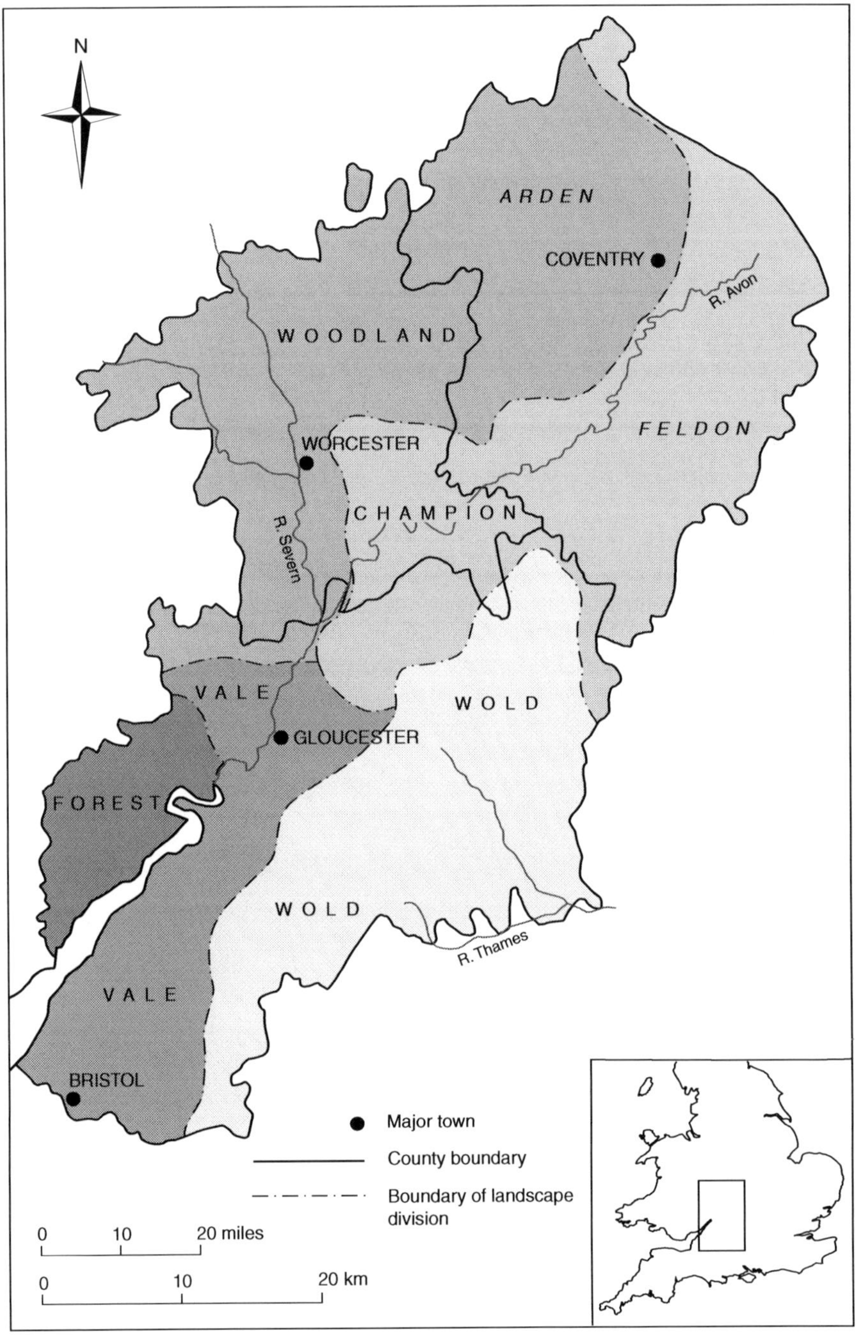

The woodland landscape included areas covered with trees, but its main characteristics were dispersed settlements and a combination of different uses of land with some arable and often extensive pastures on heaths and hill sides, and in 'wood pasture'. The 'vale' in Gloucestershire along the river Severn estuary contained all of these types of land, but also included some wetland. The 'woodland' term can be applied to large sections of western England, from the south-western counties of Dorset and Somerset up to Cheshire in the north-west.

The open fields of the region's champion and wolds resembled those found throughout Europe, with long narrow strips (called *selions* in the documents, or lands in everyday speech). Each strip amounted to a fraction of an acre, at most a half-acre (0.2 ha), which could be ploughed in a day, and they were grouped, 20 to 50 together to form a furlong (*quarantena*), running in parallel up to a headland (*forera*) on which the ploughs turned. A strip also served as a unit of land holding, usually intermingled with others. Sometimes when the location of strips in one holding was identified by reference to the adjacent strips we find that the same neighbour's name recurs – for example John Smith of Compton Murdak in Warwickshire in the late thirteenth century held 10 acres (4 ha.) in eighteen parcels or strips, and in eleven of these his neighbour was Gilbert le Grys (Dyer, 2000: 61–2). This could have been a survival of a regular ordering of the field system in the remote past, or of a more recent reallocation of land. A similar document from the same village mentions strips of the lord's demesne as lying next to peasants' strips, showing that the demesne was scattered over the fields, just like those of the tenants (see Figure 2.2). On the other hand, demesnes could be arranged into blocks set apart from the common fields of the villagers, but sometimes observing the same rotation. The boundaries between strips were marked by stakes or stones at the ends, or narrow unploughed balks. The boundary of the whole field was surrounded by a hedge or fence which was made into a continuous barrier when crops had been planted, but which had gaps ('sherds') which were opened when animals were allowed to graze on the stubble after the harvest. Most villages in the region worked two fields, though three-field systems are found in north-east Warwickshire. Winter-sown crops (wheat and rye) and crops sown in the spring (barley, dredge, oats and legumes) were planted in different parts of the same field if a two-course rotation was followed, but under a three-course system one field was planted with winter-sown crops and a second field with spring-sown crops in roughly equal quantities. If the land was divided into two fields, half of the land was fenced off each year to protect the growing crops, which gave the livestock more grazing than in a three-field system.

The amount of permanent grazing varied greatly. In some villages more than 80 per cent of the land was ploughed, and grassland was limited to the meadow after the hay making, and to small areas on roadside verges. In others an area was set aside for pasture, for example on a steeply sloping hillside where cultivation was impossible. Alternatively a ring of villages might be able to share in a large area of heathland or upland grazing where they had rights to intercommoning (Dyer, 1996: 121). Common rights of the same kind might be enjoyed at a great distance from the village, where the main village territory was linked with a distant woodland, from which also came supplies of timber for building (Hooke, 1985:78–88). Many villages in south Warwickshire and on the Cotswolds obtained their firewood and wood for hedging and fencing from small groves included within their territory, which consisted of coppiced 'underwood' (bushes) and a few mature trees (Wager, 1998: 12–16).

Figure 2.2 The openfields in Compton Verney *c.* 1300

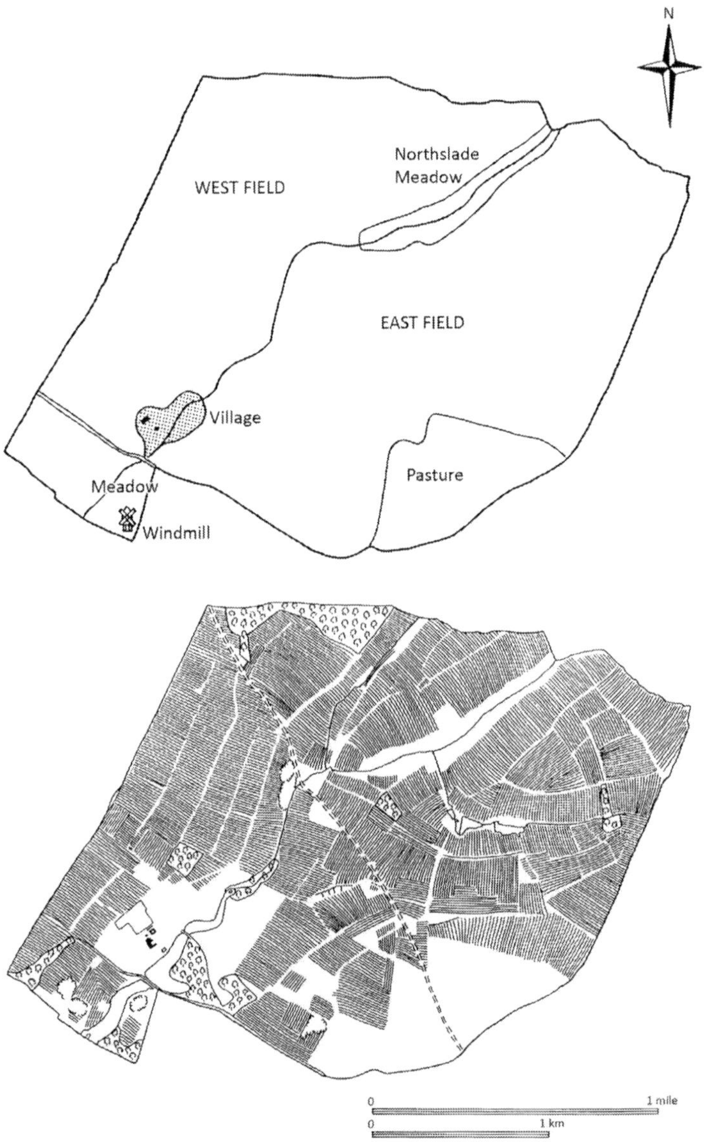

Evidence for the open fields of a typical midland village, Compton Verney (once called Compton Murdak) in Warwickshire. The upper part of the Fig shows the two fields, with areas of meadow and pasture, and the site of the village. This is a conjectural plan of its appearance in c.1300, based on documents and other evidence.The lower part is a map of the same place, in the late 1940s, with the ridge and furrow which reflects the pattern of strips. The village has been replaced by a large country house and its landscape garden with lakes, and woodland has been planted.

A holding would be described in the documents in terms of the yardland, which varied in size but usually consisted of about 30 acres (12 ha.) of arable. This was often divided into half-yardlands or quarter-yardlands (the last often being called nooks). From a few detailed descriptions (of the kind from Compton Murdak quoted above) we know that the yardland or a fraction of that unit consisted of many small parcels or strips in the open fields. The word yardland in documents was often followed by the phrase 'with appurtenances' which was the lawyers' term for the share in the meadow and use of the pasture and wood which assured each tenant a portfolio of resources.

In the woodland townships an area of arable consisted of intermixed strips with furlongs and headlands resembling champion open fields; the land was ploughed in strips, and lay in ridge and furrow. Cultivation followed agreed rotations: the orderly succession of croppings and fallowing was not based on two or three fields, but on 'seasons' in which land of all kinds, not just that cultivated in open fields, observed an agreed rotation, and often a two-course routine was followed. There were however many differences between champion and woodland open fields. The open field in the woodland landscapes occupied a smaller area of the township, between 10 per cent and 70 per cent of the total, but more often below 50 per cent (Roberts, 1973:224). The number of fields was usually greater than the two or three fields of the champion, and there could be more than ten of them; not all of the tenants had a stake in all of the fields. The open fields, together with the hedged and fenced crofts and closes that occupied much of the land were subject to grazing at the 'open time' after the harvest and during fallow years by all those with common rights. Most woodland townships had access to common pastures, heaths or wood pastures, which were usually more extensive than those available to champion peasants. A typical holding might consist of two or three crofts and closes, together with a dozen selions scattered over two or three open fields. It was sometimes customary to employ the yardland terminology normal in the champion country, but often the holding was defined in terms of acres or closes. As the holding had rights in the commons, the 'with appurtenances' phrase was often used.

II. The origins and development of fields

The earliest detailed charters, and especially those provided with boundary clauses, were compiled in the ninth and tenth centuries. They are few in number and brief, but can contain precious references to common land, usually pasture and wood. At Himbleton in Worcestershire a charter of *c.* 977 refers to the division of meadow into parcels corresponding in size to a holding of arable land, as would be the case in later centuries (Hooke, 1981:58). Another Worcestershire document of 974 for Cudleigh describes 30 acres of land lying in two fields (Hooke, 1981: 59). When a holding was split between land holders we can sometimes observe a moment at the birth of intermingling. For example two brothers at Moreton in Bredon (Worcestershire) received a holding in 990 which was divided unequally: 'the oldest shall always have three acres, and the younger the fourth' (Finberg, 1972: 490).

The boundary clauses describe the landmarks moving through an open field or around its edges, as they use words such as 'acre' and 'land' (both apparently referring in the context of defining a boundary to a strip), 'furrow' and 'headland'. At Donnington in eastern Gloucestershire, in the heart of what was later to be called champion country, a tenth-century charter boundary mentions a heath and other non-arable features, but also uses the word for 'headland' three times and *furh* twice, suggesting that in places open-field cultivation had extended up to the boundary of the estate (Grundy, 1935–6: 107–13). In the woodlands, notably in central Worcestershire, the boundaries contain words associated with arable cultivation, but less often, and the appearance of hedges and crofts as boundary marks suggest that the partially enclosed landscape known from later sources was already emerging (Hooke, 1981: 40–58).

The likelihood is that both open fields, mostly in the areas later called champion, and partly enclosed fields in the woodlands were already formed in the ninth and tenth centuries. The documents are giving us a terminus ante quem for the field systems, which may have been emerging long before the first detailed boundary clauses. Specialization of farming emerged before the development of marketing, which meant that the places near the Warwickshire Avon were provided with distant woodland appendages. This is well known from eleventh-century and later evidence, but as early as *c.* 700 Shottery in the Avon valley, which was well suited to arable, was attached to woodland 18 km. to the north at Nuthurst (the name means wood with nut trees) and even further at a distance of 25 km. at *Hellerelege* (Finberg, 1961: 87). By such long-distance connections the arable cultivators could obtain adequate grazing, fuel and building materials.

Open fields tended to be associated with nucleated villages. Exceptions to this generalization include the scatter of hamlets and small settlements among the nucleated villages known from the thirteenth century and later. Groups of houses representing between two and six holdings were set apart from larger villages on the edge of woods, near mills, on upland pastures (for shepherds), or in the fields away from the main concentrations of population (Dyer, 2002: 17–28). Some of the large villages were not closely nucleated, but consisted of straggling settlements loosely connecting two, three or four nuclei, and in some larger territories up to seven townships were engaged in the cultivation of extensive open fields, often with two or more small villages sharing a single two-field system. On the other side of the divide, although larger woodland parishes contained twenty small hamlets and isolated farmsteads, again when high quality records are available after 1200, the landscape could also contain a few quite large village settlements. Often these occasional woodland nucleations are found to be grouped around a market place, and their inhabitants included small-scale traders and craftsmen. At least three market villages of this type can be seen in the Arden of Warwickshire, at Hampton-in-Arden, Knowle and Tanworth (Dyer, 1996: 126). There might be small open fields near these villages, but they were not cultivating two or three extensive fields of the type found in the champion. Another recurrent feature, characteristic of the Severn valley in Worcestershire (Chaddesley, Hartlebury, Kempsey and Ombersley) was the location of the inhabitants in federations of small villages scattered over the territory, amounting to around ten settlements.

In view of the association between open fields and nucleated villages, if the chronology of the settlement is known, it might help to date the fields. There is a lack of closely datable pottery in the region from the second half of the first millennium AD. The distinctive wares of the period 450–650 are often found in cemeteries, mostly in the champion districts. The cemeteries and a few settlements of the same period are not sited in or very near to later villages (Ford, 1996; Carver, Hills and Scheschkewitz, 2009).

Grass-tempered ware, which may have continued in use into the ninth century, can also be found on sites away from later villages (Dyer, 2002:13; Reynolds, 2006). This might suggest that settlements moved about in the period 450–850. The earliest pottery excavated from the cores of villages in the champion areas can be dated no earlier than the tenth and more often the eleventh century, so the settlements may have begun to develop at that time. The general scarcity of dated pottery before 1000, however, might make us think of the eleventh-century wares as providing a terminus ante quem, and the villages could well have begun their growth, as they did in neighbouring counties, in the ninth and tenth centuries, or even earlier. Nonetheless although a manor house at Lower Slaughter in Gloucestershire began life as early as the seventh century and may have provided the nodal point around which the village grew, dozens of churches which also may have been associated with village formation, contain fabric of the tenth and eleventh centuries and no earlier (Kenyon and Watts, 2006).

Open fields had come into existence by the tenth century according to the documents. If they grew up together with nucleated settlements they were unlikely to have existed as early as the seventh century. This leaves the eighth and ninth centuries as their likely period of origin and early development. The explanation of the emergence of open fields must draw on the general arguments that have been well rehearsed for the landscape and agrarian history of the whole central province. New evidence from the analysis of pollen samples, combined with work on field boundaries of the Roman and medieval periods allows us to set aside the idea that open fields resulted from the assarting of woodland. The woods of the central province had largely been cleared in prehistoric times, and the pollen of the Roman and medieval periods show a small amount of tree cover which did not change a great deal (Rippon, Wainwright and Smart, 2014: 201–07). The expansion of cultivation must have been at the expense of grassland and scrub, and it was on such a scale that the continental term cerealization could well be applied. The cultivators increased their output of food grains, and pushed their ploughed land out to the edge of their territories.

The framework of boundaries within which the peasants worked – the land attached to each village, together with the area dominated by manorial lords, and the parishes from which local churches took their tithes – was subject to that process of *encellulement* again identified on the continent. This meant that the productive efforts of each peasant community, as well as the extraction of revenues by lords and the church, were concentrated in well-defined units of land which were smaller and more rigorously organised than the great estates of the seventh and eighth centuries. With cultivation squeezed into a limited space (many champion field systems occupied no more than 1000 acres or 400 ha.) the permanent pasture was eroded and grazing had to be provided for the village livestock on the fallows (Fox, 1981: 66–68). Hungry animals and growing corn are incompatible, and so there had to be a strict separation between the sown field and the fallow, which was most easily arranged by dividing the land into two fields. A single line of hedging and fences around the field that was sown each year was much more easily arranged than piecemeal enclosures for numerous patches of arable, and used less fencing wood. The reason for the extension of the ploughed area must remain controversial. An obvious explanation would be a growth in the number of people who were consuming agricultural produce, such as town dwellers, but they were relatively few before 1100, and the most likely demand came from the expanding rural population. It is said that there was no strong connection between population density and open fields because East Anglia supported a large population, while lacking open fields of the Midland type. East Anglia however was a special environment with its own institutions and social structures, and high population was only one factor in the region's agrarian and landscape development. In the champion areas of the west Midlands a relatively high population prevailed, and the woodlands were markedly more thinly inhabited (Darby, 1976: 90; Darby and Terrett, 1954: 24, 240, 284).

While the creation of the fields is not recorded, their internal subdivisions are described in the early charters, with references to acres, furrows and headlands. By the tenth century heavy ploughs were being used, and the land was being ploughed into ridges for technical reasons, above all to aid drainage on clay soils. When it was necessary to subdivide holdings, as in the case of the land being allocated to two brothers mentioned above (at Moreton in Bredon), the ridges (called acres) could become tenurial units, and the course was set for the intermixture of selions which was a normal feature of the open fields by the thirteenth century. The furlongs in which the strips were grouped are not fully recorded in the early charters, but they are listed by name in documents from the thirteenth century and later. Following a suggestion by Fox, the names are a clue to the growth of fields by the gradual accretion of new parcels of land, as references to settlements (for example Tunstall means 'site of a settlement') could be taken to mean that a furlong had taken over land previously occupied by a farmstead (Fox, 1981: 89). The embryonic open fields may have emerged in the eighth century, and they probably grew quickly to fill a village territory by the tenth, but in some communities the fields could have expanded through accretion into the twelfth century.

The decision to adopt open fields in a particular locality, or rather if we follow a more evolutionary model, the series of decision which led to open fields, remains an elusive episode in rural history (Dodgshon, 1980). Soils may have been a factor. For example part of the champion coincided with areas of sticky clay soils in the Warwickshire Feldon, the land north of the Avon in Worcestershire, and the north-west of Gloucestershire. Open-field villages, however, included those clustered along the river Avon and its tributaries, with their light alluvial soils, and nucleated villages with open fields occupied the Cotswolds, which tested cultivators' skills with steep slopes and thin stony soils, but where heavy clay was quite scarce. To pursue the problem further, the clay soils which seem to have encouraged the development of open fields were being cultivated in irregular and enclosed fields in the woodlands of north-west Worcestershire, for example (Dyer, 1991: 10–11). The choice of crops gives a clue to environmental differences, as the champion areas in the period after 1200 when documents survive, grew a range of crops, including a proportion of peas and beans often in excess of 20 per cent of the total; but conditions of soils and climate encouraged the cultivators in parts of the north Worcestershire and Warwickshire woodlands to grow large quantities of oats, and relatively few legumes.

The contribution of lords and peasants to the origins of the open fields has been much debated. Lords had a major role in the process, because their demesne was normally the largest single land holding in the field system, and they had considerable power over their tenants. The hand of lordship is normally hidden from our view, partly because big changes followed oral discussions and verbal orders, and in the case of open fields the crucial decisions were made before written records. A suggestion that the monks of Coventry Cathedral Priory were instrumental in the adoption of three fields for a group of eastern Warwickshire villages has been hesitantly advanced (Roberts, 1973: 222). However, some of the Priory's champion villages had two fields, and they held others in the forest of Arden which worked irregular field systems, so it appears that choices were made locally on the basis of environment or local customs, which makes it possible that the villagers had a voice in the process. The normal assumption is that both lords and peasants contributed to the decisions that began and developed open fields, and that assumption will lie behind the following consideration of the rationale of fields.

III. The rationale of open-fields: a social perspective

The discussion of the rationale of open fields since the late nineteenth century tends to assume the need for peasant communities, and to some extent their lords, to achieve a degree of self-sufficiency. The fields were designed to grow a combination of crops which fulfilled the needs of households, that is bread corn, drink corn, pottage corn and fodder. Wheat and rye were primarily bread corns, though occasionally wheat was malted for brewing. Barley was a wonderfully versatile crop, which could be baked into bread, malted for brewing, or boiled in pottage. Oats and legumes were in the west Midland region either pottage or fodder corns, or both. There is no record of oat bread or oat cakes, as was the case in the counties to the north and west, though paupers were given bread baked from peas and beans. Straw was an important by-product of the corn fields, sometimes cut separately after the upper part of the plant had been harvested with sickles. It was fed to animals, used as bedding for both animals and people, provided the main roofing material, sometimes served as a floor covering, and was burnt as a fuel.

The division of the fields into two was a strategy to allow mixed arable and pasture farming even when arable extended over most of the land. Animals played a vital role in supporting cultivation by providing traction and manure. Their milk and meat (and eggs from poultry) contributed to a more varied and balanced diet. Hides and wool could be made into clothing, shoes and harness.

Arable and pasture in the open fields were clearly not equally weighted, and the population consumed cereals and legumes in much greater quantity than dairy produce and meat. Dependence on corn made the peasantry vulnerable to harvest fluctuations, and the open fields were designed to minimize the risk of bad harvests. The scattered strips gave the cultivators a better chance of obtaining some crops, for example from the drier parts of the field system in wet years, or from low lying heavy land in times of drought. Among the combination of crops at least one type would yield well, and in the west Midlands a good proportion of the fields were sown with mixtures, especially dredge (oats and barley), but also maslin (wheat and rye). The rotation from one field or one season to another did not just ensure that there was ample grazing on the fallows, it also helped to prevent diseases accumulating in soil that was repetitively cropped. If the rotation included legumes, as was often the case in the well-documented later Middle Ages, the crop that was grown on the strips previously planted with peas, beans or vetch could benefit from the nitrogen that those plants fixed in the soil. The field which was being cropped had been scattered with dung from the animals that roamed over it, and had received cartloads of manure carried from the village (Jones, 2004; Dyer, 2000: 66–67). The fallow period gave ample time for the land to be given an additional ploughing in the summer, which killed the weeds and helped to create a better seed bed.

The attitude of the cultivators towards the avoidance of risk and the protection of the plough soil from the damaging effects of overcropping can be appreciated from their reluctance, as a growing population demanded more food in the thirteenth century, to make the apparently obvious transition from two fields to three. Those working in a two-field village must have been aware of the existence of an alternative method of cultivation, and a few two-field villages (again in north-east Warwickshire with its pocket of three-field townships) are thought to have made the change around 1300. Most did not, partly because they knew that would be a troublesome upheaval, but mainly because they feared that yields would fall with less frequent fallowing, and the opportunities for animals to feed would decline, thus reducing the amount of manure (Fox, 1986). In two-field villages even a small expansion in pastoral husbandry threatened the stability of the arrangements, and the number of livestock allowed to each individual had to be controlled by stints – maximum quotas of cattle, horses and sheep.

The social and landholding structure also served to ensure self-sufficiency. Holdings were listed by lords when recording rents and services, and typically a village would contain groups of peasants with holdings of equal size expressed as standard units of a yardland. Continuing to use Compton Murdak as an example, in 1279 it had twenty-two tenants with yardlands, each containing 40 acres (16 ha.) of arable, four with half yardlands, and eighteen cottagers. The distribution of land at Compton was much the same in 1086, and the units of landholding presumably originated in some unrecorded allocation of land before 1000 AD. Yardlands in the west Midlands varied in size between 20 and 60 acres (8 and 24 ha.) which must reflect underlying assumptions about the quality of the soil and the different areas needed to cater for the needs of a family. The first reference to a 30-acre yardland appears obliquely in a famous description of the obligations of peasants written in the west Midlands in the early eleventh century, but probably based on an earlier document (Harvey, 1993). At some time in the early Middle Ages an understanding of subsistence needs led to calculations of holding size that were then perpetuated for centuries. There must have been temptations as population grew to divide the holdings into half-yardlands, which would have allowed peasant families with yardlands to look after the welfare of two sons, or provide a daughter with a landed endowment for future marriage. Lords too would gain more revenue from a large number of tenants. The divisions were often made, judging from the large number of half-yardlands recorded around 1300. Inheritance strategies were however governed by the same cautions that preserved the two-field system: peasants were concerned that dividing holdings would result in the impoverishment of those who held them, because a half-yardland might yield enough to feed a family in a normal year, but it was not large enough to provide a surplus or even a cushion against adversity. Lords also were concerned that their tenants should have the resources to perform services, and could afford to keep draught animals and pay their rents.

The large numbers of cottagers and smallholders, who could represent 40 per cent of the landholding population, might be taken as evidence of social irresponsibility, which allowed the creation and proliferation of holdings capable only of supporting a precarious existence. Their presence, however, represents another type of self-sufficiency, as the agriculture of the village would not function without flexible supplies of labour. Among the yardlanders, who would in mid-life have plenty of labour from sons, daughters and servants, there would be elderly or disabled people, or recently established tenants with children too young to work, or widows and widowers who were in acute need of help with the routines of agriculture. Most households would expect to employ workers in the harvest. Self-sufficiency in labour was a feature of compact villages in which large and small holdings were sited in close proximity. The larger holding produced a surplus of cereals beyond the need of the household, but the cottagers who often cultivated no more than an acre or two needed to acquire some of that grain. By the thirteenth century cash wages could be paid and used to buy food, but even then vestiges survived of an older method of rewards in the form of meals for workers, and a share of the corn given to harvesters at the end of a day's toil. The leaders of village communities and the lord welcomed the presence of cottagers, but they may have regarded with caution an uncontrolled growth in numbers with its associated risk of creating an impoverished underclass who could bring stress into the community.

A weakness of the peasant economy, even among the elite yardlanders, was its lack of capital for buildings, equipment, animals and technical improvements. Yardlanders would own a plough, but rarely more than three oxen, when a full team required the pulling power of eight. A well-known co-operative practice, for which there is surprisingly little direct evidence, was the pooling of oxen and implements to enable the fields to be cultivated with heavy ploughs pulled by between four and eight oxen (Langdon, 1986: 62–74). This method of overcoming shortages by a self-sufficient sharing of assets is best recorded when the villagers assembled to perform ploughing services at a 'boon' on the lord's demesne. If the twenty-two yardlanders of Compton Murdak contributed two or three oxen each they could have mustered seven full ploughs. On other days, when they were not working for the lord, they could apply the same ox-power to their own holdings. In all 220 acres of tenant land (assuming that half of each 40-acre holding was fallow, and half of the cultivable land was ploughed in the spring) would be ploughed in the weeks after Michaelmas (29 September), in order to plant wheat. This area could be ploughed by seven ploughs in 63 days at a rate of a half acre each day. Implements and equipment could also be shared. For example Richard Geffe of Himbleton (Worcs) was said in 1315 to have 'detained' (that is, borrowed but not returned) an iron chain (part of the equipment for a plough or cart) that belonged to another tenant (WCL, E7).

Sharing and borrowing seem to display the open-field village in a positive light as a type of co-operative community governed by rules of fairness, trust and reciprocity (Hilton, 1975: 48–51). In some villages, so that everyone had an equal chance of access to resources, parcels of meadow land were distributed each year by lot, so that all tenants with a right to meadow had an equal chance of gaining the best plots. Many of these aspects of self-sufficiency – the variety of crops, mixed farming, rotations, the complementary functions of those with large and small holdings, co-operation and sharing – could be found in all settlements and landscapes, including those of the woodlands. The peculiarity of the champion village and the two-field version which prevailed in the west Midlands lay in the delicacy of the mechanism and the pressure on all participants to fit closely into the collective project. The scattering of strips in particular, but also the balance between arable and grazing land, ensured the commitment of every peasant cultivator to the smooth working of the interlocking arrangements (Dahlman, 1980). This could mean that the modern politicians' claim of social inclusiveness: 'we are all in this together' really applied to the open-field communities. Farming in the woodlands allowed more flexibilities, safety valves and individualism.

IV. The rationale of profit

Many studies of open fields understandably emphasise the formal structures depicted in surveys, or which are apparent from field plans and physical remains. The critical examination of the ideal of self-sufficiency which now follows is based on dynamic sources which reflect change, and in particular the by-laws and records of their enforcement. It is informed by development in the study of peasant society over the last thirty years which have undermined (though not overthrown) the concept of peasant subsistence farming, and has demonstrated the effects of market forces (Schofield, 2003: 131–56). To take the example of village social structure presented above, Compton Murdak's twenty-two tenants each holding a yardland were not as equal as the 1279 survey implies. If we are given a glimpse of individual wealth (in tax lists, for example) they show that some yardlanders had more possessions than others. Court records from throughout the region reveal that parcels of land might be sublet, which meant that a standard holding of 40 acres for which a tenant was formally responsible, might really have been fragmented and diminished in size, and a more enterprising neighbour could have accumulated more than 40 acres (Harvey 1977: 307–11). The same dynamic entrepreneur might be keeping animals in greater numbers than the rules allowed, and was gaining more money than his neighbours by sales of wool or cheese. Perhaps the labour of the cottagers was being exploited by the better-off at a time when workers were plentiful and consequently cheap, as we know that around 1279 labourers were being paid as little as 1d. per day, which would make it difficult for a family to buy enough food, especially in bad harvest years.

Even in the early days of the open fields markets existed and the peasants sold some of their produce. In the eleventh century the west Midland region contained 12,000 townspeople who relied on others to grow their food. From their beginnings the champion villages expected to produce at least a modest surplus, and in the thirteenth century, with town dwellers accounting for about a fifth of the population, widespread industrial growth, and the development in a large-scale export trade in wool, the rationale of the field systems was increasingly turned to generating profits. The market was active within the village, in labour, land and livestock; there was also a lively trade in ale, as well as foodstuffs. Better-off peasants in particular looked outside the village to sell grain, livestock and dairy products, and to buy food (such as fish), clothing, utensils, fuel, and building materials. They hired specialist craftsmen to build their houses and make carts (Hilton, 1975: 87–90; Dyer, 2012).

New demands were being made of the open fields, and this set up tensions within village society, and changed the way that lords managed their demesnes. Charters and deeds produced before 1350 suggest pressure points in the farming system, by their mention of limits on the number of animals that the new landholder could keep on the common, presumable because of anxieties that the pastures might be overgrazed (Hilton, 1966: 119–20). Throughout the thirteenth century the extension of enclosure and cultivation on the common pasture caused unrest and the destruction of newly-erected fences and hedges, again provoked by a fear of the loss of grazing, but these frictions are found most often in woodland areas (Dyer, 2006). The best opportunity to investigate the stresses and strains in the open-field villages comes from the by-laws which were recorded as formal legislative pronouncements in the manorial court records ('It is ordered by common assent of the whole vill that no-one should[...]') or which are mentioned because individuals were presented for wrong-doing before the court, sometimes with the explicit statement that their actions were 'against the by-law'. Often an offence is simply described and we have to presume that there was either a formal by-law, or a long standing and commonly understood custom. References to enforcement are reassuring evidence that the laws were not just rhetorical gestures that had no practical effect.

This exploration of by-laws will begin with those recorded before 1350. These early rules were explicitly influenced by the lords of the manor. The lord was often mentioned, and a few of them were solely directed to defending the lord's interest, such as forbidding trespass by grazing animals on the lord's demesne. In the case of laws dealing with more general agricultural problems the lord had an interest because before 1380 or 1400 directly cultivated demesnes formed part of the field system. On the estate of Worcester Cathedral Priory the proceedings of the manor courts are recorded in sequence as the steward (or sometimes a monk) and his clerk travelled from manor to manor. The rolls for July 1337 show that within a few days each of a dozen manors adopted the same (or very similar) by-laws, which must mean that the steward dictated the wording (WCL, E13). Normally the conventional phrase that the law was issued with the assent of the community probably reflected reality, and one imagines that the villagers, or a group of them, had met in advance of the assembly of the lord's court and asked that the steward approve their idea for legislation (Ault, 1972: 15–19, 64–72). By-laws originating within the community are likely to reflect the concerns and interests of the better-off peasants who wielded most influence as jurors or in other official positions.

A recurrent theme in early legislation was the regulation of the harvest. At the climax of the agricultural year dozens of workers converged on the fields including those bringing in the crops on the lord's demesne (for wages or performing labour service). In larger numbers were the peasants, their families, servants and employees who were reaping on the tenant holdings. To add to the activity in the field, tithes were being collected, and gleaners were scouring the stubble for ears of corn that had escaped the attention of the harvesters. The open fields were public, accessible spaces. A much-repeated by-law was concerned to restrict gleaning to the genuine poor and to insist that the able-bodied should seek employment as harvesters. Through this by-law we hear the voices of the yardlanders and the lord's officials, who wished to maximize the supply of labour, so that the harvest could be completed quickly and at low cost. They accepted that the poor should be able to collect corn that would otherwise be wasted, but they were intolerant of idlers claiming undeserved charity. Another law stipulated that harvest workers were not to be rewarded in the field with sheaves of corn, expressing the suspicion that strangers would infiltrate the groups of harvesters and steal part of the crop: if genuine workers were carrying sheaves from the field, no-one could tell the difference between them and the criminals. These evil doers might be receiving encouragement from within the village, judging from by-laws forbidding the sheltering of strangers. Suspicion extended beyond wage-earners, who would have been recruited from among the cottagers, landless and outsiders migrating in search of work, because the carting of corn was forbidden at night, or on feast days or Sundays, occasions when the loading of the carts would not be under public scrutiny. Another rule sought to prevent the making of 'secret' stiles or gates at the ends of closes (the plots in which houses stood in the village) which would give hidden access to the open field. Those who owned carts or who managed closes would have been tenants; only those with 15 acres or more would normally own carts. Within the village apparently respectable landholders were believed by their neighbours to be capable of stealing corn. All of this reveals at least a potential for dishonesty in the village, and a great deal of social anxiety. Strangers, the undeserving poor, untrustworthy wage-earners and sinister neighbours all threatened the security of the crops, and might reduce the amount of corn being taken into the honest tenants' barns whether for consumption in the household or sale.

The other theme of the pre-1350 by-laws was maintaining the balance between arable and pasture, though this was not such a dominant problem as in the succeeding two centuries. This was sometimes connected to the stressful period of the harvest, as laws were concerned with preventing livestock being brought on to the stubble before the harvest was completed, and being grazed 'in the midst of the corn' thus endangering the standing corn, or the sheaves piled ready for carting. At Thornbury in Gloucestershire in 1338 sheep were restricted in their access to the stubble, because it was feared that their close grazing would deprive the larger animals of feed (SRO D641/1/4C/1). Tenants were also ordered not to graze their animals on the lords' separate land. Stints were enforced, for example at Teddington (Worcestershire) in 1336 Thomas Aluart and John le Reve were found each to be keeping twelve sheep beyond 'the common limit', though we are not told how many were allowed. We know that a yardland in a variety of other villages could graze between 30 and 60 sheep. Pigs were always attracting criticism because their rooting damaged any land they entered, and so it was required that they should have rings fitted to the noses. All of these regulations suggest that the friction, though perhaps not yet acute, arose from the aim of some tenants to increase the numbers of livestock. The motive of these troublesome individuals is likely to have been the higher profits which they hoped to gain from wool, dairy produce, meat and surplus animals in the market. For the tenant of a half-yardland, the arable crops would feed the household, but rent money and any other cash income would come from the profits of pastoralism. It goes without saying that at the apex of village society the most entrepreneurial producers – those with more than a yardland who might flout the stint by keeping flocks of 100 sheep – were not just expanding their flocks and herds, but distorting the conventions in order to adjust the balance between cultivation and pastoralism.

The by-laws and the presentments to the courts of those who offended against the rules by illicit gleaning, carting at night (one even reaped by night), allowed their animals to stray on to the lord's land, or overburdened the common were highlighting particular pressure points in the operation of the fields. One imagines that much administration was needed through village meetings to decide the dates by which sown fields were fenced, or to fix the date of the beginning of the harvest, but these routines ran smoothly enough to escape mention in the courts or the documents. In only one case, at Cropthorne (Worcestershire) in 1337 was a by-law announced that 'no-one should begin to harvest before the agreed time' (WCL E13).

The annual routines do not appear in our sources because they were commonplace, but major innovations are also invisible because they were decided by word of mouth. An exception appears in the court records for Shipston-on-Stour in 1342 which shows the whole community had united in opposition to a leading tenant, William Robins (WCL E13). He had been given permission by the lord to plant pulse (beans and peas) and barley on a 'plot of fallow' in the open fields. His neighbours were ordered not to destroy the crops, as evidently they expected to put their animals to pasture on the land

and had taken direct action. This has to be seen against the background of an especially assertive and rebellious community which had very antagonistic relations with their lord. The basis of the incident, however was a tendency of the period throughout the region to make an inhok, also called 'hitching', a procedure for enclosing part of the fallow field in order to plant spring-sown crops, especially legumes. This device made the two-course rotation more intensive without the drastic step of changing to a three-field or four-field system, The inhoks increased the amount of fodder available for animals. There are many references to this modification of the open fields, and particularly in a group of open-field villages between Bredon Hill and the river Avon in Worcestershire, first recorded in 1348 but clearly by then an established practice (Field, 2004: 5). The villages evidently shared the pasture on the fallows over a large area, an example of intercommoning, not on a heath or area of upland, but over an expanse of adjacent open fields. Once historians would have interpreted inhoking as a move to increase output at a time of high population, but we now see it as a response to market demand and in particular as a means of expanding the pastoral dimension of peasant agriculture. Unlike the Robins case these adaptations of the rotation near Bredon Hill demonstrate the famous flexibility of the open fields, which with the consent of the community could increase output and change the balance of crops.

The by-laws of the period 1350–1540 (which become especially numerous after 1400) make some reference to the regulation of the harvest, but their main concern was to protect crops and pastures from an advancing army of livestock. At Cleeve Prior almost a half of the by-laws were concerned with the control of animals, keeping them out of the harvest field, insisting that larger animals be tethered, and that pigs be controlled. The Elmley Castle community attempted to be precise about the quantity of grain that should be harvested before the animals could be introduced on to the stubble – once it was after twenty selions had been harvested, in another year the limit was raised to forty. But still the animals came, and on these and other manors a group of officials, the 'wardens of the harvest' were appointed to report offenders to the manor court. Other related problems which applied throughout the year, not just in the harvest, included the refusal of individual tenants to put their animals in the care of the common herdsman, or to respect the function of the pinfold in which straying animals were placed. Another complaint related to the damage caused when animals were carelessly driven to pasture over ploughed or enclosed land (in some villages the offence was called 'staff driving').

At Broadway in Worcestershire in particular the concern of the community was to prevent overburdening of the common pasture, and the stint which limited each yardland to sixty sheep was repeated a number of times, presumably because it was not being observed (TNA SC2/210/28). Numbers were exceeded partly because tenants were taking in flocks that belonged to outsiders, and in effect selling access to the common pasture by taking money from people who had no right to keep animals on the common. Some villages set out a sensible rule to limit excessive numbers – no-one could keep more animals on the common in the summer than they could feed on their own resources in the winter.

As was the case before 1350 some problems which would be expected to cause trouble for the open fields, such as the fencing of the planted fields, are not very often mentioned. It was perceived as a major problem at Romsley in Worcestershire, and the small villages which made up the 'federal' structure of Kempsey in Worcestershire (BAH 346210–346867; WA ref 705:4 BA 54). These were in woodland landscapes, and evidently their small open-field areas posed fencing problems, or at least the inhabitants did not give repairs a priority. Hedges were sometimes damaged by those seeking firewood, and again this was a subject of complaint more in the woodland than in the champion villages. The poor were prominent among those raiding the hedges, and the offenders included the same types of people who gleaned in the harvest field. Most of the post-1350 by-laws, however, were seeking to discipline a wide cross-section of the community, such as the half-yardlanders and yardlanders, whose aim was to squeeze as many opportunities for grazing from the management of the fields.

Some of the by-laws of the open field villages hint at major changes going on in the background. When tenants were not cultivating their land, as in 1459 at Hampton Lucy (Warwickshire) where the order was to 'plough all their selions in the common fields', or at the same village tenants were told to keep one field fallow and plant the other, these were not just individual delinquencies, but relate to restructuring of the fields and their operations (WA ref 009:1 BA 2636/164). Inhoks or hitching, well established before 1350, continued throughout the region. In the fifteenth and sixteenth centuries we find that two-field villages were pursuing the logic of planting on the fallow by dividing each field in half to make four fields, which would diminish the amount of fallow grazing, but ensure that a good proportion of the land was growing peas and beans for fodder (Yelling, 1969: 32–33). In practice the amount of grazing was increased because part of the open fields, in a widespread trend was being converted to leys, that is grassland, either for a few years in a long-term rotation or permanently. Piece-meal and small-scale enclosure of open-field strips, following some degree of consolidation of scattered holdings, was practised most often in woodland field systems, but it can be found on the edges of open fields, such as those at Broadway.

At the same time as these radical changes were being made to the fields the village was going through a restructuring in the distribution of land, as a minority acquired much larger land holdings than had existed before 1350 (Harvey, 1979: 288–89). Holdings of two, three or even four yardlands did not just increase the amount of arable land available to individual tenants, but also allowed these wealthy individuals to graze more than a dozen cattle and as many as 120, 180 or 240 sheep. In practice they did not cultivate all of their land and kept many sheep above their customary limit, with occasional reports of flocks of 300. Their wealth and prominence in the community gave them much influence on the manor court, and that institution was unsurprisingly ineffective in enforcing the rules.

These changes endangered the open fields and the whole principle of common rights on which the agrarian economy had depended for centuries. In a minority of cases villages collapsed and their fields were converted into commercially profitable pastures, but most villages were able to survive, retaining their much modified and adapted fields until new and overwhelming political and economic forces swept them away in the eighteenth and early nineteenth centuries. At that time the open fields, which we have seen could be modified to allow peasants to keep abreast of economic changes, were maligned as barbaric and primitive.

V. Conclusion

Commercial development and response to market forces in the period 1200–1540 did not immediately destroy the open fields. From the thirteenth century, even under the pressures of the period after the Black Death, the peasantry were able to obtain their own subsistence from the fields, and gain crops and animal products for sale. From an early period the fields were managed to yield a surplus, and this became a growing dimension through the later Middle Ages. The balance between arable and pasture was a particular point of pressure within the village, which was managed by by-laws and enforcement measures. Major concerns were the enforcement of stints, overburdening of common grazing, and the sale of common grazing to outsiders. Two-field farming was resilient and adaptable, and the introduction of inhoks, and ultimately four fields gave open fields a new lease of life, enabling villagers to avoid wholesale enclosure for centuries after the Middle Ages.

Abbreviations for archive repositories

BAH	Birmingham Archives and Heritage
SRO	Staffordshire Record Office
WA	Worcestershire Archives
WCL	Worcester Cathedral Library

Bibliography

Ault, W. O. (1972), *Open-field farming in medieval England*, London.

Carver, M., Hills, C. and Scheschkewitz, J. (2009), *Wasperton: a Roman, British and Anglo-Saxon community in central England*, Woodbridge.

Dahlman, C. J. (1980), *The open field system and beyond*, Cambridge.

Darby, H. C. (1977), *Domesday England*, Cambridge.

Darby, H. C. and Terrett, I. B. (1954), *The Domesday geography of Midland England*, Cambridge.

Dodgshon, R. A. (1980), *The origin of British field systems: an interpretation*, London.

Dyer, C. (1991), *Hanbury: settlement and society in a woodland landscape*, Department of English Local History Occasional Papers, 4th series, 4, Leicester.

Dyer, C. (1996), 'Rural settlement in medieval Warwickshire', *Transactions of the Birmingham and Warwickshire Archaeological Society*, 100, pp. 117–32.

Dyer, C. (2000), 'Compton Verney: landscape and people in the Middle Ages' in: R. Bearman ed., *Compton Verney. A history of the house and its owners*, Stratford-upon-Avon, pp. 49–94.

Dyer, C. (2002), 'Villages and non-villages in the medieval Cotswolds', *Transactions of the Bristol and Gloucestershire Archaeological Society*, 120, pp. 11–35.

Dyer, C. (2006), 'Conflict in the landscape: the enclosure movement in England, 1220–1349', *Landscape History*, 28, pp. 21–33.

Dyer, C. (2012), 'Did peasants need markets and towns? The experience of late medieval England' in: M. Davies and J. Galloway ed., *London and beyond. Essays in honour of Derek Keene*, London, pp. 25–47.

Field, R. K. ed. (2004), *Court rolls of Elmley Castle, Worcestershire, 1347–1564*, Worcestershire Historical Society, 20, Worcester.

Finberg, H. P. R. (1961), *The early charters of the west Midlands*, Leicester.

Finberg, H. P. R. (1972), 'Anglo-Saxon England to 1042' in H. P. R. Finberg ed., *The agrarian history of England and Wales*, I, part 2, Cambridge, pp. 385–525.

Ford, W. J. (1996), 'Anglo-Saxon cemeteries along the Avon valley', *Transactions of the Birmingham and Warwickshire Archaeological Society*, 100, pp. 59–98.

Fox, H. S. A. (1981), 'Approaches to the adoption of the Midland system' in: T. Rowley ed., *The origins of open-field agriculture*, London, pp. 64–111.

Fox, H. S. A. (1986), 'The alleged transformation from two-field to three-field systems in medieval England', *Economic History Review*, 2nd ser, 39, pp. 526–48.

Gover, J. E. B., Mawer, A. and Stenton, F. M. (1936), *The place-names of Warwickshire*, Cambridge.

Grundy, G. B. (1935–6), *Saxon charters and field names of Gloucestershire*, Gloucester

Harley, J. B. (1958), 'Population trends and agricultural development from the Warwickshire hundred rolls of 1279', *Economic History Review*, 2nd ser, 11, pp. 8–18.

Harvey, B. F. (1977), *Westminster Abbey and its estates in the Middle Ages*, Oxford.

Harvey, P. D. A. (1993), 'Rectitudines singularum personarum and Gerefa', *English Historical Review*, 108, pp. 1–22.

Hilton, R. H. (1966), *A medieval society. The west Midlands at the end of the thirteenth century*, London.

Hilton, R. H. (1975), *The English peasantry in the later Middle Ages*, Oxford.

Hooke, D. (1981), 'Open-field agriculture: the evidence from the pre-Conquest charters of the west Midlands' in: T. Rowley ed., *The origins of open-field agriculture*, London, pp. 39–63.

Hooke, D. (1985), *The Anglo-Saxon landscape: the kingdom of the Hwicce*, Manchester.

Jones, R. L. C. (2004), 'Signatures in the soil: the use of pottery in the identification of medieval farming regimes', *Archaeological Journal*, 16, pp. 159–88.

Kenyon, D. and Watts, M. (2006), 'An Anglo-Saxon enclosure at Copsehill Road, Lower Slaughter: excavations in 1999', *Transactions of the Bristol and Gloucestershire Archaeological Society*, 124, pp. 73–109.

Langdon, J. (1986), *Horses, oxen and technological innovation: the use of draught animals in English farming from 1066–1500*, Cambridge.

Renes, H. (2010), 'Grainlands. The landscape of open fields in a European perspective', *Landscape History*, 31, pp. 37–70.

Reynolds, A. (2006), 'The early medieval period' in: N. Holbrook and J. Jurica eds, *Twenty-five years of archaeology in Gloucestershire*, Gloucester, pp. 133–60.

Rippon, S., Wainwright, A. and Smart, C. (2014), 'Farming regions in medieval England: the archaeobotanical and zooarchaeological evidence', *Medieval Archaeology*, 58, pp. 159–255.

Roberts, B. K.(1973), 'Field systems of the west Midlands' in: A. R. H. Baker and R. A. Butlin ed., *Studies of Field Systems in the British Isles*, Cambridge, pp. 188–231.

Schofield, P. (2003), *Peasant and community in medieval England, 1200–1500*, Basingstoke.

Wager, S.(1998), *Woods, wolds and groves: the woods of medieval Warwickshire*, British Archaeological Reports, British Series, 269, Oxford.

Yelling, J. (1969) 'The combination and rotation of crops in east Worcestershire, 1540–1660', *Agricultural History Review*, 17, pp. 24–43.

3 Open fields in Scandinavia, *c.* 900 - *c.* 1850

Carl-Johan Gadd, University of Gothenburg, Sweden

This chapter begins by looking briefly at what is known about the development of Scandinavian regular-fallow systems in the Middle Ages. It then discusses in more detail the character of field systems in Denmark and Sweden, starting with the situation in the seventeenth century, a time when the supply of source material is good in both countries. Different systems of strip subdivision are then examined, followed by a discussion of the transition from continuous cropping to the three-field system in southern Sweden, something which serves to throw important light on how subdivided fields first came into being. After a short discussion of subsequent changes in the organization of open-field systems, and the effects of different forms of landholding on their character and operation, the difficult question of the causes of strip subdivision is addressed.

I. The Middle Ages

The farms of Scandinavia were, as a rule, still held in severalty around the year 900, each comprising a ring-fenced area containing both meadow and arable. The latter was, at least in principle, sown and cropped every year (Hoff, 1997: 175; Pedersen and Widgren, 1998: 329–30). There was thus already a division between the infield land (Sw. *inägor*), containing both arable and meadow; and the outlying land beyond the farm (Sw. *utmark*), which comprised pasture and, in the more forested regions, woodland. To the east of Jutland this basic division would continue to characterize agriculture until the great land reforms that took place in Denmark from the 1760s, and in Sweden and Norway mainly after 1800, when hamlets and villages ceased to operate as the key organizational structures in agriculture.

Although pollen evidence from the southernmost parts of Skåne indicates that some form of fallowing was being practised by the eighth century, field systems were probably still simple in character, retaining the one-field system typical of continuous cropping and without physical barriers between the arable and the meadows. As long as the arable continued to occupy a relatively small proportion of the land area there was little reason to develop regulated fallowing, and the more complicated field systems with which this was associated (Hoff, 1997: 142–50; Pedersen and Widgren, 1998: 325–26). But over time, the area under arable cultivation expanded as a result of land clearance; and since the amount of manure did not, as a rule, increase at the same pace as the cropped area, it gradually became necessary to institute regular fallows. When, as a result, a large part of the land area came to be taken up by fallow, it needed to be grazed more systematically, and this also became the case with the mown meadows. In addition, the increasing cultivation of autumn-sown rye required the erection of fencing to keep grazing livestock off parts of the arable in the autumn, as well as to prevent them from straying into the hay meadows during the summer. This, however, would have involved individual farmers in substantial costs, in terms of both labour and fencing materials, especially in the plains, where a shortage of timber was becoming serious by the Middle Ages (cf. Hybel and

Peasants and their fields. The rationale of open-field agriculture, ed. by Christopher Dyer, Erik Thoen and Tom Williamson, Turnhout, 2017 (*CORN* Publication Series, 16), pp. 49-75

BREPOLS ❀ PUBLISHERS DOI: 10.1484/M.CORN-EB.5.114268

Poulsen, 2007: 3–23). In order to minimize such costs, several farms would collaborate in creating a shared fencing system. Land was redistributed so that smaller pieces of cultivated land were merged into larger fields which were fenced in common, but in which a number of different property holders held parcels. Once this change had been achieved, the system of regular fallowing every second or third year could be applied more thoroughly. Two- or three-field rotations had come into existence in the Danish-Scanian area by the thirteenth century, and a similar chronology has been observed in the central-Swedish plains (Widgren, 1997a: 180; Pedersen and Widgren, 1998: 331–32, 385–87; Myrdal, 1999: 101–02). From about the year 1000, villages and hamlets developed in the Danish-Scanian region, and perhaps a little later in the Swedish plain regions (Porsmose, 1988: 227; Pedersen and Widgren, 1998: 430; Riddersporre, 2008: 33; Ericsson, 2012: 163). It does not seem, however, that the subdivision into parcels or small strips which characterized Scandinavian fields in the seventeenth century had occurred to any great extent before the late Middle Ages (see below).

The gradual transition to biennial and triennial crop rotations in the Scandinavian plains after A.D. 800 was associated with significant developments in plough technology. Before the introduction of the one-year fallow typical of the new crop rotations, long leys – lasting several years – had been employed, which were broken up using a spade when the land was ready to be cultivated once again. The wooden or small iron shares of the plough implements of that time were incapable of breaking a well-established fallow. The new biennial or triennial rotations thus necessitated the use of new, heavier implements, with heavier shares. In the period between 850 and 1200 the mouldboard plough was gradually adopted in Denmark, the Skåne plains, in large parts of Norway and in western Sweden. In east-central Sweden, in contrast, ards continued to be used, although after *c.* 1000 their iron shares were made larger and heavier so that they, too, were capable of breaking a fallow (Hoff, 1997: 171–72; Myrdal, 1997: 150, 153, 156–59; Myrdal, 1999: 63; Myrdal, 2011: 83–84).

II. The early modern period

The large-scale land surveys made in the seventeenth century make it possible to reconstruct the boundaries of different kinds of Swedish and Danish field systems (figure 3.1).[1] In broad terms, over most of the Jutland (*Jylland*) peninsula field-grass systems prevailed. But in eastern Jutland, and extending onto the Danish islands and into the Skåne plains, three- and two-field systems could be found. This was the most cultivated part of Scandinavia, with arable covering around 50 per cent of the land area. In the woodlands extending from northern Skåne up to the region of the great Swedish lakes much less land was under cultivation – as little as 2 per cent of the surface area – and continuous cropping (*ensäde*) was prevalent. In the lowlands at the latitudes of the great lakes, and along the Baltic coast of Sweden, another zone in which two- and three-field systems

[1] In Sweden, an extensive mapping project was carried out *c.* 1630–1655 and was resumed in the 1680s. In Denmark, the results of detailed land surveys were collected in a sort of nation-wide cadastre (*matrikel*) in the 1680s which modern scholars have been able to transform into maps. For Sweden, see Höglund 2008; for Denmark, see Frandsen, 1983a.

Figure 3.1 Field systems of Sweden and Denmark, *c.* 1650 - *c.* 1680

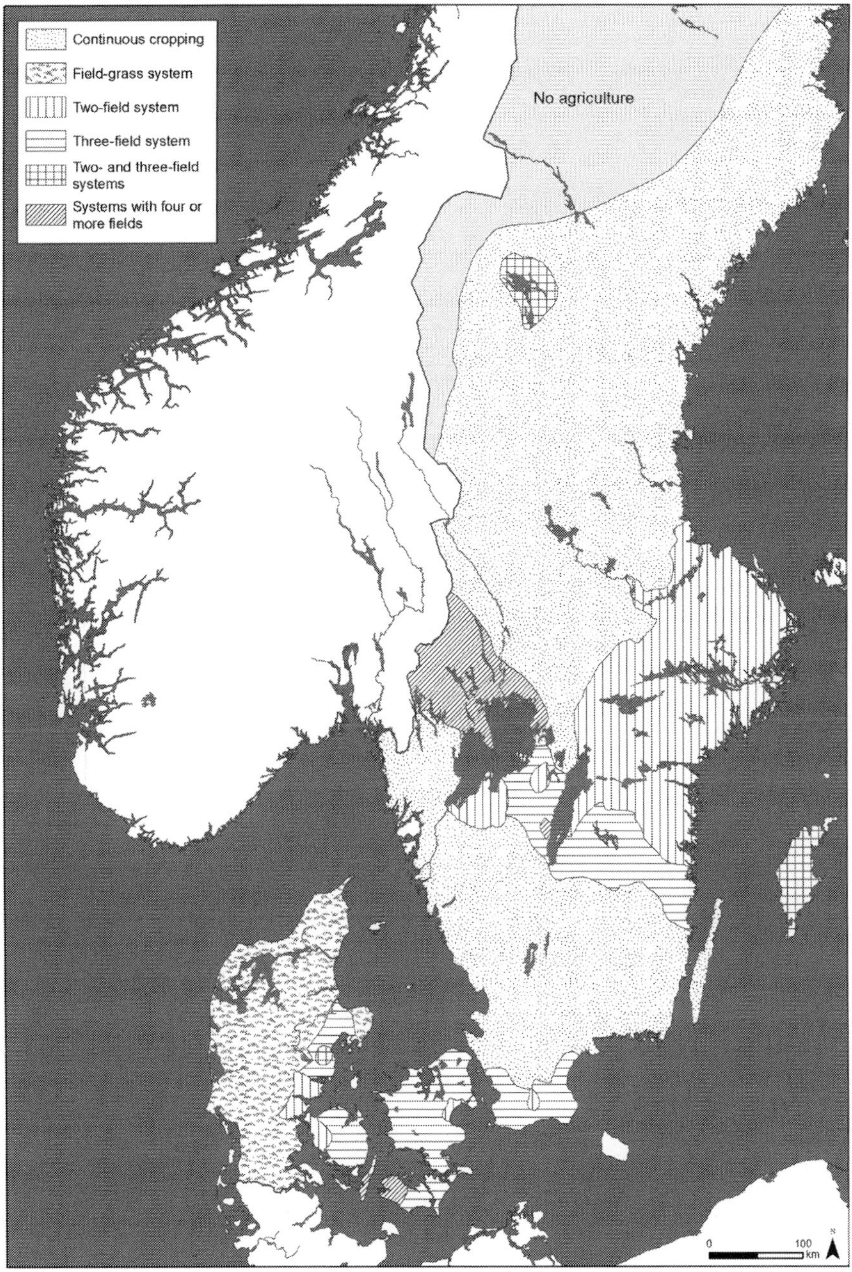

Sources: Frandsen, 1983a, map on front Frandsen 1999: 147; Widgren, 1995: 83; Myrdal, 2011: 85.

predominated could be found. Here, however, the proportion of the land under cultivation was much lower than in Denmark and Skåne, often around 10 to 25 per cent of the land surface. West and north of this lowland region continuous cropping was used, as it was across most of Norway, although zones of four- and five-field systems also existed, as they did in parts of western Sweden (for cultivation rates, see Frandsen, 1988a: 17; Gadd, 2000: 26–42; for field systems, see Frandsen, 1983a: map on front endpaper and passim; Frandsen, 1983b; Myrdal, 1999: 291; Gadd, 2000: 128; Lunden, 2004: 171). In large parts of Norway and in the Swedish interior north of the 63rd parallel arable agriculture was absent, mainly for climatic reasons.

All the principal systems of agriculture revealed by the seventeenth-century surveys had an open-field structure, in the sense that individual properties were intermingled in the fields as narrow strips, or as more block-shaped parcels, with no fences between them. In most cases, the open fields were also common fields: that is, individuals produced their own grain on their own parcels of arable, or hay on their own portion of meadow, but all peasants living in the village or hamlet had the right to graze their livestock over the fields after the harvest, or the meadows after they had been mown.

In the main, Scandinavian open-field systems followed the same pattern as their continental and English counterparts. It should be observed, however, that on the Scandinavian Peninsula north of the Skåne plains, settlement agglomerations were often smaller and the extent of cultivation less than was usually the case in Western Europe, and meadows and pastures played a more important part in the agrarian economy, something that explains the greater significance of animal production in Scandinavian agriculture, and the comparatively large number of draught animals per arable area (Gadd, 2000: 170; Gadd, 2011: 123).

Field-grass farming

Most of Jutland was farmed under a field-grass system of farming (Da.: *græsmarksbrug*). While the arable land was often divided from the outlying pastures by some kind of physical barrier it was not subdivided internally by fencing. The fields were cultivated under a system of long rotations, lasting at least six and often as many as nine or ten years, with fallows occupying around half this time: in other words, limited areas were cultivated and sown for a number of years and thereafter laid to grazing. As a rule, one continuous area, generally containing strips belonging to all the farms in the village, was cultivated and manured each year, while another was put to grass for a number of years, in a kind of long fallow system (cf. Boserup, 1965: 15–16). Often this field-grass system was combined with the use of continuous cropping (Da: *alsædebrug*) on parts of the land closest to the village (Frandsen, 1983a: 96–97, 254–55, 260, 266–67; Frandsen, 1988b: 118).

Continuous cropping

In most woodland and some transitional areas of Sweden, and in most of Norway, such a system of continuous cropping (Sw. *ensäde*) constituted the principal form of arable farming, and continued to do so into the eighteenth or even nineteenth century. This system was also found in small areas of the Danish islands and, as noted, was used together with the field-grass system in Jutland (Myrdal 2011: 85; Gadd 2000: 128;

Figure 3.2 Part of western Falbygden, a relatively densely-populated continuous-cropping area, *c*. 1640

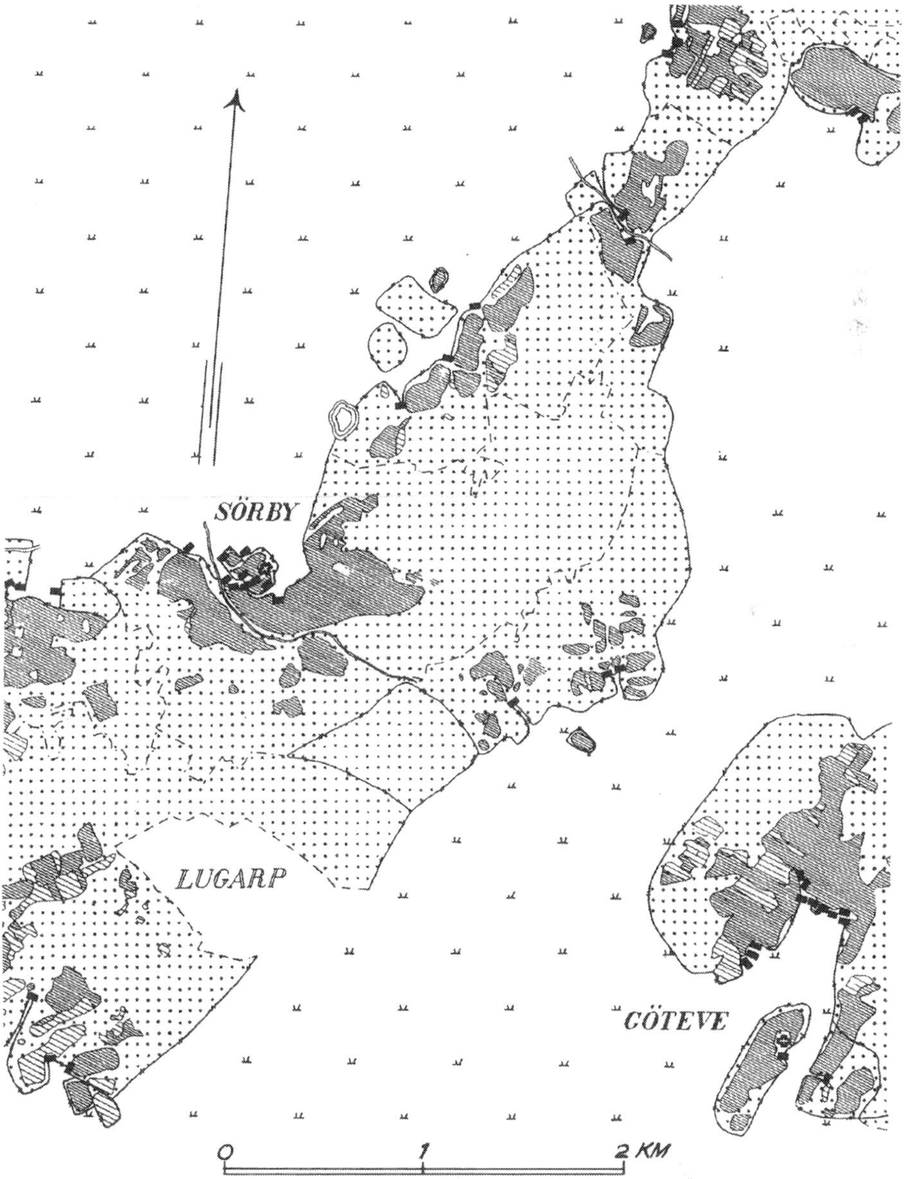

Legend: Slanted lines=arable; dotted areas=meadow; grass indications=common pasture. Black rectangles indicate cadastral homesteads.

Source: Lindgren, 1939, map.

Lunden, 2004: 171; Frandsen, 1983a: map on front endpaper). Continuous cropping was characterized by having no or almost no fallow. Typically, all or almost all of the arable was sown each year. The arable and meadows (the latter usually extensive) lay within the same fenced infield area. Since the meadow was not separated by a fence from the arable, it could not be used for grazing immediately after hay-making. Not until after the grain harvest had been completed in the autumn could the infield as a whole be opened up for grazing.

The continuous-cropping system was typical of areas in which arable farming was of relatively limited importance, compared to cattle raising and by-employments such as charcoal burning. Furthermore, in woodland areas swiddens – short-term intakes from the woodland – often complemented the permanently cropped arable. However, the continuous-cropping system was also used in transitional areas between woodland and plain, and in Danish coastal areas. Generally, it was typical in regions where meadows and grazing land were extensive, compared to the area under cultivation, and where supplies of manure were thus relatively abundant (Frandsen, 1988b: 296; Gadd, 2005: 65–67). In the most thinly populated regions each infield area was rather small and normally contained the arable and meadow of an isolated cadastral homestead (*jordebokshemman*, which was, in the seventeenth century, often subdivided into two or three discrete holdings);[2] or of a small settlement agglomeration comprising a few homesteads, reminiscent of the agrarian structures which had been found more widely across Scandinavia in the ninth or tenth centuries. Where continuous cropping was employed in more densely populated zones in the seventeenth century, in contrast, the infield areas were often large and contained the arable and meadow belonging to several dozen farms, often located at considerable distances from each other. The settlements were spread out along the boundary between the infields and the outlying land (Figure 3.2; Lindgren, 1939: 108). As a rule, the arable land belonging to each peasant lay relatively close to the site of his farm. These large, coherent infields were a form of inter-commoning (Campbell, 1928: 28; Campbell, 1933; Lindgren, 1939: 83–4). Since grazing animals were regularly pastured on the arable fields after the crop had been gathered, communal decisions and usages regarding the timing of cropping and other activities were necessary which were, in essence, the same as those found in two- and three-field systems, even though the cooperating farms, rather than being grouped into nucleated villages, might lie several kilometres apart, scattered around the margins of the infield area (Widgren, 1997a: 17; Gadd 2005: 68).

Short-fallow systems using regulated fallowing

Two- and three-field systems, and other short-fallow systems[3] with regulated fallowing, all had in common the fact that the arable was divided into two, three or sometimes more

[2] Swedish seventeenth-century surveyor's maps show the fields and meadows of cadastral farms or homesteads. These had, as a rule, been actual family farms in the mid-sixteenth century, when the cadasters were originally drawn up. In many cases, but with great regional variations, the homesteads became subdivided in the following centuries, so that one homestead contained several family farms.

[3] Cf. Boserup, 1965: 16, 51-52 and *passim*.

fields[4] that were each separately fenced for protection against grazing animals. Within each field every peasant farmer had a number of unfenced parcels or strips. In the plains, as a rule, the fences that surrounded the individual fields and meadows had by the seventeenth century developed into quite intricate systems which allowed livestock to be turned onto any part of the infield that was not currently occupied by growing crops or un-cut hay. The outlying land beyond the infield was, in contrast, fed by livestock throughout the entire grazing season.

Across most of the Danish-Scanian regular-fallow region the three-field system was used. Only in eastern Jutland, western Funen and a relatively small area in the interior of Skåne, were two-field systems the norm. The three-field system thus prevailed on both sides of the Sound. Mainly a fallow-barley-rye rotation was used, rather than the fallow-rye-barley rotation that was normal further south on the Continent, and on the most southerly of the Danish islands. An important feature of the Danish-Scanian rotation system – perhaps a consequence of a shortage of pasture – was that the fallow was untilled and used for grazing for the whole of the year, and was not ploughed up until the following spring, when the land was prepared for sowing the barley (Frandsen, 1983b: 300; Dahl, 1989: 159–60; Pedersen and Widgren, 1998: 386).

In seventeenth-century Denmark and Skåne most villages contained between ten and twenty farms, although villages with between thirty and forty and, in rare cases, as many as fifty could also be found. Figure 3.3 shows a typical three-field village on the Skåne plains in 1703. The total arable area cultivated by the village amounted to *c.* 240 ha. Eleven peasant-farmers lived in the village, each farming around 22 ha. divided into about 50 strips. In both the Scanian plains, and in Denmark, some less fertile portions of the cultivated land were not cultivated as frequently as the two years in three that was the norm under the system. It therefore seems likely that in Lilla Uppåkra each farm would have had only around 13 ha. under cultivation in any one year, scattered across a little less than thirty separate strips (Gadd, 2000: 125). However, the number of strips per peasant farmer in Lilla Uppåkra seems to have been somewhat smaller than was usual in Denmark, where it was quite normal for a peasant farmer to hold between sixty and seventy strips, and holdings of a hundred strips were by no means unknown (Dahl, 1942: 16; Porsmose, 1988: 282; Dombernowsky, 1988: 230).

In the central-Swedish lowlands two-field systems predominated, while three-field systems were found in the adjacent northern parts of the South Swedish Highlands (Figure1). Figure 3.4 shows a typical two-field village in Östergötland county as it was in 1694. The settlements of east-central Sweden were small, and this hamlet contained only five homesteads and one *utjord*, i.e. a collection of strips that belonged to a farm located in another village. The total arable area was 62 ha., or 12 ha. per peasant farmer, and the number of strips per farmer was about thirty-five. Each year a farmer tilled and sowed about 6 ha., on average, distributed over about seventeen strips (Gadd: 2000, 120–21). The arable of the

[4] Sometimes the number of fields was in fact higher, for topographical or other reasons, than is implied by the rotation system itself. In this chapter the term'three-field system' is used for a three-year rotation with three or more separately fenced fields. The corresponding is applicable to two-field system, etc. Cf. Frandsen, 1983a: 253; Gadd, 2005: 72–73.

Figure 3.3 Lilla Uppåkra, SW Skåne, a village in the Danish-Scanian three-field zone, 1703

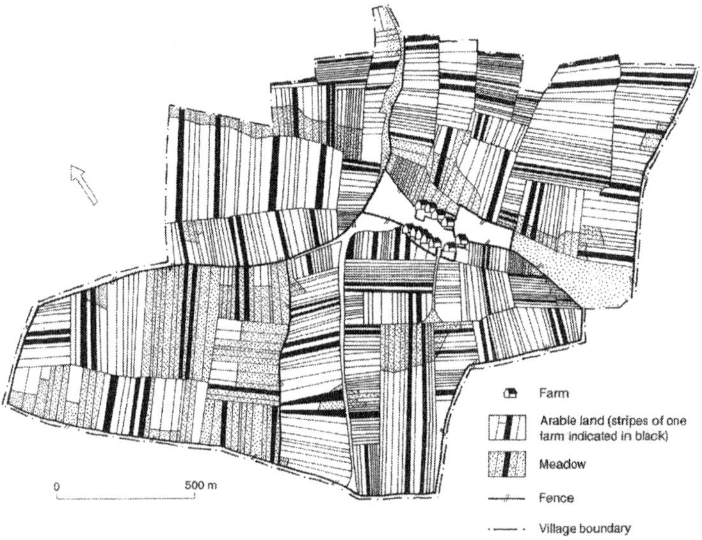

Source: Gadd, 2000: 123; Gadd, 2005b: 41.

village had, at some point in the past, been laid out according to the principle of *solskifte* (sun-division; see below), which gave the distribution of properties a very regular pattern.

Seventeenth- and eighteenth-century eastern Scandinavia was characterized by its strong emphasis on the cultivation of rye, a crop which was sown in late summer (August).[5] Consequently, during the whole of the grazing season some part of the arable land carried a standing crop and thus had to be protected from grazing livestock. This necessitated a relatively complex system of fencing in which the two arable fields and the meadow were securely divided from each other. The animals could be let in to graze the meadow after the hay had been gathered in July, and later moved onto the stubble field after harvest in August and September, while the newly sown rye in the other field remained protected behind fences.

The predominance of rye cultivation in east-central Sweden in the seventeenth century was the culmination of a long process of evolution that had taken place during the preceding centuries. In the thirteenth century barley had been almost the only crop grown in the region. The transition to rye cultivation began mainly after 1350, and involved a radical change to the timing of activities across the farming year, and in the character of the fallow grazing.

[5] Around 1800, about 50 per cent of the grain sown in east-central Sweden was rye, and another 5–10 per cent wheat, which was also sown in late summer (Gadd, 2000: 134). Since these kinds of corn were sown more thinly, yielded more and commanded higher prices than the spring-sown kinds, their importance was in fact greater than the percentages indicate.

Figure 3.4 Östra Nederlösa, province of Östergötland, a hamlet in the East-Central Swedish two-field zone, a°1694

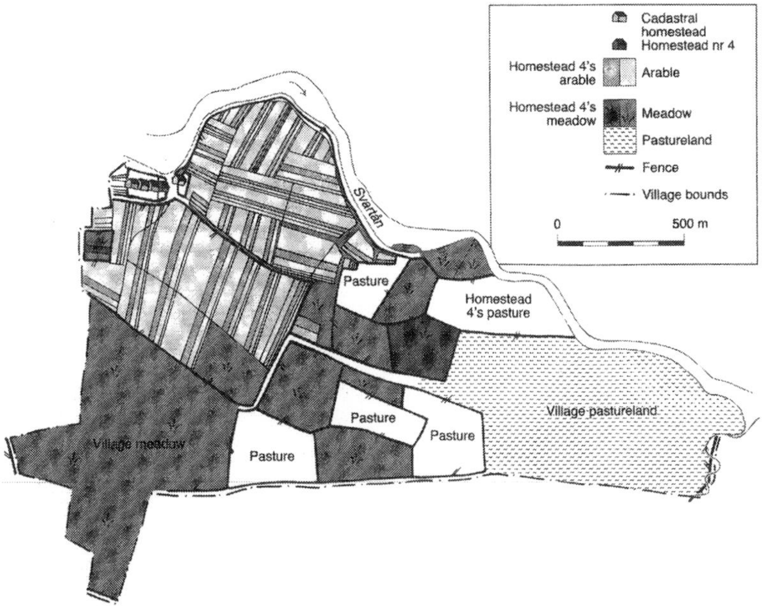

This compilation of a surveyor's map shows the hamlet's arable, meadows and pastureland. The regularity of the strips, ordered according to the solskifte principle are striking when compared to the apparent randomness that characterizes the fields of Åsle in Western Sweden (figure 3.5). In reality, the number of strips was greater than the map indicates, since the surveyor has mapped the strips of three of the homesteads as continuous pieces of land

Source: Gadd, 2000: 121; Gadd, 2011: 127.

The preparation for the sowing of barley started with autumn ploughing, prior to sowing the following spring, and thus left the fallow grassy and useful for grazing for the whole summer. The preparation for winter rye, in contrast, included three successive ploughings in late spring and summer before sowing in August. The result was a more or less grass-less, 'brown' fallow (Myrdal and Söderberg, 1991: 417; Myrdal, 1999: 238–39, 291; Gadd, 2000: 143).

As we have seen, rye cultivation in southern Scandinavia generally stimulated a transition to a three-field system. Why, then, was the two-field system retained in east-central Sweden? The answer is, as explained by a Swedish agronomist in the 1840s, that the growing season at these northern latitudes is so short that it was not possible to prepare the soil for new sowing without an intervening fallow year (Nathorst, 1847: 185–86; Gadd, 1983: 214).

A radically different form of two-field system was found south of Lake Vänern in western Sweden. This was adapted to spring-sown crops, oats and barley, which were the kinds of corn that predominated in the western parts of the Scandinavian peninsula. Autumn grain was sown, but only in small quantities (*c.* 6 per cent of the seed sown in this area 1800; Gadd, 1983: 324) and so late in the season that the grain crops of the current year had already been gathered in. There was thus little need to protect a standing crop in early autumn, and this helps to explain why the fencing system remained very simple in this region: in essence, just one single, long fence divided all the land of a village – both arable and grassland – into two parts. The grasslands on the same side of the fence as the current year's fallow would serve as grazing ground, the grasslands on the same side as the cropped field would serve as meadow. In late autumn, when rye was to be sown, the grazing animals were simply moved from the fallow-and-grazing side of the fence to the side that contained the stubble after the newly cropped harvest, and the meadows that had been mowed for hay. Since the grassland served as summer grazing the one year, and as meadow the other, the boundary between infields and outlying land – normally an important line in Scandinavian open-field systems – had effectively been dissolved (Gadd, 1983: 206; Gadd, 2005: 70; Höglin, 1994: 25).

Inter-commoning

One way to reduce the burden of constructing and repairing fences around and between different parts of the infields was to adopt the system of inter-commoning (Da.: *vangelag;* Sw.: *gärdeslag, vångalag*). Two or more villages or hamlets with arable or meadow fields lying adjacent coordinated their use – fallowing the same year, sowing spring grain the same year, etc. – so that they did not need to maintain fences between them. This led to the development of large fenced areas that contained the arable land belonging to several different settlements: a single village could in this way come to collaborate with two, three or more neighbouring villages. In the most cultivated parts of Scandinavia – the Danish islands and the Skåne plains – as many as nine villages might have shares in the same *vangelag* (Frandsen, 1983b: 308). North of the Scanian plains inter-commoning was to be found both in continuous-cropping and in two- and three-field regions, although inter-commoning in the Central-Swedish two- and three-field regions rarely included fields from as many villages or hamlets as was usual in the Danish and Scanian plains (Lindgren, 1939: 87–88; Helmfrid, 1962: 128–32). Once established, inter-commoning had a preservative effect on the field system, since the established cooperation between several villages made any significant changes in agricultural practices extremely difficult to implement. This seems to be one of the main reasons why, in the period after the Middle Ages, the boundaries between areas characterized by different kinds of field system displayed a remarkable degree of stability, even in cases where natural conditions were similar on both sides of the boundary (Frandsen, 1983a: 268; Frandsen, 1983b: 308, 311–12).

Systems of strip distribution: regular systems

In the two regions where arable farming had the greatest economic importance, the Danish-Scanian area and the eastern part of the Central Swedish plains, two systems for regular strip division were found, namely *bolskifte* (*bol* division) and *solskifte* (sun-division). In both cases the existence of land assessment units (*bol*, *attung* and *markland*) was fundamental. These units were measures of wealth, not of area. In principle they were similar to the Roman *mansus* or the English hide (Ericsson, 2012: 23–24, 343–51). *Bol* division was widespread in Denmark, especially in Sealand, and on the Skåne plains. In this system, each village was divided into a number of *bol* which in broad terms corresponded with its size. A farm's rights in the fields, the meadows and the resources of the outlying land were determined by its bol value. Lilla Uppåkra, shown in Figure 3.3, with its eleven peasant-farmers, was divided into 4½ bol. In seventeenth century Denmark, however, villages of about 10 bol, distributed amongst twenty cadastral farms,[6] were not unusual. As a rule, all full bol units of a village held parcels of equal size in each furlong of the arable, and thus the same proportion of the village's total arable land. Each bol was often made up of farms of different size (a half, a quarter or an eighth of a bol): a farm constituting a quarter of a bol would thus hold a quarter of all the parcels of land belonging to that bol. Such size differences are likely to have existed at the time that the system of bol division was first established. Furthermore, farms often consisted of parts of more than one bol, and would therefore hold more than one strip in each furlong. The bol thus seems to have been created as a reference unit in an allocation system that was internal to the village, and that facilitated an equitable distribution of arable land – relative to tax or rent demands – between farms of different sizes (Porsmose, 1988: 235–36; Nordholm, 1967: 90; Sporrong, 1973: 10–14; Frandsen, 1983a: 59–61; Hahnemann, 1997; Riddersporre, 2008).

Sun-division (*solskifte*) was characterized by an extremely regular division of the arable fields. Each homestead in the village or hamlet had one strip in each furlong, and these always lay in the same sequence. The word '*sol*' (sun) in '*solskifte*' alludes to a way of expressing the relative position of the strips in terms of points of the compass, where the south and east was 'closer to the sun', north and west was 'more distant from the sun' (Ericsson, 2012: 241–42). Sun-division existed in two main forms, one predominant in east-central and south-east Sweden, the other in Denmark and Skåne. The former mirrored an ownership structure in which the homesteads of each hamlet were usually of different sizes and belonged to owners of different social categories (viz. freeholders, noblemen, the Church or, especially after the sixteenth-century Reformation, the Crown). Typically, the system included the regulation of house-plots, the width of each reflecting that of the associated strips in each furlong, and thus its total arable area, as well as its share in the hamlet's meadow and outlying land (Sporrong, 1994: 33). The system was closely associated with the village assessment units, called the *attung* or the *markland*, depending on the region. For example, a hamlet in the province of Östergötland in *c.* 1640 would often consist of four homesteads of about two *attung* each, giving a total assessment of eight *attung* (Ericsson, 2007a: 29–45). An expert on the Swedish sun-division system has described it as 'a technical solution to an equity problem' which created 'order and a comprehensive view' of the total land assets of the hamlet in question. It appears that sun-division on this east-central Swedish model, i.e. taking account of the size differences between the farms, was more difficult to implement in comparatively large hamlets or villages. Such agglomerations, which often consisted of around ten homesteads, were generally divided into two subgroups (sing. *holme,* pl. *holmar*), whose arable was divided into relatively large blocks and systematically distributed over the village fields. Thereafter each block was divided and allocated to the farms in the respective subgroups according to the rules of sun-division. This two-step procedure broadly resembles the system of *bol* division (Ericsson, 2012: 242, 245–47; Hemfrid, 1962: 154–60).

[6] The Danish *matrikel* of the 1680s show 'farms' (*gårde*). Apparently, these were cadastral farms since they were often subdivided between a number of different holders (*brugere*). Cf. Frandsen, 1983a: 49, 50; 1983b: 296, 298).

The second type of sun-division, found in Denmark and Skåne, was associated with villages belonging to a single owner: here the land was divided between tenant farms in equally sized holdings. The basic principle of having a regular sequence of holdings, recurring in each furlong, was the same as in east-central Sweden. Where the Danish-Scanian system was different was that in each furlong the strips were all of equal width. While the east-Swedish form of sun-division reflected differences in total landholding that had existed between the farms before the system had been established, the Danish-Scanian 'equalizing' sun-divisions had evidently made a clean sweep of all differences in farm size that may have existed before the new pattern was initiated (Hahnemann, 2007).

How old were the *bol* and sun-wise subdivision systems? We know that the assessment units themselves – the *bol*, *attung* and *markland* – existed before the regular subdivisions of the arable. The *bol* seems to go back to around the year 1000, the *attung* to about 1100, and the *markland* to the thirteenth century (Christensen, 1983: 27; Ericsson 2012: 335; Myrdal, 2008: 130). In all three cases their primary purpose seem to have been the assessment of farms, either for the distribution of collective taxes to the Crown, or for the levy of rent from tenants to landowners, or both. Relatively soon these assessment systems proved useful in other ways, especially when it came to expressing owned shares of a hamlet or village. Both freeholders and large-scale landowners benefited from the systematic assessment of the land, for it facilitated the execution of land transfer by purchase or inheritance (Sporrong, 2008b: 242–47).

The evidence suggests that the first stages in the development of these assessment systems was strongly linked with significant socio-economic changes that occurred during the Middle Ages. Large farms, which in the tenth and eleventh centuries had been worked either with slaves or using other forms of dependent labour, were, in the period *c.* 1150–1350, transformed into small manors. Their demesnes were farmed using labour rents extracted both from semi-free ('serf') or otherwise dependent smallholders, many of them the descendants of former slaves; and from personally free larger-scale tenant farmers. There also existed, especially in Sweden, a stratum of freehold farmers who, in their capacity of manning warships and paying taxes, were patronized by royal power. With the late-medieval population decline the socio-economic structure changed further, as the number of demesnes declined and the crofter-like semi-free smallholders more or less disappeared, to be replaced by a class of personally free tenant farmers (Porsmose, 1988: 264, 311–12; Ericsson, 2007: 113; Widgren, 2014).

Both the *bolskifte* and the east-Swedish *solskifte* appear to have played a key role when, as a result of these social processes, farmland was, from the thirteenth century, divided up between peasant farms. This applies both to the internal distribution between tenant farms, and to the settlement of boundaries between tenant farms (owned by nobility, the Church or – later on – the Crown), and freeholds. It is doubtful, however, whether these systems of land division always coincided with major changes in manorial structure, since the latter seem to have occurred mainly before 1500. It has been pointed out that the Skåne regional law (drawn up *c.* 1210) mentions fields in which parcels seem to be block-shaped (have a *blockstruktur*) rather than being long and narrow. Similar observations have been made in a study of deeds of transfer (*diplom*) from Uppland (east-central Sweden) that were drawn up as late as in the fifteenth century (Frandsen 1983a: 9; Rahmqvist 2013; Sporrong, 2008a: 245). A significant proportion of the sun-divisions systems in east-central

Figure 3.5 The arable fields of Åsle, a village in the three-field zone of eastern Falbygden, *c.* 1640

Note: the house symbols indicate cadastral homesteads.
Source: Sporrong, 1970: 28.

Sweden thus seem to have been established as late as the sixteenth century, when they replaced a structure of block-shaped parcels (Sporrong, 2008a: 17; Sporrong, 2008b: 247). This relatively late dating of the east-Swedish sun-divisions is important because it shows that they were introduced in a society where the two-field system and nucleated villages or hamlets were already in existence. The Danish, 'equalizing' sun-divisions also seem to have been introduced at a relatively late date, extending from the late Middle Ages into the seventeenth century (Porsmose, 1988: 358–59).

Systems of strip distribution: irregular strip divisions in western Sweden

Both the continuous-cropping regions and the two- and three-field systems of western Sweden were characterized by a much more erratic layout of holdings (Figure 3.5). There was no visible order to the distribution of strips and individual farms often had significantly more arable in one of the fields than in the others (Sporrong, 1973: 19; Widgren, 1999: 49; Gadd 2000: 120; Bonow, 2005: 66–85; Bonow 2008). These patterns of division give the impression of immaturity, of having remained at an earlier stage of development than the bol- and sun-divisions just discussed (cf. Sporrong, 2008a: 7; Sporrong, 2008b: 251), although their *chronological* age is not necessarily any greater. Perhaps the main immediate reason for the lack of systematic distribution of strips was the lack of basis for land distribution, since assessment units of the *bol* or *attung* type were absent from the regions in question. This in turn was the result of taxes being more individualized in character – borne by individual farms or homesteads, rather than by communities – at an early stage (cf. Ericsson 2012: 339). It is also possible that the high importance of livestock production in western Sweden (see, e.g., Myrdal and Söderberg, 1991: 126–32, 179) made significant reforms of the distribution of the arable within field systems less important for peasant farmers and other landowners (Sporrong, 1973: 21; Gadd, 2000: 120).

Field subdivisions in continuous cropping systems

A characteristic feature of Swedish hamlets and villages with fields in continuous cropping was that the division of arable into strips was seldom, if ever, as extensive as in two- and three-field areas. True, in the relatively large west-Swedish continuous-cropping village of Karleby (Figure 3.6), with its twenty-four homesteads, the average number of parcels per homestead was as many as eleven, but they often lay so close to one another, near to the holder's house-plot, that in many cases they were close to forming coherent pieces of arable. A similar picture of relatively few strips per homestead in the continuous-cropping area is presented in studies of northern Småland (Vestbö, 1996: 31–38; Vestbö-Franzén, 2004: 90–92; Vestbö-Franzén, 2008: 196–97). This said, some Scandinavian villages and hamlets could be found where, notwithstanding their continuous-cropping system, land was more minutely divided. Examples existed in parts of Bohuslän, which is today a West-Swedish province but which until 1658 belonged to Norway. In a particularly well researched village, surveyed in 1728, eleven homesteads had their arable distributed over a total of 200 strips, i.e. there were eighteen per homestead, none of which lay adjacent or in close proximity in the manner described at Karleby. The reason for this relatively extensive strip division is debated, but an important factor was probably the character of the Norwegian inheritance system (Widgren, 1997b: 9–10, 39–42, 116–18).

The permanence of strip configurations

In many cases, the particularly rich character of the Swedish source material, in the form of surveyors' maps, allows us to track the configuration of strips within particular field systems over time. In some cases, we can follow a village or hamlet for about 200 years before the strip-consolidation reforms of the nineteenth century, similar in many respects to the enclosures in eighteenth- and nineteenth-century England, radically changed the picture. An examination of the pattern and configurations of scattered strips during the period *c.* 1640–1830 has been made for a region with a predominance of irregular strip

Figure 3.6 The arable fields of Karleby, one of the largest villages in the continuous-cropping zone of western Falbygden, *c.* 1640

The parcels belonging to homesteads nr 2 (upper part of map) and 23 (lower part of map) are shaded. The house symbols indicate cadastral homesteads.
Source: Sporrong, 1970: 26.

division in western Sweden. It is interesting to observe that virtually no changes occurred during this long period, nor were there any transfers of ownership of single strips between homesteads (Bonow, 2005: 72–84, 172). This is likely to have been a general pattern, shared by most of Sweden. An important reason for such apparent stasis was a law that prevailed from the sixteenth century, forbidding the transfer of land from one cadastral homestead to another (Bäärnhielm, 1995: 25). It was possible to divide a farm among heirs (although it was, at least in the eighteenth century, more common for one heir to buy the others out); it was also possible to sell and buy parts of homesteads. When a peasant-farmer bought, for example, a quarter or an eighth of a farmstead he took over the corresponding shares of each strip in the fields and meadows that belonged to it, and the corresponding rights in the outlying land. Division of farms therefore, in practice, increased the number of strips in total (Gadd, 2000: 119).

The transition from continuous cropping to the three-field system: some examples from seventeenth-century Sweden

On the whole, there are few documented examples in either Denmark or Sweden of a transition from one kind of field system to another during the period 1500–1800 (Myrdal, 1999: 290; Gadd, 2000: 128–29; Frandsen, 1983a: 257). One important exception, however, is the changeover from continuous cropping to a three- or sometimes a four-course rotation that occurred in the northern and north-eastern parts of the south-Swedish highlands. This process is interesting, since it involved a transition from a system characterized by single farms and small agglomerations with fields that show a low degree of strip division, to one with larger settlement agglomerations and more extensive sub-division. In some cases the sources actually show us how this transition was accomplished.

Ådel Vestbö-Franzén (2004) studied the transition from continuous cropping to a three-field system that took place in the north of Jönköping County, a territory extending over some 11,500 square kilometre in the northern part of the Southern Swedish Highlands. To a large extent, the shift in field system occurred in the period between 1550 and 1630. The environment of Jönköping County was mainly woodland, but this characteristic was gradually weakening due to continuous deforestation. As a result, the cultivation of autumn-sown rye gradually had to be moved from swiddens to the permanent arable of the infields. Previously, only spring grain had been sown there, which is why it had been possible to use the infields' total area for grazing after harvest. But when autumn-sown rye began to be grown there, the need to protect the crop with fences encouraged the adoption of a three-field system (Vestbö-Franzén, 2004: 150, 164, 226). This involved the clearing of new areas of permanent arable, a change from more-or-less rectangular parcels to longer, narrower and smaller strips, and a process of settlement agglomeration, as homestead sites were gathered into larger hamlets or villages. In the seventeenth-century examples, the transition often began with the introduction of a three-year rotation in the separately fenced infields of isolated homesteads (Vestbö-Franzén, 2004: 80–82). The establishment of a regular three-field system involved several previously isolated homesteads co-operating in the cultivation of a new common field, in the establishment of an accompanying fencing system, and in the relocation of the previously dispersed homesteads to a new hamlet site. Of particular interest are a number of maps which were made when such a transition was actually in progress. These examples also show, however, that the establishment of a new field system and the relocation of buildings to a new site only demanded to a limited extent simultaneous action on the part of several farmers. The introduction of the three-field system was often gradual, in the sense that a farm would, during a transitional period, have some of its arable within the new fields and at the same time retain some of its old, enclosed continuously-cropped fields. Neither did the relocation of buildings to the new settlement site necessarily take place simultaneously. In individual cases such a move could await, for example, a change of head of household (Vestbö, 1996: 76; Vestbö-Franzén, 2004: 82). In Jönköping county it was mostly in areas that were originally characterized by small hamlets and isolated farms that the transition to the three-field system took place. Larger hamlets and villages often retained their continuous-cropping system. Evidently, the change was easier to implement when the number of participants was small (Vestbö-Franzén, 2004: 235–36).

Another important study based on the evidence of land surveyor's maps from the 1630s and 1640s is that by Gunnar Lindgren (1939). Lindgren studied Falbygden, a *c.* 1,100 square kilometer plateau region in central Västergötland which, in the 1640s, might be characterized as a transitional area between plains and woodland (Lindgren, 1939: 61). In the 1640s, the Falbygden region was divided between a three-field zone in the east and a continuous-cropping zone in the west. The dividing-line between the two was sharp (the distance between the two villages Åsle and Karleby in figures 3.5 and 3.6, with their very different strip configurations, is a mere 4 km., as the crow flies), although it did not follow any natural-geographic boundaries. Neither geology nor soil showed any obvious differences between the two zones (Lindgren, 1939: 93). In the three-field zone most of the meadow was fenced off from the arable fields, thus forming a fourth fenced area. In the continuous-cropping zone, on the other hand, all the arable and most of the meadow lay within the same ring fence. When we first meet Falbygden in the survey maps of the 1630s and 1640s the arable fields of both zones were, as a rule, part of inter-commoning systems. To the east this was of the relatively complicated kind in which each village, because of the three-field system, collaborated on fences with up to three other groups of villages. In the west, inter-commoning was of the more uncomplicated kind typical of continuous cropping districts, in which a large area containing meadow and scattered areas of arable was fenced collectively by several farms and hamlets. In many cases these inter-commoning areas were very large, and included up to a hundred homesteads, sometimes more (Lindgren, 1939: 82–83). Lindgren considered that the situation in *c.* 1640, as depicted on the various maps examined, was the result of the three-field system having advanced westwards in previous centuries, at the expense of the continuous-cropping system. Later research had confirmed his views on this point: but while Lindgren thought that the expansion had been a feature of the late-medieval period, more recent work suggests that it may have taken place after 1500, even as late as *c.* 1600 (Widgren, 1995: 86; Jansson, 1998: 223; Vestbö-Franzén, 2004: 21). The settlement pattern within the two zones of Falbygden displayed marked differences. The average size of the agglomerations (villages and hamlets) was much larger in the three-field than in the continuous-cropping zone. The main reason for this difference, according to Lindgren, was that the need for co-operation over the management of fields and, in particular, fences encouraged larger concentrations of farms. Farms had been relocated to villages placed at the point where its three fields met, in order to minimize the costs of travel and transport (Lindgren, 1939: 106; Höglin, 1994: 16).

The arable holding of each homestead in the Falbygden three-field zone was 30 per cent larger than in the continuous-cropping zone. Accordingly, if all arable except the fallow fields was sown, the sown area per homestead was of equal size in the two parts of Falbygden. Evidently, when the three-field system had been introduced, more land had been brought into cultivation, in order to compensate for the reduction in sown area resulting from the adoption of a regular fallow. As a result, the area used as pasture was smaller in the three-field zone. The transition from continuous cropping to the three-field system thus involved a change in the main focus of farming from stock breeding to arable farming (Lindgren, 1939: 74, 94). Lindgren sketches a rough outline of how the transition was accomplished. In the new dispensation, the arable would be divided into three fields, but because each homestead had formerly possessed the relatively coherent pieces of farmland typical of the continuous-cropping system, several homesteads would have found all their arable in only one of the new fields. Since the peasants in question could not let all of their farmland lie fallow every third year, a redistribution of land had to be undertaken. This could conveniently be made by dividing the parcels belonging to the homesteads in question into three parts, two of which were exchanged for equally-sized pieces belonging to homesteads with their arable mainly situated in the other fields. A procedure like this would provide an explanation for both the shape of strips and their distribution across the three fields (Lindgren, 1939: 91–92). To be sure, this model can be discussed and perhaps improved on several points, but what seems especially worthy of note is the rather un-bureaucratic process that is proposed: peasants exchange land with one another in a way that implies little overall planning, other than the creation of the three fields themselves.

After the 1640s, field systems in the Falbygden area changed remarkably little until the old agricultural structure was entirely dissolved as a consequence of the radical consolidations of strips ('enclosures'; *enskifte* and *laga skifte*) that occurred in the nineteenth century. What did happen in the intervening period was that within the continuous-cropping fields of western Falbygden the system of fallowing developed in several ways. As a response to land reclamation, the area occupied by fallows increased from 5 per cent in the 1640s to about 25 per cent by the early nineteenth century in the most cultivated continuous-cropping parishes. In addition, fallowing was made more regular: in the 1640s the leys had been left untilled for an unspecified number of years before they were brought into cultivation again, but by 1800 one-year fallowing was practised every fourth, fifth or sixth year. The strips lying fallow were tilled several times each summer with ard and harrow in order to keep them clean of weeds, and they were manured before being sown with barley. In other words, a system of regular fallowing had developed within the framework of the continuous-cropping system, that is, without separately fenced fields. Such a development was not unique to Falbygden; something similar has been noted in a continuous-cropping zone in the interior of Skåne (Gadd, 1983: 208–10; Gadd 2005: 66–67; Dahl, 1989: 100–01).

Considering the great changes in field systems that had taken place before the early seventeenth century, it is remarkable that no major alterations in their basic structure occurred during the period 1640–1800, with the exception of some continuation in the westward movement of the boundary between continuous cropping and the three-field system. As we have seen, already in the 1640s large inter-commoning systems existed on both sides of the boundary. Since inter-commoning, as a rule, impeded significant changes in field systems, this appears to be an important part of the explanation behind the small degree of change in the continuous-cropping area.

The overall picture of the transition to the three-field system in northern Småland and Falbygden shows both similarities and contradictions. In both areas, the transition involved a change from relatively few and large parcels to more numerous, smaller and narrower strips. In northern Småland, however, the transition occurred mainly in areas of isolated homesteads or small settlement agglomerations which, even after the transition, only counted three or four homesteads: it seems to have been more difficult to implement in hamlets covering, e.g., six or eight homesteads. In eastern Falbygden, in contrast, the transition took place in areas which, judging by the size of the villages in 1640, covered 20 homesteads or more. Hopefully, future research will provide a clearer answer to the question of what made the transition to the three-field system in eastern Falbygden possible.

III. Changes within open-field systems

The consolidation of strips

From the 1750s a number of laws were passed in Sweden which were intended to reduce the number of strips in the open fields. The legislation on *storskifte* of 1757 decreed that if a landowner in a village or hamlet applied for reform, surveyors would be called in who would bring about a consolidation of strips that would apply to the entire village. The existing division of the arable into jointly fenced fields was not changed, which makes the *storskifte* a reform within the framework of existing open-field systems. The *storskifte* strip consolidations had a particularly large impact in the plains of Skåne and east-central Sweden (Olai, 1987; Helmfrid, 1961; Bäck, 1984: 191–92). On the Skåne plains the number of strips fell from about 50 per farm to between 10 and 15; in one particularly well-researched parish in east-central Sweden, the number of strips per farm declined from fifty to between five and six (Gadd, 2000: 272–82; Olai, 1983: 68–73). It was mostly freehold peasant-farmers who applied for reform.[7]

Seeing that Swedish peasants, in the Middle Ages and the sixteenth century, took part in, and readily accepted, reforms that involved more extensive strip subdivision it is perhaps surprising that they, on the whole, supported the *storskifte* strip consolidation reforms after *c.* 1750. We might note a number of factors which may help explain this change in attitude. Firstly, the average size of strips had continued to decline as a result of the subdivision of farms which took place in the centuries before 1750. In addition, the number of strips per farm tended to increase due to the substantial land reclamation that characterized Sweden in this period.[8] While, as a rule, reclamation before the middle

[7] A study of 400 storskifte commissions, mostly concerning villages or hamlets, shows that peasant-farmers applied for storskifte in two thirds of the cases, persons of standing in the remaining third (Olai, 1987: 48, 73; Gadd, 2011: 151).

[8] In the eighteenth century the arable area increased by more than 70 per cent, during the period 1700–1870 by more than 300 per cent (Gadd, 2000: 232, 379: Gadd, 2011: 158–59).

of the eighteenth century had led to subdivision of farms, after 1750 it more often led to an increase in the area of arable per farm (Gadd, 2000: 233–34). There was probably a limit to how extensive strip division could be, and still remain accepted as comfortable, and it is conceivable that Swedish farms had approached or exceeded this limit by the eighteenth century. Secondly, an acceptance of consolidation may have been encouraged by technical changes that ensured the more rapid completion of routine agricultural tasks carried out on each strip, making the time spent traveling between strips relatively more important. In particular, the scythe had become the main harvesting implement on the Swedish plains between *c.* 1600 and 1750, doubling the speed of harvesting (Gadd, 2011: 145–46).

New crops

In the seventeenth century it became more common to sow peas and, if the local climate permitted, beans within temporary enclosures (large enough to contain the strips of all the village members) which were fenced in from the fallow: flax and, after 1800, potatoes might also be cultivated in this manner. As a result, in the eighteenth century the area actually fallowed was, in general, much smaller than the 50 or 33 per cent of the arable that was traditional in two- or three-field systems (Myrdal, 1999: 290–91; Gadd, 1998: 191). In the nineteenth century it also became usual to cultivate clover and grasses for animal fodder within such temporary enclosures from the fallow. This practice was found not least in the central-Swedish two-field areas. The one-year period of cultivation was hardly ideal for sown grasses, since a sown ley will produce for three or four years after one sowing, but it was difficult to introduce significantly longer crop rotations in open-field villages. Among other things, any attempt to do so would have involved in extraordinarily difficult negotiations between the peasant-farmers of the village. Clearly, there was a limit to the adaptability of the open-field system. The desire to introduce crop rotations better adapted to the cultivation of grasses and clover became an important reason for implementing the more radical strip-consolidation ('enclosure') reforms (*enskifte* and *laga skifte*) of the nineteenth century, which resulted in a structure of dispersed farms sited on coherent blocks of land (Gadd, 2000: 207–08; Gadd, 2011: 152–54).

IV. Did tenures and other dimensions of land-holding impact on field division?

While the 'equalizing' sun-division systems found on Danish manors were no doubt the result of top-down decisions, the sun-divisions found in east-central Sweden do not show any connection with the predominance of nobility-owned land, and it is in general not possible to identify any particular power factor behind their implementation. Sun-divisions were as common in regions in which freeholders predominated as they were in those with many noble landowners (Sporrong, 1985: 190; 212; Sporrong, 2008 a:16). Furthermore, the actual implementation of a system of sun-division does not seem to have required any great degree of specialized knowledge. A document that certifies the implementation of a *solskifte* in 1414 shows that the division was carried out under the guidance of a deputy district judge who worked with twelve men from the local peasantry. Special surveyors were thus not needed (Ericsson, 2012: 242–44). As regards field systems more generally, there are no signs that these developed as a result of top-down decisions. In Denmark, where a single magnate sometimes owned several villages, it has been shown that no connection existed between the extent of the manors and the character of field systems. Rather, the peasant farmers themselves had adapted their farming methods to the local natural circumstances (Frandsen, 1983a: 255). Similarly, in Jönköping county in northern Småland – where noble landowners owned about 40 per cent of the cultivated land by 1700 (Gadd 2000: 43; Forsell, 1834: 8) – there is no evidence that they exerted a significant influence on the field systems used on their tenant farms. Each noble landowner, as a rule, had tenant farms dispersed over a large area, while demesnes were relatively small and unimportant. Most of the tenant farms lay in villages or hamlets that also contained farms owned by other social categories, the most important among them being freehold peasants. The nobility's tenants used the same field system as the other peasant-farmers of the village in which they lived. In cases where the field system was changed from continuous cropping to the three-field system it was, to all appearances, the local peasant-farmers who took the decision (Vestbö, 1996:77–78; Vestbö-Franzén 2004: 213, 223).

V. Why were fields subdivided into strips?

All seventeenth-century Scandinavian field systems treated here were – if two or more farmers had their arable in a commonly fenced field – open-field systems, according to the definition made earlier in this publication. As we have seen, subdivision into strips was most extensive (in terms of the number of strips per holding) in the Danish and Scanian plains; somewhat less so in the central-Swedish two-and three-field regions; and probably least extensive in the Swedish continuous-cropping regions. In previously-Norwegian Bohuslän, on the other hand, in locations where relatively large hamlets had developed in continuous-cropping regions, their fields seem to have been more subdivided than in Swedish hamlets practising the same cropping system (Widgren, 1997b).

It has sometimes been suggested that the division into strips would have facilitated co-aration, that is, when peasants co-operated through forming teams of oxen that belonged to a number of different owners. The idea has been criticized on the basis that even where joint ploughing has appeared in connection with subdivided fields, there is no proof that the two are causally connected (Dodgshon, 1980: 30–34). Here we can add the observation that in Scandinavia subdivided fields existed without joint ploughing, as a rule, being present at all. Most Swedish and Danish peasant-farmers seem to have owned the beasts of burden necessary for their own current needs (Gadd, 2000: 168–70; Dombernowsky, 1988: 377). An important part of the reason for the large number of draught animals relative to the extent of the arable is the short growing season, which in practice meant that each farmer had to command sufficient animals to till his own fields (Gadd, 2000: 170; Gadd, 2011, 158).

McCloskey has argued that by holding strips or parcels in different parts of the village farmland, the individual peasant reduced his risk of total crop failure, because crop yields usually varied more across the village's total acreage than they would have done across a single farm with consolidated arable (McCloskey 1975a; McCloskey 1975b). As Fenoaltea has pointed out, moreover, across a large area the soil becomes ready for tilling and sowing and crops ripen at somewhat different times during a spring or autumn. By having strips located in different parts of the village's farmland the individual farmer and his household could extend the period of productive agricultural work (Fenoaltea, 1976; Fenoaltea, 1977). Both these theories seem to have an explanatory value. But how vital was the subdivision into strips in achieving these assumed benefits? Were they first-hand reasons for such subdivision, or were they relatively unimportant advantages that helped the farmers to accept or even appreciate a subdivision that had been carried out for other reasons?

In this context, I would like to draw attention to another observation made by McCloskey: that the cost for pre-industrial peasants of subdivided fields was in fact relatively small. The disadvantage most usually highlighted by scholars is the time which would have been wasted in moving from one strip to another performing farming tasks. However, as McCloskey notes, often a strip represented a day's work or more, so that in most cases the only travel required was from village to strip and back again (McCloskey, 1975a: 79–80). Even in cases where more than one strip was worked during a day, they often lay in relative proximity to each other – the result of the tendency of land in the same part of village fields to become ready to be worked at about the same time. In view of this, it appears that we may not need to look for any long-run advantages arising from subdivision in order to explain the existence of subdivided fields. The reason for more extensive subdivision may be sought in the temporary advantage of facilitating redistribution of land connected with the transition to the new field system, or when land possession among farms of a village needed to be made more equitable in relation to rent and tax burdens. If the negative effects of subdivision were small, it should have made little difference to the peasant if he had thirty smaller parcels instead of ten larger ones.

In this context, we might profitably look again at the research carried out into the transition from continuous cropping to a three-field system in Falbygden and northern Småland. Here, the increased subdivision of land into strips which occurred at the point of transition seems to have been primarily a result of the reallocation of the arable over the three new fields. Evidently, subdivision was not regarded as a drawback, at least not one big enough to motivate actions to reduce it. If, on the other hand, sub-division in itself had benefits, these were not important enough for the peasant-farmers of nearby continuous-cropping villages or hamlets to reduce the size of strips, increase their number and spread them more evenly over the arable.

When peasants in Denmark and central Sweden took part in the establishment of bol- and sun-division systems an important aim seems to have been to create equity between, on the one hand, the pressure of tax and rent, and on the other the size of landholdings. The result was a more extensive subdivision into strips (cf. Sporrong 2008a: 17–21; Sporrong, 2008b: 245–46). This subdivision appears to have taken place within existing two-and three-field systems. Thus, it does not seem to have been the result of a change in the character of the field system itself, as appears to have been the case in Falbygden and northern Småland. We do not know whether the fragmentation of land was seen as an advantage in itself but, if it was, this advantage does not seem to have been very strong, considering that within the sun-division zone some hamlets never adopted this reform and instead retained their larger, block-shaped parcels.

Evidently, the factors that gave rise to more extensive subdivision were not always the same. In the regions of northern Småland and Falbygden such a change seems to have been the result of the transition from annual cropping to a system of regulated fallowing and separately fenced fields; elsewhere, it was the consequence of a desire for increased taxpaying equity among peasants. In the Middle Ages, and throughout most of the early-modern period, the resulting subdivision seems to have been tolerated, maybe even appreciated, but it appears to have required an additional, triggering or driving factor to bring it about.

As we have seen, judging by their ready acceptance of the *storskifte* consolidation of strips, it seems that Swedish peasant farmers had acquired a more negative attitude to the subdivision of fields by the middle of the eighteenth century. Some possible reasons for the change in attitude has already been mentioned (see pp. 67-68). To this we might add a number of further factors, relating to important social and economic developments. If strip subdivision had indeed been accepted because of risk minimization or because it extended the work seasons for individual households, then the importance of these advantages should have decreased with time. With progressively better communications, cereals could be transported at lower cost from surplus areas to deficit areas. In Sweden an increase in production per farm, and a stagnant tax burden after *c.* 1700, served to increase the peasant-farmers' savings (or rather, hoarding; Gadd, 2000: 64, 195–97, 342–47). This hoarded wealth could, if the need arose, be exchanged for grain. Furthermore, the increase, especially after *c.* 1750, in the relative number of landless and semi-landless workers made the labour supply for peasant-farmers more flexible. Instead of being dependent on a steady workforce that successively worked the strips belonging to the household, farmers could now hire temporary workers during the harvest season or at other bottlenecks in the farming year, individuals who could work for a number of different employers, depending on the need.

Another factor to be considered is the availability of professional and essentially impartial land surveyors (Bäck, 1992; Gadd, 2000: 281–82, 295). These were able to grade land, according to soil quality, piece by piece, before redistribution: farmers who received land of poorer quality could be compensated by receiving it in larger quantities. The services of professional land surveyors had not been available to medieval and sixteenth-century peasants. The simplest way of attaining a more equitable land distribution, or of distributing arable holdings over several fields, had thus been to divide it up into numerous, smaller strips, using the relatively simple administrative and surveying methods that were then available. The grading of the land carried out by the eighteenth-century land surveyors involved a solution to a transaction-cost problem that had not been available to medieval or most early-modern peasants.

Bibliography

Bäärnhielm, M. (1995), 'Vad lagboken berättar' in M. Widgren ed., *Äganderätten i lantbrukets historia*, Stockholm, pp. 17-36.

Bäck, K. (1984), *Bondeopposition och bondeinflytande under frihetstiden: centralmakten och östgötaböndernas reaktioner i näringspolitiska frågor*, Stockholm.

Bäck, K. (1992), *Början till slutet: laga skiftet och torpbebyggelsen i Östergötland 1827–65*, Borensberg.

Bonow, M. (2005), *Gård, gräns, giftermål: familjestrategiers betydelse för markens och landskapets utformning i Norra Åsarps socken, Västergötland ca 1640–1880*, Stockholm.

Bonow, : (2008), 'Det fria stångskiftet i Västergötland' in: A. Ericsson ed., *Jordvärderingssystem från medeltiden till 1600-talet*, Stockholm, pp. 203–218.

Boserup, E. (1965), *The conditions of agricultural growth: the economics of agrarian change under population pressure*, London.

Campbell, Å. (1928), *Skånska bygder under förra hälften av 1700-talet: etnografisk studie över den skånska allmogens äldre odlingar, hägnader och byggnader*, Uppsala.

Campbell, Å. (1933), 'Vångalaget i Skåne. En boskapsskötselns samfällighet', *Skånes hembygdsförbunds årsbok*, Lund.

Christensen, C.-A. (1983), 'Begrebet bol. Et vitnesbyrd om vikingetidens storbondesamfund', *Historisk Tidsskrift*, 4, pp. 1–33.

Dahl, S. (1942), *Torna och Bara: studier i Skånes bebyggelse- och näringsgeografi före 1860,* Lund.

Dahl, S. (1989), *Studier i äldre skånska odlingssystem*, Stockholm.

Dombernowsky, L (1988), 'Ca. 1720–1810' in: C. Bjørn ed., *Det danske landbrugs historie. 2, 1536–1810*, Odense, pp. 211–394.

Ericsson, A. (2007), *Attungen–ett medeltida fastighetsmått: en agrarhistorisk undersökning baserad på attungsbelägg i SDhk till år 1376 och Folke Dovrings kasuistik*, Uppsala.

Ericsson, A. (2012), *Terra mediaevalis: jordvärderingssystem i medeltidens Sverige,* Uppsala.

Fenoaltea, S. (1976), 'Risk, transaction costs, and the organization of medieval agriculture' *Explorations in Economic History*, 13, 2, pp. 129–51.

Fenoaltea, S. (1977), 'Fenoaltea on open fields: a reply', *Explorations in Economic History*, 14, 4, pp. 405–10.

Forsell, C. af (1834), *Socken-statistik öfver Swerige*, Stockholm.

Frandsen, K.-E. (1983a), *Vang og tægt: studier over dyrkningssystemer og agrarstrukturer i Danmarks landsbyer 1682–83*, København.

Frandsen, K.-E. (1983b), 'Danish field systems in the seventeenth century', *Scandinavian Journal of History*, 8, pp. 293–317.

Frandsen, K.-E. (1988a), '1536–*c.* 1720' in C. Bjørn ed., *Det danske landbrugs historie. 2, 1536–1810*, Odense, pp. 13–156.

Frandsen, K.-E. (1988b), 'The field systems of southern Scandinavia in the 17th century. A comparative analysis', *Geografiska annaler. Series B, Human Geography, 70,* pp. 117–21.

Frandsen, K.-E. (1999), 'Studies of field systems and agrarian structures in some European regions' in B. Liljewall ed. *Agrarian systems in early modern Europe. Technology, tools, trade,* Stockholm, pp. 146–96.

Gadd, C.-J. (1983), *Järn och potatis: jordbruk, teknik och social omvandling i Skaraborgs län 1750–1860*, Göteborg.

Gadd, C-.J. (2000), *Det svenska jordbrukets historia. Vol. III, Den agrara revolutionen: 1700–1870*, Stockholm.

Gadd, C.-J. (2005), 'Odlingssystemens förändring under 1700-och 1800-talen' in: U. Jansson and E. Mårald ed., *Bruka, odla, hävda: odlingssystem och uthålligt jordbruk under 400 år*, Stockholm, pp. 63–91.

Gadd, C-J. (2011) 'The agricultural revolution in Sweden, 1700–1870' in: J. Myrdal and M. Morell ed., *The agrarian history of Sweden: from 4000 BC to AD 2000*, Lund, pp. 118–64.

Hahnemann, S. (1997), 'Um solskift ganger á by. Om gammel jorddelen og et forsøg på en nyvurdering af de agrarhistoriske begreber "bolskifte" og "solskifte"', *Historisk Tidsskrift,* 97, 2, pp. 237–71.

Helmfrid, S. (1961), *The storskifte, enskifte and laga skifte in Sweden: general features*, Stockholm.

Helmfrid, S. (1962), *Östergötland 'Västanstång': Studien über die ältere Agrarlandschaft und ihre Genese*, Stockholm.

Hoff, A. (1997), *Lov og landskab: landskabslovenes bidrag til forståelsen af landbrugs- og landskabsudviklingen i Danmark ca. 900–1250*, Århus.

Höglin, S. (1994), 'Byar och landskap i Skaraborg' in: G. Berggren and A. Boqvist ed., *Mylla mule människa: det västgötska kulturlandskapet, några brottstycken*, Skara, pp. 11–45.

Höglund, M. (2008), 'Inledning', in M. Höglund ed., *1600-talets jordbrukslandskap: en introduktion till de äldre geometriska kartorna*, Stockholm, pp. 9–15.

Hybel, N. and Poulsen, B. (2007), *The Danish resources c. 1000–1550: growth and recession*, Leiden.

Jansson, U. (1998), *Odlingssystem i Väneromrädet: en studie av tidigmodernt jordbruk i Västsverige*, Stockholm.

Lindgren, G. (1939), *Falbygden och dess närmaste omgivning vid 1600-talets mitt: en kulturgeografisk studie*, Uppsala.

Lunden, K. (2004), 'Recession and new expansion 1350–1814' in: R. Almås ed., *Norwegian agricultural history*, Trondheim, pp. 144–232.

McCloskey, D. (1975a), 'The persistence of English common fields' in: W. N. Parker and E. L. Jones ed., *European peasants and their markets: essays in agrarian economic history*, Princeton, N. J., pp. 73–119.

McCloskey, D. (1975b), 'The economics of enclosure. A market analysis' in: W. N. Parker and E. L. Jones ed., *European peasants and their markets: essays in agrarian economic history*, Princeton, N. J., pp. 123–60.

Myrdal, J. (1997),'The agricultural transformation of Sweden, 1000–1300' in: J. Langdon and G. Astill ed., *Medieval farming and technology: the impact of agricultural change in Northwest Europe*, Leiden, pp. 147–71.

Myrdal, J. (1999), *Det svenska jordbrukets historia, Vol. II, Jordbruket under feodalismen: 1000–1700,* Stockholm.

Myrdal, J. (2011), 'Farming and feudalism: 1000–1700' in: J. Myrdal and M. Morell ed., *The agrarian history of Sweden: from 4000 BC to AD 2000*, Stockholm, pp. 72–117.

Myrdal, J. and Söderberg, J. (1991), *Kontinuitetens dynamik: agrar ekonomi i 1500-talets Sverige*, Stockholm.

Nathorst, J. T. (1848), (Untitled speech before the Second Swedish Public Agricultural Meeting, printed) in Arrhenius, J. ed., *Berättelse öfver förhandlingarne vid det andra allmänna svenska landtbruksmötet i Stockholm 1847*, Stockholm.

Nordholm, G. (1967), *Studier i Skånes äldre ekonomiska geografi,* Lund.

Olai, B. (1983), *Storskiftet i Ekebyborna: svensk jordbruksutveckling avspeglad i en östgötasocken*, Uppsala.

Olai, B. (1987), *'-till vinnande af ett redigt storskifte-': en komparativ studie av storskiftet i fem härader*, Uppsala.

Pedersen, E. A. and Widgren, M. (1998), 'Del 2. Järnålder' in: S. Welinder, E. A. Pedersen and M. Widgren., *Det svenska jordbrukets historia. Del 1, Jordbrukets första femtusen år: 4000 f. Kr.-1000 e. Kr.*, Stockholm, pp. 237–459.

Porsmose, E. (1988), 'Middelalder o. 1000–1536' in: C. Bjørn ed., *Det danske landbrugs historie. 1, Oldtid og middelalder,* Odense, pp. 205–417.

Rahmqvist, S. (2013), 'Jordens läge i byn under medeltiden' in: M. Lennersand, Å. Karlsson and H. Klackenberg eds, *Fragment ur arkiven: festskrift till Jan Brunius*, Stockholm, pp. 350–58.

Riddersporre, M. (2008), 'Det skånska bolet' in: A. Ericsson ed., *Jordvärderingssystem från medeltiden till 1600-talet,* Stockholm, pp. 23–38.

Sporrong, U. (1973), 'Bysamhället' in: K. Norborg ed., *Svensk landsbygd. Kulturgeografiska studier av markanvändning, bebyggelse och miljö*, Stockholm, pp. 7–44.

Sporrong, U. (1985), *Mälarbygd: agrar bebyggelse och odling ur ett historisk-geografiskt perspektiv*, Stockholm.

Sporrong, U. (1994), 'The old agrarian landscape before 1750' in: L. Wastenson and S. Helmfrid ed., *National atlas of Sweden. Landscape and settlements*, Stockholm, pp. 30–35.

Sporrong, U. (2008a),'Europas periferi kontra de agrikulturella regionerna – en tillbakablick' in: A. Ericsson ed., *Jordvärderingssystem från medeltiden till 1600-talet,* Stockholm, pp. 11–22.

Sporrong, U. (2008b), 'Jordägande och jordvärderingssystem från medeltiden till 1600-talet – ett försök till sammanfattning' in: A. Ericsson ed., *Jordvärderingssystem från medeltiden till 1600-talet*, Stockholm, pp. 241–61.

Vestbö, Å. (1996), *Odlingslandskap i förändring: agrara processer och rumslig organisation i norra Småland från senmedeltid till 1650*, Stockholm.

Vestbö-Franzén, Å. (2004), *Råg och rön: om mat, människor och landskapsförändringar i norra Småland ca 1550–1700*, Stockholm.

Vestbö-Franzén, Å. (2008), 'Byamålen i norra Småland: något om oregelbundna regleringar och regelbunden oordning med utgångspunkt i kartor från 1640 till storskiftet' in: A. Ericsson ed., *Jordvärderingssystem från medeltiden till 1600-talet*, Stockholm, pp. 177–202.

Widgren, M. (1995), 'Var medeltidens jordbrukssystem uthålliga? Vägar till ett svar' in: G. Dahlbäck ed., *Miljö och livskvalitet under vikingatid och medeltid,* Stockholm, pp. 77–98.

Widgren, M. (1997a), 'Fields and field systems in Scandinavia during the Middle Ages' in: J. Langdon and G. Astill ed., *Medieval farming and technology: the impact of agricultural change in Northwest Europe*, Leiden, pp. 173–92.

Widgren, M. (1997b), *Bysamfällighet och tegskifte i Bohuslän 1300–1750*, Uddevalla.

Widgren, M. (1999), 'Västsvenskt odlingslandskap under järnåldern och medeltid – några aktuella forskningsfrågor' in: L. A. Palm ed., *Problem i västsvensk medeltid*, Göteborg, pp. 46–61.

Widgren, M. (2014), 'Hur drevs den vikingatida-medeltida storgården? Några frågor från Lägerbovada, Ydre' in: *O. Karsvall and K. Jupiter ed., Medeltida storgårdar: 15 uppsatser om ett tvärvetenskapligt forskningsproblem*, Uppsala, pp. 59–72.

4 The open-field landscape in two Swedish provinces on the fringe of possible cultivation

Hans ANTONSON, Lund University and KMF forum AB, Sweden

I. Introduction

In his *French Rural History* of 1931 Marc Bloch (1966: 36) wrote about 'open fields with long furlongs', commenting that 'the most striking feature of the ploughlands is the uninterrupted vista they present. This is to say there will be no fences of any kind'. It would have been interesting to know if he ever identified these characteristics in Norway. However, he makes no reference to the Scandinavian conditions, or to the literature concerning Scandinavian open-field systems – somewhat surprising, perhaps, given that he wrote the book while visiting the Institute for Comparative Research in Human Culture in Oslo, the capital of Norway. The reason for mentioning Norway is because the provinces studied in this article were, until 1645, a part of Norway (itself at the time a Danish Province), having been claimed by both Norway and Sweden during a series of wars beginning in 1519. They were Swedish in ecclesiastical terms, however, forming part of the Uppsala archdiocese (Antonson, 2009).

With few exceptions (e.g. Roeck Hansen, 2002; Antonson, 2014), studies of Swedish open-field systems (Sw. *tegskifte*; Dk. *vangebrug*) have concentrated on the southern parts of the country (e.g. Hannerberg, 1977; Sporrong, 1985; Dahl, 1989; Höglin, 1990; Widgren, 1997), where there were many similarities to arrangements found in, for example, England (Göransson, 1961; Thirsk, 1964; Baker, 1973; Butlin, 1973; Dahlman 1980; Dodgshon, 1980; Rowley, 1981; McCloskey, 1989). Here, however, I present a case study from the Provinces of Jämtland and Härjedalen in mid-Sweden (Figure. 4.1), where farming conditions were very different from those in the south. The aim is to describe and analyse the character of open-field landscapes in a northern part of Europe where agriculture was based on animal husbandry, due to the harsh climate. This is done through the retrogressive method, using historical maps and an important eighteenth-century by-law. The article focuses on a number of fundamental questions concerning the similarities of the region's fields with, and their differences to, those found in more southerly areas of Sweden, and in western Europe more generally; the origins of open fields in this northern region; and the relationship between their character, and that of the wider environment and economy. The chapter starts with a brief description of the characteristics of open-field systems, with some asides regarding Sweden. This is followed by a description of the study area, and the conditions here during medieval and early modern times, which is in turn followed by a review of the source materials used and methodology employed, and then by the detailed case study itself. The chapter ends with a discussion and conclusion.

Peasants and their fields. The rationale of open-field agriculture, ed. by Christopher DYER, Erik THOEN and Tom WILLIAMSON, Turnhout, 2017 (*CORN* Publication Series, 16), pp. 77–97

BREPOLS PUBLISHERS DOI: 10.1484/M.CORN-EB.5.114269

Figure 4.1 The location of the principal areas and places mentioned in the text

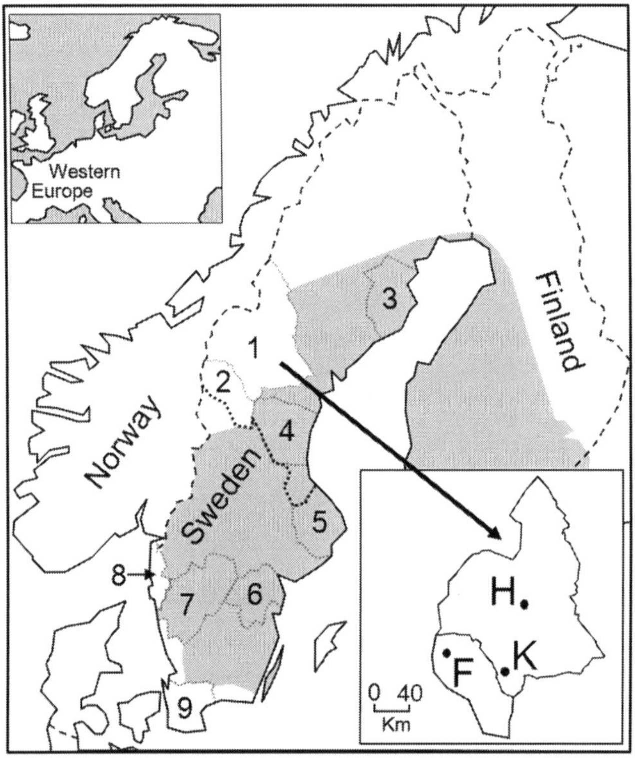

Provinces: 1=Jämtland, 2= Härjedalen, 3=Västerbotten, 4= Hälsingland, 5=Uppland, 6=Östergötland, 7=Västergötland, 8=Bohuslän, 9=Scania (Sw. Skåne)

Settlements: H=Hägred hamlet, K=Klövsjö church-village, F=Funäsdalen Village

The black dotted line marks the southern boundary of Norrland, Sweden's northernmost region. The area shaded in grey represents the extent of the kingdom of Sweden in the fifteenth century

II. Characteristics of open-field systems

Hopcroft (2003: 135) made an attempt to summarize the physical layout of the open-field system during the late Middle Ages and found four elements, namely: '1) communal arable land for growing the primary food crops and for communal grazing of herds; 2) communal meadows for the production of hay; 3) wasteland or commons, used by the community as a whole for gathering firewood and grazing; and 4) private gardens attached to houses in the village'. Here, emphasis is put on both tangible and intangible features as well as on the system *per se*–as a whole, composed of several parts. Gadd has described how, in the period between 1700 and 1850:

> In the plains and transitional districts, the villages and hamlets used open-field farming [...]. Each field was fenced so that the growing crops could be protected from grazing animals. In each growing season, all the strips [i.e. parcel, sub-divided fields] in one of the fields were left fallow, and the livestock were turned out on it to graze the grass that rapidly took over. Although the tilling and harvesting of the strips was undertaken by each individual peasant household, and the harvest thus belonged to the individual peasant, the grass that grew on the fallow land after the harvest was a fodder resource that belonged to all the villagers in common' (Gadd 2011: 124).

Here the arable is described as private, although used for periods in a collective manner. Each strip was individually owned and lay intermingled with those of other farmers. The strips were grazed by all the livestock of the hamlet or village during the fallow season. Another characteristic feature was the openness of the landscape, that is, the absence of fencing between the strips. But at the same time, one or several common fences surrounded large bundles of arable strips, which together made up an individual field, or enclosure (Sw. *gärde* or *vång*).

Despite broad similarities, the western European literature does acknowledge that there were several exceptions to this general picture of the essential features of an open-field system, and many variations in their character and layout. Baker and Butlin (1973a) argue in the concluding section of their edited volume that there were differences in both time and space in the British Isles, and that the systems found in some districts were more mature than others. They identify at least three origins for the pattern of ownership in intermingled strips, namely communal assarting, partitioning of holdings and remodelling of holdings. They also state that the distribution of strips could be highly regulated (sometimes by *Solskifte* or sun-division), irregular or uneven. In some places, such as parts of Northumberland, regular distribution of strips across the fields ensured and equality of landholding, both in terms of quantity and quality; elsewhere, as in some places in Scotland, the land was reallocated annually. They describe (1973a: 644) how 'open fields must *almost* [my emphasis] always have been cultivated in common even if they were not pastured in common'. Differences could also encompass common grazing in the field, which was sometimes practised by neighbours and sometimes by the whole community: they refer to examples from Newton Longville in Buckinghamshire where farmers pastured their own arable strips on the field after harvest, while in parts of Northumberland furlongs (strips grouped into blocks) were allowed to revert to pasture for a period of time. Not all open fields were entirely open, in the sense that some had enclosures within them. Similar variations existed elsewhere in Europe. Prince (1977) described how in some of the villages in Vosges, France, there was no communal grazing on the arable during the fallow season. Campbell (1981: 114) identified six basic elements and fourteen functional attributes within different common field systems which made him propose 'five principal types of field system plus several sub-types'.

In Sweden there are at least five types of open-field system with ownership in non-fenced, intermingled strips. The first three are found in the south Swedish plains. Towards the east we find the *Solskifte* system in, for example, the provinces of Östergötland and Uppland (Sporrong, 1985). Another parcelling system appears in the central parts of the Västergötland Province, named by scholars as the *Fritt stångskifte* (literally, the 'free rod division'). Its origins appear to be relatively late, around the early 1600s. In both cases, the strips acted as a representation of the share which the farm had in the land and resources of the village or hamlet, and thus served as the basis for sharing the rights and obligations (Bonow, 2008). The *Bolskifte* system (Bol division) is found in the former Danish Province of Scania (Göransson, 1961; Riddersporre, 2008). In the western parts of the Province of Bohuslän, as well as in the Province of Hälsingland, we find the arable divided into long narrow strips as a result of farm subdivision due to partible inheritance (Widgren, 1997; Antonson, 2014). And in the Province of Västerbotten the long and narrow strip fields are found on the margin of the younger, more block-like fields of each hamlet, and were established on land added to the core of the arable following coastal reclamation (Roeck Hansen, 2002).

In at least three of these open-field systems there was considerable variation in the number of individual fields, and in the proportion of land laid to fallow each year – that is, in both field systems and course-rotation systems. In the province of Östergötland, a regular two-course-rotation (two fields, one lying fallow each year) was practised (Sporrong, 1985), but in large parts of the Province of Västergötland a three-course-rotation (three fields, one lying fallow) was generally found (Gadd, 2011). In the province of Uppland some hamlets operated a two-course-rotation system in shared fields, meaning that neighbouring hamlets did not have to build a fence along their common boundary. Every second year the farmers in a hamlet grazed their cattle jointly in the fields with one neighbouring hamlet, and the next year with those of another (Höglin, 1990). Some of the communities in the same province practised a two-course-rotation, however, using four fields instead of the more usual two (Sporrong, 1985). In Västergötland a four-course-rotation system (four fields and 25 per cent fallow) was sometimes employed (Hannerberg, 1941; Myrdal and Söderberg, 1991). In Scania varieties of two-and three-course-rotation occurred (Dahl, 1989). Jansson (1998) found that in parts of Västergötland the hamlets located beside the northbound ox-drifts leading towards the Bergslagen mining areas had different field layouts than those of other settlements in the locality.

III. The study area

The Provinces of Jämtland and Härjedalen are located in the Swedish interior around latitude 63° N (Figure 4.1). Both lie within the two-thirds of the country that is named *Norrland*, literally 'the north lands'. Both Provinces are crossed by a number of rivers and streams, in particular the Indalsälven and the Ljungan, and there are numerous large lakes. The region is separated from Norway to the west by a range of mountains. Even today the arable appears as a number of small islands in an ocean of forests – even in the more intensely cultivated districts – and can in no way be compared to the landscape of the south Swedish plains. During the seventeenth and eighteenth centuries the arable was full of clearance cairns and the land was described as rich in rocks, stones and boulders, Hagström (1751: 2) noting that 'Here one see more cairns or stone piles cast together than elsewhere in Sweden'.

Jämtland has, compared with other provinces located in *Norrland's* interior, relatively good conditions for agriculture, partly because of the occurrence of clayey till and partly because warm winds find their way from the west through the mountains. However, compared with the conditions in south Sweden, Jämtland can be depicted as marginal and climatically vulnerable (Parry, 1978). In a royal taxation document from 1808, which proposed a more moderate level of taxation for the area, it was said that:

> While considering this county's northernmost location compared to Uppland [...] and the more southernmost located counties, here one has a smoother cold and longer winter, where the frost is forced deeper, and consequently weakens the soil. [...] The short summer makes that tilling and hay making for cattle feed – which often occur at far distant sites – requires more people and work than in the south. It should not be concealed that the land here is very troubled with cold [water] springs and hidden waters, requiring digging and maintenance of many deep ditches. [Moreover] the quality of fields and meadows, etc. cannot nearly be compared to the south' (Enagrius, 1808: 27).

Both Provinces displayed a variety of settlement forms in the seventeenth and eighteenth centuries: single farmsteads, hamlets and villages could all be found. A hamlet can be defined as a settlement agglomeration of no less than three and no more than twenty farms, while a village is one with no less than ten or twenty farms (Uhlig and Lienau, 1972), although no distinction is made in Sweden between these differently sized settlement categories: all are termed, quite simply, as a *by*, as long as there is a measure of co-operation between farms within a village, or between neighbouring settlements on infield or outfield land (Sporrong, 1990; Antonson 2004). In theory this could mean that a single farmstead might be termed a *by*, if its inhabitants co-operated in the management of the outfield with another single farmstead. Co-operation on the outfields between all three settlement categories is a well-documented phenomenon, evident on innumerable historical maps from different parts of Sweden, and has a variety of names, including *marker* ('land) or *skogelag* ('forest team, group of forest owners'). This phenomenon is known from the provinces of Närke (Jansson, 2003), Jämtland (Antonson, 2004), Småland (Tollin, 1999) and Västergötland (Bonow, 2005). In 1747 the surveyor Gerhard Edman wrote on a map of Hölje hamlet in Lit parish (NAS, map number LSA Y31–17:1) that Hölje has 'jointly owned forest with the neighbouring Åsmundsgården' hamlet, and that Jöns Andersson in Nästgården 'is a stakeholder in the forest, but not regarding the infields'. In other words, forests were collectively managed. Despite being a single farmstead, Edman regarded Nästgården as a *by*, just like the larger settlements of Hölje and Åsmundsgården. The three settlement categories have parallels in certain parts of Norway (e.g. Ambrosiani, 1974; Øye, 2000).

In Jämtland, a hamlet or village can take a variety of spatial forms, in terms of the distribution of its constituent farmsteads. The hamlet of Handog in 1708, for example, had six farms evenly spread out at four different locations within the infields (Figure 4.3d). Over time both the population and the number of individual settlements increased, but only very gradually, and the average hamlet, to judge from sixteenth-century accounts, comprised only 2.7 farms. In 1566, 48 per cent were located in hamlets containing between one and three, rising by 1700 to 54 per cent (Antonson, 2004). Although the majority of settlements thus consisted of very few farms, it should be stressed that during the 1600s some larger settlements – larger hamlets as well as small villages – could also be found (e.g. Figure 4.4). It has been shown that this slow increase in the number of farms was due to partible inheritance (Antonson, 2004; Antonson, 2014: see also Figure 4.2). The figures strongly imply that in medieval times the majority of settlements had consisted

Figure 4.2 The infields of Hägra hamlet, Rödön parish, from a map of 1693 by Christoffer Stenklyft

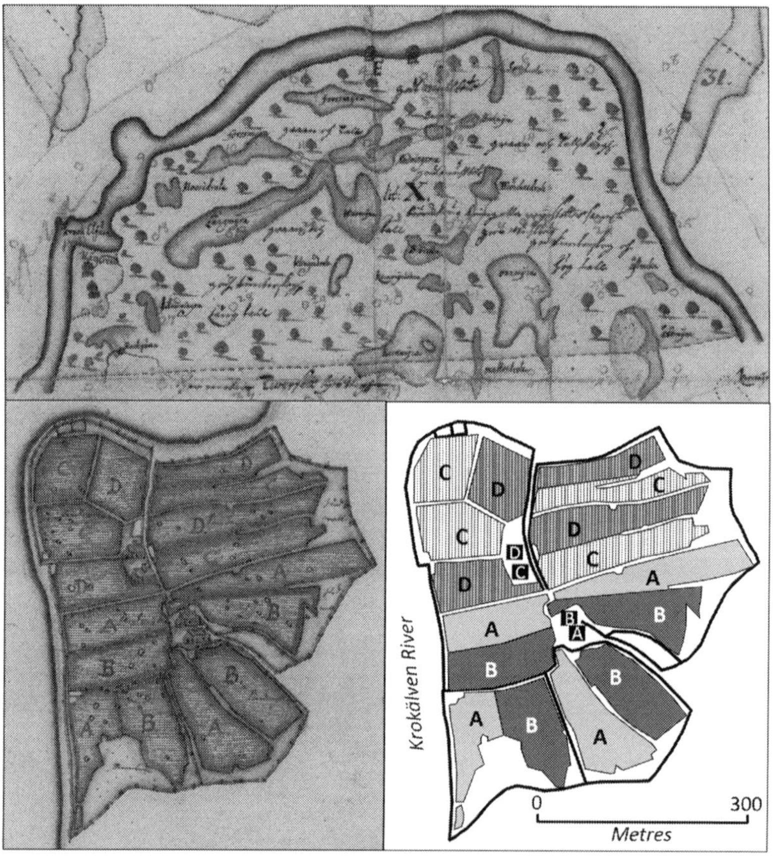

The hamlet occupies a promontory jutting out into the river Krokälven (part of the river Indalsälven). The upper map show the vast communal outfields, comprising areas of forest and meadow. The infields are depicted in the lower left-hand corner: the relative size of infields, and outfields, is clearly shown. On the lower left is an extract from this map, showing the infields in more detail. This is redrawn lower right, with the individual farms shown as black squares, the fences as thick black lines and the arable strips as shaded areas. The farms are clustered in two locations, separated by arable land, and their strips are concentrated in particular areas. Farms C and D thus have their land in the northern part of the arable, while A and B have theirs in the southern part.

As can be seen in the detail of the original map (lower left), the surveyor was careful to provide an accurate representation of the fencing, in elevation. He has clearly shown the kind of withy fence, with diagonal braces tied together between paired poles, which was widely used in Sweden at this time.

Source: NAS, map number LMA 23–Rödö–4

of single farmsteads. We should also note that while there is evidence for an economically strong aristocratic elite in Jämtland during the high Middle Age, they were not numerically significant (Hanson et al., 2005) and in the period after the Black Death in the mid-fourteenth century, they disappear from the sources. By the sixteenth century the region's farmers were free-holders (Antonson, 2009).

The earliest maps of the region show that the arable land of hamlets and villages took the form of block-shaped parcels, and/or irregular strips, mostly due to the undulating character of the terrain, and the many impediments to cultivation it posed. The strips were relatively large and ownership was intermingled (Figure 4.2 and 4.3). During the eighteenth century surveyors from the south came to Jämtland to carry out the Storskifte partition (enclosure) land reforms, following the legislation of of 1749: storskifte as a verb means 'to divide in large blocks', but as a noun means 'large parcel of land'. Some of the surveyors were astounded to find that the hamlets and villages *'since ancient times had large parcels of land'* (Antonson, 1995: 22).

From the late Middle Ages the arable was tilled using a *högplog* (North Sea plough) with a mouldboard. It was a light plough, drawn by a single horse and easy to manoeuvre. The average area of sown fields on a group of 63 farms, from eight central parishes in the region, was a mere 1.63 ha. (taking into account the fact that half the land lay fallow each year under the two-course rotation); the area of the parcels cultivated ranges from 1.54 to 6.72 ha. This can be compared to the situation in southern Sweden, where seventeenth-century maps show that the average farm area in the Seminghundra district of the Province of Uppland was 12 ha., of which 6 ha. were sown annually; that in the Kumla district of the Province of Närke it was 6 ha., of which 3 ha. were sown annually; and that in eastern Falbygden in the Province of Västergötland it was 5 ha., of which 3.4 ha. were sown annually (Lagerstedt, 1942; Lindgren, 1939; Hannerberg, 1941). Furthermore, the arable area as a whole was very small in Jämtland, compared to the amount of outfield (forest and pastures), as shown in Figure 4.2. The actual ratio of land lying in infields, to that of the outfields, is unfortunately difficult to ascertain because land surveyors in the seventeenth and eighteenth century concentrated only on the area of the former, and only noted the area of the meadows within the outfields.

A detailed study of maps surveyed between the 1640s and 1899 reveals that Jämtland farms practised either a two-or three-course-rotation system (Antonson, 1993). The amount of fallow (50 per cent or 33 per cent) often corresponded to the number of individual open-fields, as in the Provinces of Östergötland and Västergötland, but as in southern Sweden exceptions were numerous: there are thus examples where a surveyor has drawn one large field, within a single boundary, even though the farmers in the associated hamlet were yearly fallowing half their land (Figure 4.3b).

IV. Methodology and source material

Two key sources exist for the study of Jämtland's field systems: by-laws drawn up in the eighteenth century, and a range of maps surveyed in the seventeenth and eighteenth centuries. A by-law is a contract between all the farmers in a hamlet or village concerning activities such as hunting, fishing, fencing and the common departure to the summer pastures. In Sweden the oldest surviving by-law dates from the 1600s but earlier agreements of a similar kind probably existed, although they were not committed to writing (Erixon, 1978). A royal ordinance of 1742 suggested that villages should adopt a *mönsterbyordning* ('ideal by-law') which could be adapted to local conditions. In the Västerbotten Province the ideal by-law had a great influence on the new local by-laws drawn up from the mid-1700s, but these also contain traces of local traditions and customary law (Isaksson, 1967). How far a particular by-law mirrors older situations, however, depends on whether radical changes had occurred in the hamlets or villages in question before it was introduced.

Sweden has many large-scale maps from the seventeenth and eighteenth centuries (Sporrong and Wänström, 1990; Baigent, 1990). They are distinguished by their wealth of details concerning each individual farm. Normally the maps have two parts: the map itself, and an accompanying administrative text, either written on the map or on other sheets. The text includes the name of the hamlet and the number of farms it contains; a brief description of the outfields; the farmers' names and the amount of tax they pay; the quantity of land sown; and the volume of meadow hay harvested. The maps contain a variety of information, depending on the individual surveyor, relating to fences, fields, soil type, roads, impediments to cultivation, meadows and administrative boundaries. More than 3,000 examples of these large-scale maps from the region have now been studied and some are discussed in greater detail below. The maps belong to two different archives, namely the Land Survey Board archive (LSA), which is the national collection, formerly located in the town of Gävle; and the County Cadastral Authority archives (LMA), which is the local collection, formerly located at the counties land survey offices. Today all are deposited at The National Archives of Sweden (NAS) in Stockholm, but are also digitally accessible via the internet at the National Land Survey of Sweden's homepage. It is important to emphasize that Jämtland has relatively few examples of the oldest types of map found in Sweden: the geometrical maps of the 1640s and the geometrical maps from the period between *c.* 1690 to *c.* 1730: although the former fail to depict the individual strips within the infields, the latter are often more detailed, and show the landscape before the great land reforms which began in 1749. There are only 17 places in Jämtland which have an example of both kinds of map deposited at the LMA, and 108 places with such maps at the LSA. This is far less than in the provinces along the Bothnian coast, or in the southern part of Sweden.

The method applied here is the retrogressive approach, not to be confused with the *retrospective* approach (for a discussion of the concept, see Baker, 1968; Baker and Butlin 1973b; Antonson, 2014). Marc Bloch was the pioneer advocate of the *retrogressive* approach, which he himself called '*la méthode* régressive' (Bloch 1931: XIV). The method begins by studying a feature from a time period close to our own, perhaps surviving in today's landscape or shown on a nineteenth-century map, and then travels back in time via a number of several sources (such as a sixteenth-century map) in order to understand aspects of past landscapes, such as the settlement pattern in a particular area in the fourteenth century, the details of which are not provided by contemporary written sources. The retrospective approach, in contrast, can be said to work in the opposite direction.

V. Case studies

My first example concerns the origins of the pattern of arable strips, as exemplified by the Hägra hamlet in the parish of Rödön (Figure 4.2). The hamlet is located in the central rural district of Jämtland, called Storsjöbygden. Hägra's four farms were mapped in 1698 by the surveyor Christopher Stenklyft, probably to unravel the ownership situation concerning two farms belonging to the Crown (the other two farms were owned by freeholders). The map does not specify which farm is which, providing only the names of the owners of the various strips, but on the accompanying text the surveyor does number the land belonging to each of the farms, and the same system of numbering is used in the 1818 Storskifte partition map (NAS, map number LMA 23–Rödö–4). This also shows the location of the farms and their associated strips, both before and after the Storskifte land reform. The southernmost is farm A, followed by B and C, with D in the north. Farms A and B were the larger two, paying tax for 3 acres each, compared with 2 acres for farm C and D. The redrawn map (Figure 4.2) shows that the four farms were located at two sites in the centre of the arable which consisted of sixteen strips, spread over eleven patches of arable. Some of the strips were clearly demarcated with narrow headlands while some were defined with a hatched line. The strips were relatively large and the headlands, together with the central location of the four farms, ensured that each and every farmer had access to their own strips without having to cross someone else's land.

The strips are arranged in two groups, a northern and a southern. Within each group there is a marked degree of regularity, with two of the farms owning every second strip. This suggests that the strips belonging to A and B originally formed a single undivided holding, which we might refer to as farm 1. The same reasoning applies to farms C and D, the undivided ancestor of which we can call farm 2. It is not inconceivable that the single farmsteads 1 and 2 were themselves once a single property, divided at an early date into two discrete blocks, a northerly and a southerly. It was the second phase of subdivision which produced the intermingled strips. This seems to have occurred at a relatively late date. A ransack protocol (a special court protocol) from 1615 describes four farms at Hägra, but cadastral accounts from the period 1545–1571 refer to between one and three farms. It is unclear when the first farm in Hägra was divided into two farms, but this may have happened during the late Middle Ages. Nor do the surviving sources tell us how the Crown came into possession of two of the farms in the hamlet but, since Jämtland was hit hard by the Black Death in 1350, and plague continued into the 1400s, leaving many farms still desolate in the 1600s (Antonson, 2009), it is possible that they were deserted, and were only re-occupied as *skattevrak* (literally, 'taxation wreck'). In other words, they may have come into the possession of the Crown during the period when no-one had been present to pay the taxes owed on them. Some historical maps show more intricate examples of farm division, such is Tullingsås hamlet in Ströms parish (NAS, map number, LSA Y37–17:1), but while it is possible to interrogate the evidence in a similar logical manner, in order to reveal the origins of the pattern, the larger number of farms involved makes it hard to use them as instructive illustrations.

As mentioned above, there are a number of places–single farmsteads, hamlets or villages–which deviate from the general picture, in that the relative proportion of sown land and fallow is not reflected in the number of separate fields. Some hamlets

Figure 4.3 a–d A collage of hamlet infields with parcelled fields in the Province of Jämtland

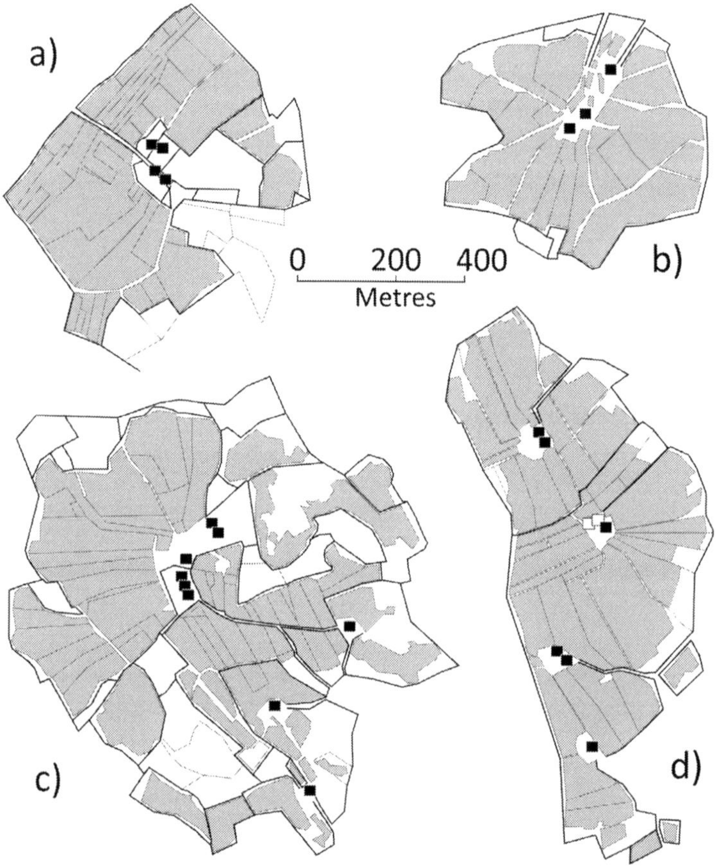

a) Forsta in Lockne parish, 1750 (LMA, map number LSA Y32–9:1); **b**) Hölje in Lit parish, 1747 (LMA, map number LSA Y31–17:1); **c**) Dvärsätt in Rödön parish, 1746, (LMA, map number LSA Y47–5:1); **d**) Handog in Lit parish, 1708 (LMA, map number LSA Y31–14:1)

The arable strips take a variety of mainly irregular shapes. Some strips are demarcated with a border, some with a headland. In some places the dry meadow, as well as the arable, is divided into strips.

The settlements associated with these various types of field range from agglomerated to dispersed. Most of the individual farms are accompanied by a toft with a garden (Sw. kålgård.)

The thick black line represents fences, grey-shaded areas are arable land, and the thin black lines are the boundaries of individual strips. The black squares are the sites of occupied farms, the white squares represent uninhabited farms which are managed by individuals living in a neighbouring hamlet.

thus practised a two-course rotation, with 50 per cent of their land fallow each year, yet all the arable strips lies within one single enclosure. A handful of such examples exist both in remote and central locations, including the hamlets of Stocke (mapped in 1697), parts of Mjälle (1696–1703) in the parish of Frösö; Södra Sölvbacken (1699) in Oviken parish; Hölje (1747) and Söre (1703) in Lit parish. Håv (1693) in Frösö and Lockne Parsonage in the parish of Lockne (1701), together with Undersåker parsonage in the parish of Undersåker (1702), are all single farmsteads and thus do not have intermingled strips. They do, however, display the same pattern, in that all their land lay within a single field enclosure.

The example of Hägra hamlet can again be used to examine this phenomenon. The redrawn map (Figure 4.2) shows that all the arable and the headlands were contained within one single outer fence. In the south, farms A and B each have two of their strips within two separate field enclosures, meaning that their strips lay within three fields. The surveyor, Stenklyft, noted that the arable was fallowed, but not how much each year. However, all the surrounding single farmsteads, hamlets and villages for which we have the relevant information –about twenty in all– were fallowing half their land each year in the decades around 1700. The presence of the two southerly fields makes sense if we assume that the farmers of the Hägra hamlet were also fallowing half their land: the two fields would represent the first and the second year of cultivation respectively in a two-course-rotation system. The remaining twelve strips lay within one enclosure, without subdividing fences to prevent cattle grazing on the fallows from entering areas under crop. This must mean that there were no grazing on any part of the field when one section lay under crops.

Not all hamlets and villages, we should note, had their subdivided arable demarcated with headlands. While some were like Hägra, for instance Hölje in Lit parish (Figure 4.3b), in others the arable strips are shown bounded with a hatched line on the map, as with Dvärsätt, in Rödön parish, Forsta in Lockne parish and Handog in Lit parish (Figure 4.3b–d), or Tullingsås in Ström parish. The implication is that some of the strips could only be accessed by crossing another farmer's land. This, however, was not a common feature.

The next example discusses, in some detail, cases where one hamlet or village is surrounded by one single enclosure, completely undivided by internal fencing. The church-village of Klövsjö in the southern part of Jämtland is a particularly interesting case because of the survival of a parallel source. The village never underwent the Storskifte partition, and a by-law dating from 14 November 1793 describes an older situation, in terms of both the physical form and the function of the agrarian landscape. Of particular note are the references made in the document to the 'ring fence' around the township: 'On the 18th of May all the cattle shall be driven from the joint enclosure [or else] under penalty of 24 ch[illings]', 'No one shall, [or else] on pain of a penalty of 16 ch, release the cattle on the common field during autumn until the harvest is in [the barn] and a day has been appointed by the bailie'. In addition, no less than four paragraphs refer to summer pastures (shielings/summer-farms), dealing with the arrangements relating joint departure to and from these, or concerning movement to a new summer pasture. In order to understand these terms and phrases it is necessary to move away from the models provided by south Swedish open-field systems, and to think in other directions.

The Klövsjö church-village was mapped in 1764 (Figure 4.4). The map shows seventeen farms, ten crofts (not shown on the map), a church and numerous small strips. The crofts

Figure 4.4 Klövsjö church-village, as mapped by J. Sundström in 1764 (NAS, map number LMA 23–Klövsjö–2)

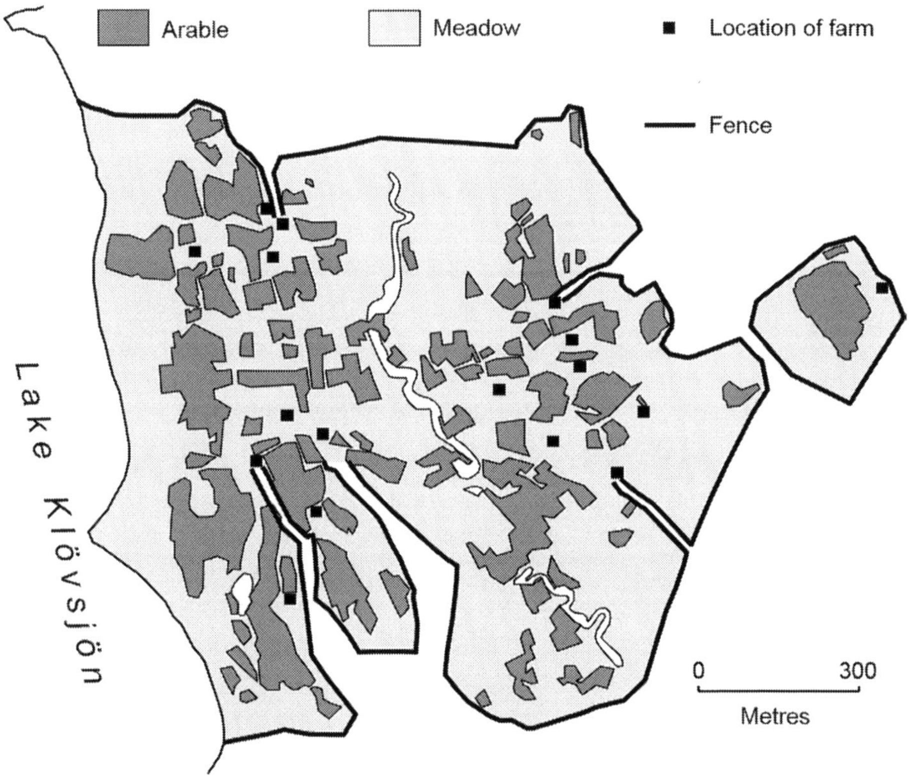

This map, the oldest large-scale map of the settlement at Klövsjö, is a so-called concept map, relating to an unfinished Storskifte partition. It is accordingly drawn in part in lead pencil rather than ink. The infields, lying within a single ring fence, contain one small enclosed field, 88 strips and 17 farms; the crofts are not shown.

were located outside the ring fence and were surrounded by small enclosures. The main church-village infield was enclosed within a single ring fence, but the many strips (eighty-eight in all) were not individually fenced, nor were there any fences separating the arable from the meadow. Erixon (1960), who studied the village, felt that the apparent absence of such fences was in fact due to a mapping error on the part of the surveyors but there are other, more convincing explanations. The by-law states that the large enclosure was held in common (no specific information is given about the arable) but this should not be taken to mean that the land was itself communally owned. The map is the original field survey, not the redrawn copy that was supposed to be sent to the king's administration. It is still partly drawn in lead pencil, it does not contain any description of who owned which pieces of land, and only one of the larger field strips is demarcated with a hatched line, and appears to have had two owners. This, however, probably means that each of the remaining field strips was individually owned, which is in line with general practice–most of the eighteenth–and nineteenth-century Jämtland maps show that the arable was individually and not collectively owned. Thus, the common activities that the by-law refers to are more likely to be concerned with the organization of grazing. Fences only served to prevent animals from entering certain land: if a field was fenced it was to keep the cattle out, not in.

The by-laws reveal that the normal routine on a farm in Jämtland was for the animals to be kept on one or more summer-farm pastures during the summer, far from the village infields. After the harvest, and when the hay was safely in the barn, the animals returned to the infields and, until the ground was snow-covered, could live off the stubble and gleanings. They were free to wander wherever they wished, regardless of who owned the land. Therefore there was no reason for the fields and meadows to be separated by fences, as they were in southern Sweden, where during the summer the farmers had to protect their harvest from grazing cattle because these had no other place to feed than on the infields (Antonson, 2004). A careful analysis of the many historic maps from the region suggests that several Jämtland villages and hamlets displayed similar traits to Klövsjö: the existence of a single large enclosure within a ring-fence can thus be paralleled in the case of Hägra hamlet, just discussed.

But field systems of this type are also recorded outside the province of Jämtland. Bjørkvik described, in the 1950s, the Norwegian village of Vik in the Province of Sogn, Norway. Prior to 1859 the village had just one fence, marking the boundary between infield and outfield. The animals grazed freely within the fence until 25 May, when sowing began (Bjørkvik, 1956). Similar arrangements could be found in the most northern part of Sweden, along the Lule River valley, in the Province of Norrbotten. In Enequist's (1937) study of the landscape along the lower part of the river the oldest geometric maps, from the mid-1600s, show that the hamlets and villages was surrounded by ring-fences. The arable belonging to each farm could in some cases be separately fenced within the ring-fence, but often such internal fences were lacking (here too, Enequist thought that their absence from the maps was due to errors made by the surveyors).

What remains unclear is the antiquity of this kind of field system. Another village, Funäsdalen, in the neighbouring Province of Härjedalen, was mapped in 1688 (Figure 4.5) by Christoffer Stenklyft. The Funäsdalen village shares many features in common with Klövsjö. One single fence surrounded the villages, with its fifteen farms, sixty-three barns, fields and meadows: evidently, the system was in place by the later seventeenth century. Given that the movement of stock to distant summer pastures was an integral part of the system, the antiquity of this practice ought to shed some light on that of the system itself, but unfortunately this is a matter that has been vigorously debated, and research findings are not consistent. Sandnes (1991) argued that the summer-farm system in Norway first developed during the 1100 and 1200s, but was initially confined to grazing and haymaking, and did not involve seasonal residence. The medieval Norwegian Frostating provincial law of 1260 mentions *'seter'*, which is the term for a Norwegian summer-farm (Hagland and Sandnes, 1994). In Ängersjö, Sweden, on the border between the two provinces of Hälsingland and Härjedalen, archaeologists have encountered building remains from the transition between the Viking period and the high Middle Ages (approximately

1000–1150 AD) which may represent the site of a summer-farm. However, there is no certain evidence for the processing of milk (Magnusson, 1989), though some structures may have been used for cooling, such as a stone dam in a brook and cellar pits beneath wooden buildings (Karlsson and Emanuelsson, 2002). In the most recent study of the subject Larsson (2009) suggests that there is no evidence for the summer-farm system before the fifteenth century. If this dating is correct, it is reasonable to assume that the collective fence around the church-township infields (as for Funäsdalen village, Hägra and Hölje hamlets) also came into existence is of similar antiquity – that is, it developed at the very end of the Middle Ages, or during the early modern period.

VI. Discussion and conclusion

It can now be said that the open-field systems within the area studied were both similar to, and different from, those found in southern Sweden (and across western Europe more generally) in terms of form, function and process (Baker 1975). In both areas, the arable land lay in strips that were open, in the sense that they were not individually fenced in; these were grouped into larger fields enclosed within a ring-fence, which was maintained by the community; and they represented the intermingled properties of a number of different owners. In both areas, moreover, each farm had in addition its own private garden within the toft. The greatest similarity, however, is that these various elements were combined into a farming *system*. Different parts were dependent upon each other. In terms of differences, we should note that the strips in the northern provinces studied here took less regular and often larger forms than in the south (they are broadly similar to those which, in Britain, were described by Dodgshon (1980: 48) as *'irregular sub-divided fields'*). This may have been due to the particular character of the physical landscape, for Jämtland was full of impediments to agriculture, such as small humps and clearance cairns, and the arable was often located on land which sloped steeply towards rivers and streams. Had the physical environment been flatter, and less cluttered with boulders and other obstacles, the strips would probably have had a different shape, closer to those found (for example) in parts of the Province of Hälsingland (Antonson, 2014).

While the operation of field systems in both the north and the south involved both individual and collective action, the specific balance between them displayed some differences. Starting with the individual, the many headlands present within the northern fields often made it possible for farmers to access their own strips without crossing those of neighbours, while the sheer size of the parcels made it possible to plough, weed, sow and harvest without trampling on another's land (Figure 4.2 and 4.3a). There are of course exceptions to this rule, as I have noted (Figure 4.3 c–d), although the scarcity of maps makes it hard to judge how common they were. While the northern fields may in this respect have been more individualized in character, all the farmers shared responsibility for building and maintaining the outer fence around the field (the ring-fence). While in many contexts plenty of material for constructing wattle fences was available in the surrounding forests, the same cannot be said about the supply of labour, for from 1350 until at least 1450 repeated outbreaks of the Black Death massively reduced the population, and at least 50 per cent of the farms were deserted (Antonson, 2009). Moreover, during the sixteenth and seventeenth centuries Jämtland was hit six times by war. Working together was labour saving.

A key difference between the north and the south was the presence in the former of the summer farm system. All the animals, regardless of who owned them, were supposed to leave the hamlet or village for the summer farms at a certain date in the spring, returning together when autumn came. During the period in which the cattle were gathered at the hamlet/village there were similarities with the open-field systems found in southern Sweden, and more widely across western Europe: the infield was grazed collectively, all the livestock being turned out together onto the stubbles and fallows of the arable, and onto the grass regrowth on the meadow strips and meadow headlands, regardless of who owned the land.

Dodgshon (1980) listed the various explanations which had been proposed for the emergence of sub-divided fields in Britain, including the development of the manorial system and feudalism, the use of a heavy plough, equal sharing (in terms of soil types, exposure, drainage; proportional adjustments between rights and duties; risk aversion), subdivision of land between co-heirs (partible inheritance), assarting (expansion of the cultivated acreage at the expense of woodland and waste), sun-division, population growth, and the transaction cost of purchasing or exchanging labour. Some of these factors clearly have no relevance in the region studied here. There were no feudal lords in either Jämtland or Härjedalen in the period from the fourteenth century; nor was a heavy wheeled plough in use here; nor was land parcelled out according to the rules of sun-division. One possible reason for the emergence of strip fields may have been population growth (cf. Campbell, 1981), leading to the repeated division of farms through partible inheritance. In Jämtland, it appear that population growth started during the late 1400s, after the plague abated, and while interrupted by periods of warfare in the sixteenth century had certainly resumed by the early 1700s (Antonson, 2004; 2009). As well as leading to the subdivision of farms and the development of holdings as intermixed strips, demographic growth also meant that the arable, and thus the infield, expanded. In the earlier maps we can see this happening at the expense of the dry meadow; the clearance of surrounding forests, in contrast, appear to have been a relatively late process, starting after the Storskifte partition (Antonson, 1993). The fact that the arable could be expanded means that, where the will and the need was present, it was possible to divide a farm into large holdings (as appear to have been the case in an early phase of the Hägra hamlet) without running the risk of creating farms too small to be viable. Nevertheless, as can be seen in Figure 4.2, the arable blocks were divided into evenly distributed strips, suggesting a concern for equitable distribution.

It seems likely that, within the region studied, systems based on a single field enclosure were once ubiquitous. In some marginal areas, as in Funäsdalen village (Figure 4.5), they persisted into the seventeenth century; in the case of Klövsjö church-village (Figure 4.4) they survived even longer, to be recorded in late eighteenth-century by-laws. In large parts of Jämtland, however, this system started to disappear in the 1600s, where farms were owned by the Crown and managed by members of the upper class, such as army officers, educated in the south of Sweden (Bergström, 1991). Systems were remodelled so that the percentage of fallow often corresponded to the number of fields (Antonson et al., 1996). It appears that Hägra hamlet (Figure 4.2) was mapped in a transitional phase, when two ways of enclosing the fields were in operation at the same time: the map shows an earlier phase, with a single undivided enclosure lying within an outer fence; and a later phase, represented by fields created within the enclosure which allowed one

Figure 4.5 The oldest large-scale map of the Funäsdalen village, drawn by Ch. Stenklyft in 1688

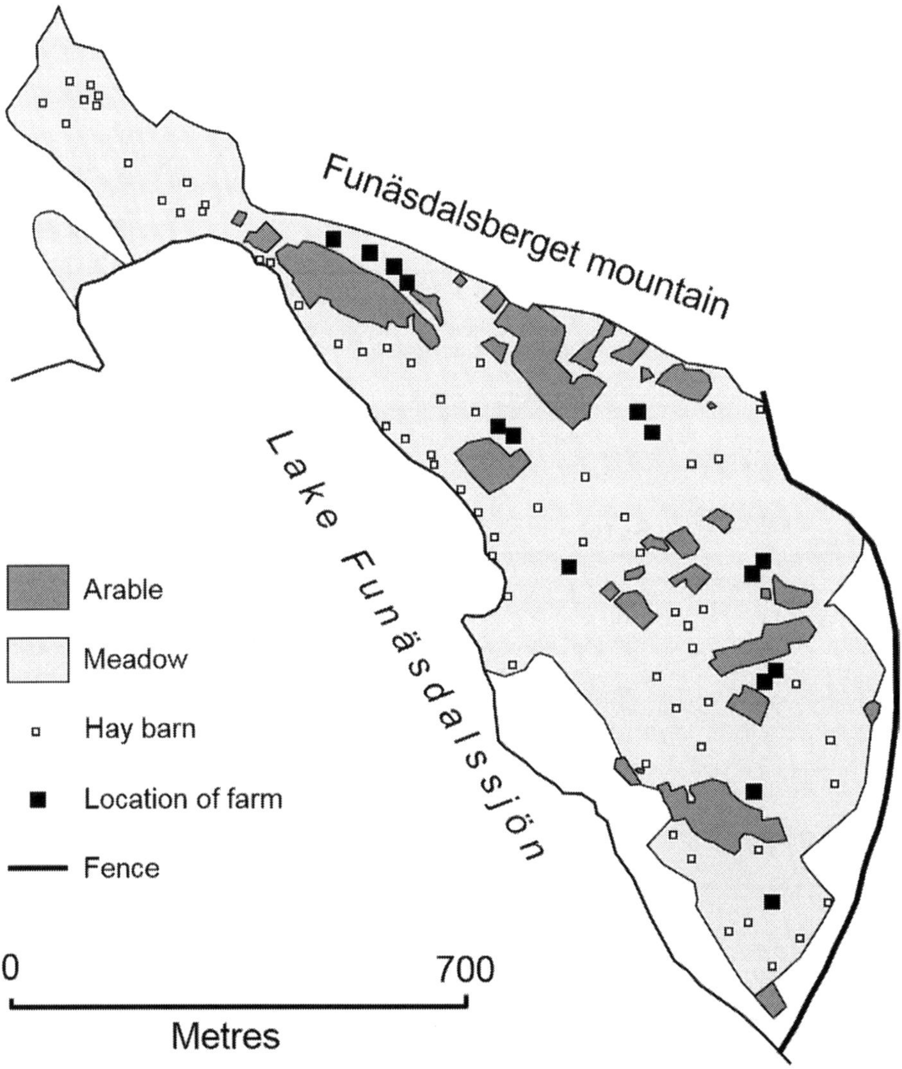

This is a so-called concept map, drawn in both ink and lead pencil. The village contains 15 farms, all located within the infields, which comprise a single area partly enclosed by a fence (to the east) and partly by features of the natural topography – the very steep mountain Funäsdalsberget and Lake Funäsdalssjön. As in figure 4.4, the strips have an irregular shape.

Source: NAS, map number LMA 23-Tännäs-3

portion of the arable to be sown and one to be fallowed, allowing livestock to graze the fallow without harming the adjacent areas under arable cultivation. In marked contrast, the field system in Klövsjö church-village at the second half of the eighteenth century was a relic of older practice, once widespread throughout Jämtland but by now a rare and perhaps unique survival.

In terms of the second research question set out at the start of this article, concerning process, it is evident that strips developed over a long time period through the division of farms by inheritance, a process which may have started as early as the late Middle Ages but which probably came to a halt during the many plague outbreaks (1350–1450), not resuming until around 1500 (Antonson, 2009). The existence of fields subdivided into strips should not, therefore, be confused with a systematic distribution of land at a single point in time, resulting from the operation of medieval provincial laws (e.g. sun-division). While subdivision of properties by inheritance was crucial in the development of the region's field systems, we should also note the crucial role played by the summer-farm system, although as we have seen the date of its inception remains unclear from the sources studied here, and earlier research on this matter has largely been based on circumstantial evidence. Its apparent origins in the fifteenth century may be an illusion, created by the fact that this is when written sources become more abundant and reliable (Larsson, 2009): the archaeological evidence suggests that it may be much older (Magnusson, 1989). Either way, we must avoid the pitfall of assuming a static, unchanging society. The summer-farm system may not always have had the same functions as those we see in the documents from the eighteenth century.

The later development of the region's field systems are clearer than their initial emergence. Beginning in 1749, when land ownership changed through the Storskifte partition reform, fewer fences appear to have been erected within the infields, according to the evidence of the historical maps. This is in line with research from the south of Sweden, where it appears that, as the arable land expanded, it eventually became more cost effective and labour efficient to enclose a single pasture, rather than to divide the area under cultivation from the wider pastures (Gadd, 2000). It remains to be answered whether this was also the case in the Jämtland and Härjedalen Provinces, where the summer farm system was to persist for nearly two centuries more. Either way, as in the rest of Sweden, the Laga skifte partition from 1827 (Pettersson, 2003) led to the death of intermingled ownership in small strips, when holdings were redistributed into compact units.

I conclude this discussion by responding to the third research question set out at the start of this article. I would argue that the landscape in the Provinces of Jämtland and Härjedalen was essentially one characterized by open-field systems, at least in the period between the sixteenth and the mid eighteenth centuries, when historical maps provide a fairly good picture of their form and function. This is a reasonable conclusion, given the many similarities with the southerly provinces, discussed above, where open fields have been more intensively investigated. But it must be acknowledged that there also are differences, some smaller and some more profound, between the systems of the two regions. Difference, however, also appears to be the hallmark of open fields across the whole of western Europe. Finally, I maintain that the retrogressive method is best used when focusing on change (Baker, 1975), and best achieved when there is a clear starting point in later sources and a clear end point or cross-section in earlier ones.

Acknowledgements

I wish to thank the two anonymous reviewers, the editors and Ph.D. Ulf Jansson who read and commented upon an earlier version of this chapter. I also wish to thank Professor emeritus Ulf Sporrong, who read part of this chapter in a different context. Finally, I would like to thank the late Roger Tanner for translating parts of this article and the editors of this book for tidying up the final text.

Bibliography

Ambrosiani, B. (1974), 'Comments on Iron Age Farms in SW-Norway', *Norwegian Archaeological Review*, 7, 1, pp. 40–43.

Antonson, H. (1993), *Jämtlands äldre agrarlandskap. Försök till en regional indelning baserad på lantmäteriakter för tiden 1650–1899*, Östersund.

Antonson, H. (1995), *Riksintressen i Jämtlands län: landskapsstudier av ett 30-tal byar baserad på lantmäteriakter från 16-, 17 – och 1800-tal,* Östersund.

Antonson, H. (2004), *Landskap och Ödesbölen. Jämtland före, under och efter den medeltida agrarkrisen [= Landscape and Ödesbölen: The Province of Jämtland, Sweden, before, during and after the Medieval agrarian crisis*, Stockholm.

Antonson, H. (2014), 'Retrospection of medieval landscape change in mid-Sweden: historical maps analysed using a retrogressive approach' in: E. Heide and K. Bek-Pedersen ed., *New focus on retrospective methods*, Helsinki, pp. 163–85.

Antonson, H, Jansson, U. and Vestbö A. (1996), 'Svenska byar utan systematiska odlingssystem' in: B. Roeck Hansen ed., *Bebyggelsehistorisk Tidskrift,* 30, pp. 21–44.

Baigent, E. (1990), 'Swedish cadastral mapping, 1628–1700: a neglected legacy', *Geographical Journal*, 156, pp. 62–69.

Baker, A. R. H. (1968), 'A note on the retrogressive and retrospective approaches in historical geography', *Erdkunde: Archiv für Wissenschaftliche Geographie*, 22, pp. 243–44.

Baker, A. R. H. (1973), 'Field systems of southeast England' in: A. R. H. Baker and R. A. Butlin ed., *Studies of field systems in the British Isles,* Cambridge, pp. 377–429.

Baker, A. R. H. and Butlin, R. A. (1973a), 'Conclusions: problems and perspectives' in: A. R. H. Baker and R. A. Butlin ed., *Studies of field systems in the British Isles,* Cambridge, pp. 619–56.

Baker, A. R. H. and Butlin, R. A. (1973b), 'Progressive and retrogressive approaches' in: A. R. H. Baker and R. A. Butlin ed., *Studies of field systems in the British Isles,* Cambridge, pp. 39–41.

Baker, A. R. H. (1975), *Historical geography and geographical change*, London.

Bergström, C. (1991), *Lantprästen: prästens funktion i det agrara samhället 1720–1800. Oland-Frösåkers kontrakt av ärkestiftet,* Stockholm.

Björkvik, H. (1956), 'The farm territories: habitation and field systems, boundaries and common ownership. Old Norwegian peasant community 2. Investigations undertaken by the Institute for Comparative Research in Human Culture, Oslo', *Scandinavian Economic History Review,* 4, 1, pp. 33–61.

Bloch, M. (1931), *Les caractères originaux de l'histoire rurale française*. Serie B, skrifter 19, Oslo.

Bloch, M. (1966), *French rural history: an essay on its basic characteristics*, trans. J. Sondheimer, London.

Bonow, M. A. (2005), *Gård, gräns, giftermål: familjestrategiers betydelse för markens och landskapets utformning i Norra Åsarps socken, Västergötland ca 1640–1880*, Stockholm.

Bonow, M. A. (2008), 'Det fria stångskiftet i Västergötland' in: A. Eriksson ed., *Jordvärderingssystem från medeltiden till 1600-talet*, Stockholm, pp. 203–18.

Butlin, R. A. (1973), 'Field systems of Northumberland and Durham' in: A. R. H. Baker and R. A. Butlin ed., *Studies of field systems in the British Isles,* Cambridge, pp. 93–144.

Campbell, B. (1981), 'Commonfield origins – the regional dimension' in: T. Rowley ed., *The origins of open-field agriculture,* London, pp. 112–29.

Dahl, S. (1989), *Studier i äldre skånska odlingssystem,* Stockholm.

Dahlman, C. (1980), *The open field system and beyond.* A property rights analysis of an economic institution, Cambridge.

Dodgshon, R. A. (1980), *The origin of British field systems: an interpretation,* London.

Emanuelsson, M., and Segerström, U. (2002), 'Medieval slash-and-burn cultivation: strategic or adapted land use in the Swedish mining district', *Environment and History,* 8, pp. 173–96.

Enagrius, C. E. ed. (1808), *Samling af landtmäteri-författningar, innehållande så wäl kongl. maj:ts nådiga förordningar, resolutioner, rescripter och instructioner, samt af kongl. maj:t fastställde delnings – grunder och skattläggnings-methoder, som ock kongl. kamar: collegii och kongl. landtmäteri-contoirets kungörelser och circulairer, rörande landtmäteriet och justeringen i Swerige och Finland, ifrån 1763 års början til 1807 års slut. Utgifwen af C. E. Enagrius. = Stockholm. 1–2. 1808–1816 =. [Del 1], Med kongl. maj:ts allernådigste privilegio, Stockholm.*

Enequist, G. (1937), *Nedre luledalens byar. En kulturgeografisk studie*, Uppsala.

Erixon, S. (1978) [1931], 'Granne är grannes broder', *Byalag och byaliv*, Nordiska museet, Stockholm, pp. 32–45.

Erixon, S. (1960), *Svenska byar utan systematisk reglering: en jämförande historisk undersökning*, Stockholm.

Gadd, C.-J. (2000), *Den agrara revolutionen: 1700–1870 in*: J. Myrdal ed., *Det svenska jordbrukets historia, vol. 3, Stockholm.*

Gadd, C.-J. (2011), 'Agriculture in industrial society, 1870–1945' in: J. Myrdal and M. Morell ed., *The agrarian history of Sweden: from 4000 BC to AD 2000, Lund, pp. 118–64.*

Göransson, S. (1961), 'Regular open-field patterns in England and Scandinavian solskifte', *Geografiska Annaler*, 43, 1, pp. 80–104.

Hagland, J. R. and Sandnes, J. (1994), *Frostatingslova.* Norrøne bokverk, Oslo.

Hagström, J. O. (1751), *Jemtlands oeconomiska beskrifning eller känning, i akt tagen på en resa om sommaren 1749*, Stockholm.

Hannerberg, D. (1977), *Kumlabygden: forntid, nutid, framtid. 4, By, gård och samhälle*, Kumla.

Hansson, A., Olson, C., Storå, J., Welinder, S. and Zetterström, Å. (2005), *Agrarkris och ödegårdar i Jämtland*, Östersund.

Höglin, S. (1990), *Gårdar och gärdesgårdar i Markim*, Stockholm.

Hopcroft, R. L. (2003), 'Open-field systems' in: J. Mokyr ed., *The Oxford Encyclopedia of Economic History*, Vol. 4, Oxford, pp. 135–40.

Isaksson, O. (1967), *Bystämma och bystadga. Organisationsformer i övre Norrlands kustbyar. Om samspelet mellan lokaltradition och central påverkan*, Umeå.

Jansson, U. (1998), *Odlingssystem i Väneromradet: en studie av tidigmodernt jordbruk i Västsverige = [Farming systems in the Lake Vänern region: a study of early modern farming in Western Sweden]*. Stockholm.

Jansson, U. (2003), 'Allmänningarnas institutionella geografi i Viby socken på 1600-talet – en plats för lokalsamhället' in U. Jansson ed., *Med landskapet i centrum: kulturgeografiska perspektiv på nutida och historiska landskap*, Stockholm, pp. 211–32.

Larsson, J. (2009), *Fäbodväsendet 1550–1920: ett centralt element i Nordsveriges jordbrukssystem*, unpublished PhD thesis, Sveriges lantbruksunivitet, Uppsala.

Magnusson, G. (1989, 'Medeltida fäbodlämningar i Ängersjö' in: O. Hemmendorff ed., *Arkeologi i fjäll, skog och bygd. Del 2: Järnålder–medeltid* (*Fornvårdaren* nr. 24), Östersund, pp. 167–74.

McCloskey, D. N. (1989), 'The open fields of England: rent, risk, and the rate of interest, 1300–1815' in D. W. Galenson ed., *Markets in history: economic studies of the past*, Cambridge.

Myrdal, J. and Söderberg, J. (1991), *Kontinuitetens dynamik: agrar ekonomi i 1500-talets Sverige*, Stockholm.

Øye, I. (2000), 'Norway in the Middle Ages: farms or hamlets – and villages too?', *RURALIA III Památky arkeologické – Suplementum*, 14, pp. 12–23.

Parry, M. L. (1978), *Climatic change, agriculture and settlement*, Folkestone.

Pettersson, R. (2003), *Ett reformverk under omprövning: skifteslagstiftningens förändringar under första hälften av 1800-talet*, Stockholm.

Prince, H. (1977), 'Regional contrasts in agrarian structures' in: H. D. Clout ed., *Themes in the historical geography of France*, London.

Riddersporre, M. (2008), 'Det skånska bolet' in: A. Eriksson ed., Jordvärderingssystem från medeltiden till 1600-talet, Stockholm, pp. 23–38.

Roeck Hansen, B. (2002), *Gårdsgärdor och tegskiftesåker: resursutnyttjande och kulturellt inflytande i det gamla landskapet Västerbotten*, Umeå.

Roeck Hansen, B. (2010), 'Early land organisation around the Gulf of Bothnia', *Journal of Northern Studies*, 1, pp. 87–96.

Rowley, T. ed. (1981), *The origins of open-field agriculture*, London.

Sandnes, J. (1991), 'Utmarksdrift og ressursutnyttelse i Norge i eldre tid' in: H. Ilsøe and B. Jørgensen ed., *Plov og pen: festskrift til Svend Gissel: 4. januar 1991*, Copenhagen, pp. 213–22.

Sporrong, U. (1985), *Mälarbygd: agrar bebyggelse och odling ur ett historisk-geografiskt perspektiv,* Stockholm.

Sporrong, U. and Wennström, H.-E. (1990), 'Maps and mapping' in: L. Wastensson ed., *National Atlas of Sweden*, Stockholm.

Thirsk, J. (1964), 'The common fields', *Past and Present,* 29, pp. 3–29.

Thulin, G. (1909), *Redogörelse för de ecklesiastiska boställena III. Jämtlands län*, Stockholm.

Tollin, C. (1999), *Rågångar, gränshallar och ägoområden: rekonstruktion av fastighets-struktur och bebyggelseutveckling i mellersta Småland under äldre medeltid,* Stockholm.

Uhlig, H. and Lienau, C., eds (1972), *Rural settlements (Die Siedlungen des ländlichen Raumes, L'habitat rural),* Giessen.

Widgren, M. (1997), *Bysamfällighet och tegskifte i Bohuslän 1300–1750,* Uddevalla.

5 Open-field farming in Finland

Petri Talvitie, University of Helsinki, Finland

I. Introduction[1]

Finland, located between the 60th and 70th parallels, is the northernmost agricultural country in the world. Climatically, Finland falls into boreal northern Europe with low average temperatures and several months of snow cover limiting vegetation periods. Due to the environmental risks, Finland was settled later than many other European countries – huge wilderness areas remained unpopulated until the sixteenth century, or even later – and traditional burn-beating (Sw. *svedjebruk*, i.e. burn-clearing of forests) occupied an important place in regional agro-systems until the nineteenth century.

In this chapter I analyse the distribution and functioning of open fields in Finland (see figure 5.1) from the high Middle Ages until the late eighteenth and early nineteenth centuries, when most examples disappeared as a consequence of state-led land consolidation reforms, the so-called *storskifte* and *nyskifte* reforms. Throughout this period Finland was an integral part of the kingdom of Sweden.[2] The history of open fields in Finland offers an interesting case study for several reasons. First of all, the north-eastern corner of the European-wide open field system divides Finland into two halves. On the southern side of the border, fields were organized – broadly speaking – along the same principles as elsewhere in European open field regions, whereas on the eastern and northern side the settlement pattern was dispersed and arable fields were held privately. As a consequence, it is possible to investigate the factors that set limits on the diffusion of open-field farming. Additionally, the system's history is well documented in written sources, because it emerged in Finland at a comparatively late date, compared with other parts of Europe. By the time the diffusion of open fields into the area began during the thirteenth and fourteenth centuries, communal field systems had already begun to disappear in England.

II. Sources and earlier research

The large-scale cadastral maps, available from the 1630s, are the most important source available for all studies of the early agrarian landscape of Finland. The same applies to Sweden, where the number of seventeenth-century maps available is almost 20,000. For Finland, the figure is much lower, somewhere between 3,000 and 4,000, although this is

[1] I would like to thank Professor Ilkka Nummela from the University of Jyväskylä for valuable comments on this text.

[2] Present-day Finland gradually came under Swedish rule from the mid-twelfth century as a result of Swedish religious and political expansion. Finland was annexed to the Russian Empire in 1809.

Peasants and their fields. The rationale of open-field agriculture, ed. by Christopher Dyer, Erik Thoen and Tom Williamson, Turnhout, 2017 (*CORN* Publication Series, 16), pp. 99-119

BREPOLS PUBLISHERS DOI: 10.1484/M.CORN-EB.5.114270

Figure 5.1 The historical districts of Finland within present-day borders

just a rough estimation, because the cadastral maps have not yet been fully inventoried.[3] Most of the maps available are from the central agricultural districts of south-western Finland; the eastern and northern parts of the country were surveyed rather sporadically. In general terms, the geographical coverage of the mapping is satisfactory, and the material allows us to obtain a relatively accurate picture of all the major field systems and agrarian settlements before the land consolidation reforms of the eighteenth century. It should be noted, however, that the large scale cadastral maps reflect the rural landscape during the early modern period, which was (or might have been) rather different from that of the medieval period, when the open-field system emerged in Scandinavian countries. It is possible to overcome this challenge, at least to some extent, by using the metrological examination and retrogressive studies developed in the field of human geography, and by combining such information with field evidence (Sporrong, 1986; Roeck Hansen, 1996).

Important evidence on the open fields of Finland is also provided by the records of the local district courts (*häradsrätter*), which were the most important arbitrators of land disputes in the countryside. The court records, available comprehensively from the mid-sixteenth century onwards, contain much information about the operation and distribution of open fields. The juridical material supplements the often static, cross-sectional image provided by the cadastral maps. The court cases demonstrate, for instance, that the peasants remodelled field layouts by exchanging individual strips. The strips were exchanged when a farmer managed to prove that his strips were of poorer quality than those of his neighbours. Strips were also sold, donated and mortgaged, although such transactions were considered to be illegal by the Swedish crown, because they reduced the tax-paying ability of peasantry. These activities, however, were not uncommon (Säihke, 1963: 8; Jokipii, 1974: 35). In addition, the cases heard in the district courts are important because they reveal the day-to-day challenges that peasants had to face in an open-field village: endless boundary disputes, repeated neglect of fencing duties, and recurrent claims concerning the trampled harvest. The village by-laws are also interesting in this respect. Unfortunately, there are only a handful of by-laws available for the medieval and early modern period. Village communities only began to draw them up on a large scale during the latter half of the eighteenth century when most of the open fields had already been consolidated (Jutikkala, 1963). Most of the source material described above is available for the early modern period. For the medieval period, the evidence is much more fragmented, consisting of a few deeds and court cases from the fourteenth and fifteenth centuries. During that time the open-field landscape was already in place in many parts of Finland, and it is therefore also necessary to carry out archaeological field-work.

Despite the rich source material, open fields in Finland have been studied to only a limited extent, and many central questions remain to be answered. Furthermore, the topic has been investigated mainly by rural historians: archaeologists, human geographers and landscape historians have not been interested to the same extent. Longitudinal analyses for Finland, similar to those executed for Sweden by historical geographers, are rare (Kilpelä, 2006: 55). The most valuable contributions were made by historian Eino Jutikkala during the 1940s and 1950s and by Swedish cultural-geographer Birgitta Roeck Hansen during the 1990s. Jutikkala concentrated especially on the implementation of *tegskifte*, a regular

[3] Cadastral maps are available online at www.vanhakartta.fi (in Finnish).

strip-system found in the central parts of eastern Sweden and in the south-western parts of Finland. He also touched upon the question of rationality, although his ideas were, for the most part, based on previous Swedish research (Jutikkala, 1950; Jutikkala, 1953; Jutikkala, 1963). Birgitta Roeck Hansen was interested in the Swedish influence on land division and cultivation practices in Finnish agrarian society. She argued, for example, that the strongest impact on landscape organization was brought about by the Swedish colonization between the twelfth and fourteenth centuries. Her studies were based on the cadastral maps, which had not been used before to the same extent (Roeck Hansen, 1996; Roeck Hansen, 1998).

III. The agricultural districts of Finland

Rural historians regularly divide the territories of present-day Finland into three clearly distinguishable agricultural districts: the livestock district of northern Finland, the burn-beating district of eastern Finland, and the arable land district of western Finland. As a rule, the western district is further subdivided into a southern arable land district and the Pohjanmaa arable land district. The classification is based on the statistical data collected in the early nineteenth century, and presumably the boundaries of the agricultural districts were altered to some extent during earlier centuries. This applies in particular to the extent of swidden cultivation. On a general level, however, the division is valid for the high Middle Ages, when open fields spread to present-day Finnish territories (Soininen, 1974).

In the northern agricultural district the growing season is very short, about 155 days a year at best, and the climate was too harsh for large-scale arable farming. As a consequence, the inhabitants had to resort to other means of making a living. These included animal husbandry, fishing and hunting. A considerable area was devoted to grassland in these regions, and the arable fields were often very small, and sown every year. The most important crop was barley: the potato was not introduced until the early nineteenth century. The fallowing of fields was unnecessary, because the farmers possessed comparatively large cattle herds, and consequently the barley fields could be kept adequately manured (Soininen, 1974).

In contrast, in the eastern district of the country, grain growing held an important place in the local agro-system, although it should be emphasized that approximately half of the grain was grown by swidden cultivation (see figure 5.2). The principal crop was burn-beating rye or root rye (in Finnish *juureinen*). The productivity of swidden agriculture was high, especially with the *huuhta* technique – the migratory cultivation of rye and turnips in the ashes of burned spruce forest. An average yield ratio (i.e. output to seed planted) for rye was twenty to one, but in good years it could reach thirty or forty to one, sometimes even more. At the same time, swidden cultivation was much more extensive than arable farming. The burn-beaten fields gave only two or three harvests, after which the areas were used as pasture for some time. It was possible to burn the land again after twenty to forty years. For the *huuhta* technique, the sustainable resting period between sowings was sixty to seventy years. Moreover, the risk of crop failure was comparatively high, and therefore the peasants also cultivated permanent arable fields (Soininen, 1974; Solantie, 2012).

Figure 5.2 The burn-beating districts of Finland in the 1830s

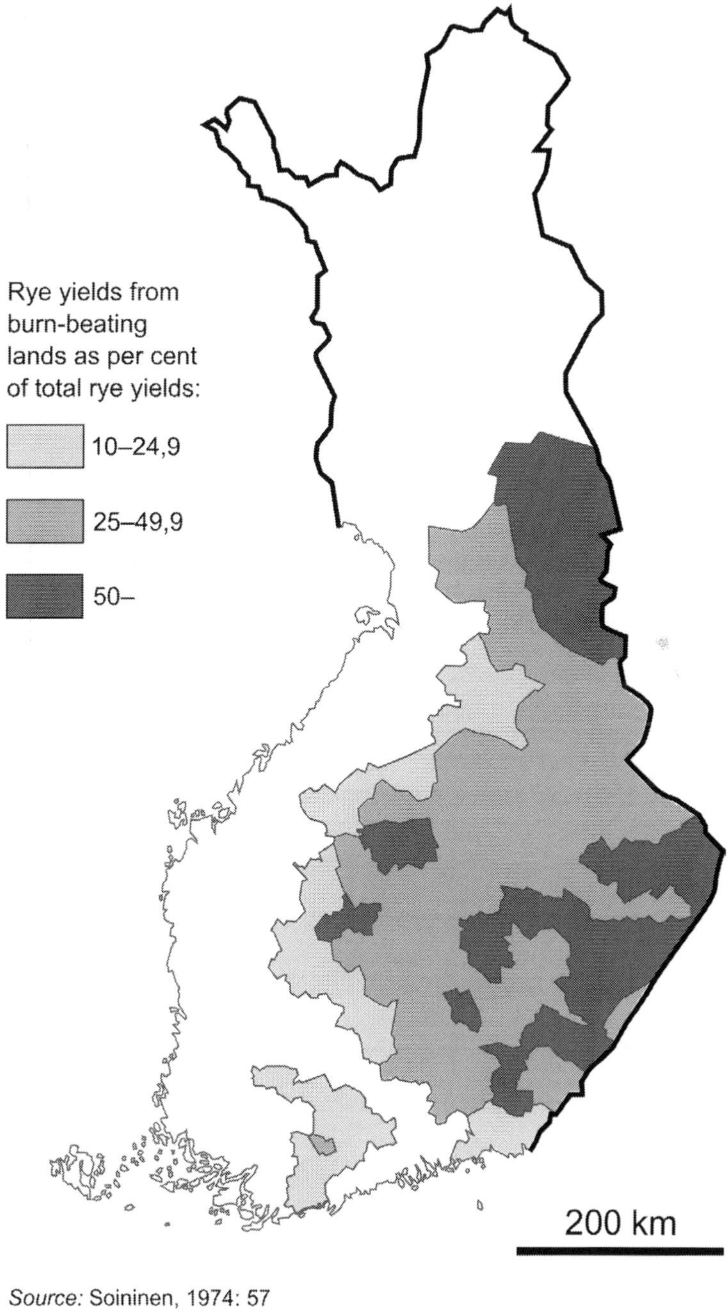

Source: Soininen, 1974: 57

Figure 5.3 The distribution of various rotation systems in Finland during the late eighteenth century

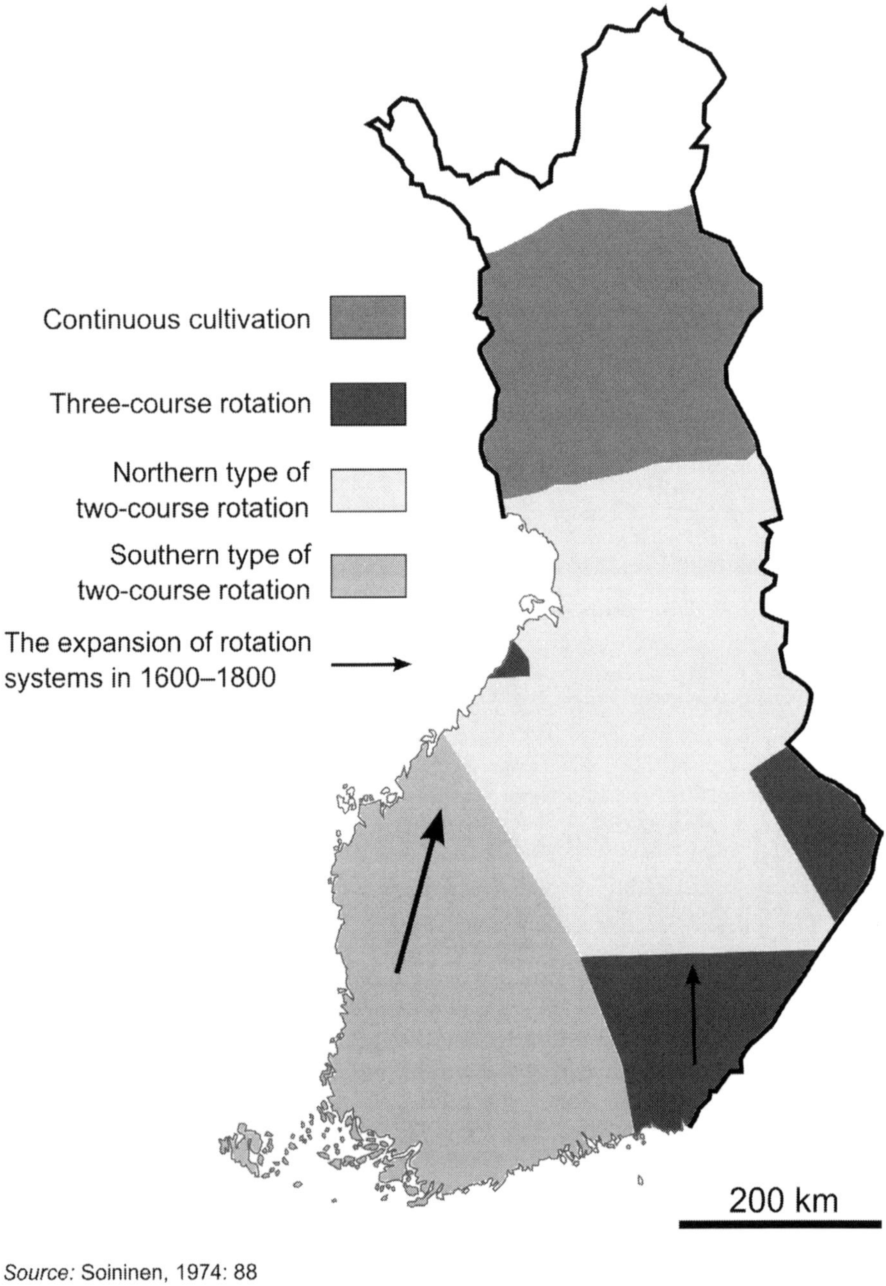

Source: Soininen, 1974: 88

In the western agricultural district, most of the yields were harvested from permanent arable fields, although peasants regularly resorted to burn-beating in western and southern Finland, with the possible exception of the Åland Islands and some of the costal parishes in Pohjanmaa. Swidden techniques were used, on the one hand, to clear new arable fields, but in the forested areas, where fields were small, swidden cultivation was an important source of livelihood in its own right. Burn-beating was especially important around the Salpausselkä moraine zone, which runs across southern Finland. In this area, peasants continued to burn forests until the late nineteenth century.

In the south-western parts of Finland, the dominant field system was a two-course crop rotation of winter rye (winter rye – fallow) (see figure 5.3). Barley was sown in a secondary crop rotation within the fallow field. The biennial system also dominated in the northern parts of the western arable land district (e.g. in northern Pohjanmaa), but here without the secondary rotation. Barley and rye were both part of the main rotation system (winter rye – fallow – barley – fallow). In a couple of parishes in northern Pohjanmaa, the farmers also used a three-course crop rotation of winter rye and barley. The boundaries of different field systems, however, were not fixed. During the eighteenth century, for instance, some of the farms in Pohjanmaa changed from three-year to two-year rotation, because yields were diminishing (Soininen, 1974; Roeck Hansen, 1996: 64).

The boundaries between different agricultural districts were determined by the climate. According to Solantie, the most important factor was the wintering of rye, but the frequency of night frost during the growth season was also crucial (Solantie, 1988; Solantie, 2012). In eastern Finland, the environmental risks were in general too high for the Scandinavian-type of arable rye used in the western parts of Finland, because the fields were often covered with a thick blanket of snow before the ground froze properly and snow mould fungi easily destroyed the sprouts. Root rye wintered considerably better than the arable rye, because it did not grow as densely, preventing the spread of snow mould. Root rye had other advantages compared to the arable rye, and it was therefore suitable for cultivation with the *huuhta* technique. In south-western Finland, the wintering risks of arable rye were less pronounced, but in order to make a secure living peasants had to rely on secondary occupations, such as animal husbandry, burn-beating, fishing as well as the selling of wood, tar and other forest products. The geography of arable and swidden cultivation also reflects differences in soil quality. In south-western Finland, clay and silt dominated the plough layer, whereas in eastern Finland, the main soil type was derived from moraine, typically covered with coniferous forest (Orrman, 2003).

IV. The character of the open fields

The different agro-systems were closely connected to systems of land division and settlement patterns. The distribution of open-field agriculture was restricted to the arable district in western Finland, and more precisely to regions where, during the nineteenth century, the prevailing arable system was a two-course rotation of winter rye. This is a rough generalization, however, because the distribution of different field systems during the earlier centuries has been studied only to a limited extent. In the eastern parts of Finland, on the other hand, the settlement pattern was dispersed. The farmsteads lay scattered throughout the landscape, and the fields were owned privately and surrounded by fences. Open-field agriculture never took root in these sparsely populated regions, where extensive swidden cultivation remained the norm. The farming of common fields was too intensive for slash-and-burn farmers, since the burning of forests required a mobile way of life. Burn-beating areas were sometimes located far away from the farm, and if resources were exhausted in a certain area, peasants abandoned their arable fields and moved to another place (Pirinen, 1982).

The private nature of land ownership in eastern Finland is clearly demonstrated by the fact that fields were often named after the family who cleared the fields (e.g. Hallikainen's field, Kopoinen's field), whereas in south-western Finland, the common fields were often named after cardinal points or landmarks (e.g. North field, South field, Church field; Virtanen, 1934). The distinction between western and eastern land-holding patterns was not absolute. Hamlets with common fields and intermingled plots could be found in eastern Finland (Pirinen, 1982: 350), and *vice versa*: single farms were common in western Finland, particularly in coastal areas and newly settled forest areas. In general, however, the boundary was clear and explicit.

While the open fields of south-western Finland displayed a measure of variation in terms of organization and plot size, in broad terms they conformed to the so-called 'regular' open-field or common field system described by, among others, Joan Thirsk, in having three basic characteristics: (1) the physical layout (fragmented land holding pattern, no permanent fences between strips); (2) the practice of common pasture on the stubble and fallow; and (3) the common regulation of farming activities. According to Hans Renes, a fourth point should be added to the definition, namely, enforced crop rotation (Thirsk, 1964; Hoffman, 1975; Renes, 2010). For south-western Finland, these characteristics could be described as follows:

1. Arable and meadow were divided into long and narrow strips among the cultivators, except in southern Pohjanmaa where most of the fields were divided into irregular blocks, as they were in northern Sweden (e.g. Hälsingland) and Dalecarlia. The block field pattern has also been found in the southern parts of Finland, but was probably rare there (Jutikkala, 1953). On average, holdings were subdivided into forty to fifty strips, but sometimes contained as many as a hundred. The boundaries of arable strips were normally marked with stones, wooden poles or small furrows, but meadows were commonly divided only before harvest, especially if there was a risk that the river would flood around them in spring and destroy the boundary markers. The outer limits of the common arable and meadow were permanently fenced in order to keep the cows out of the corn. In a few parishes in Häme and Satakunta, in contrast, all village land, including fields, meadows and commons, was customarily divided into two parts with one longitudinal fence. On the other side of the fence, the fields were laid fallow every second year and cattle could pasture there freely throughout the summer (Virtanen, 1934; Jutikkala, 1963).
2. The arable and meadow were thrown open for common pasturing after harvest and in the fallow seasons. The landless people were also traditionally allowed to let their cows graze on the common fields at these times. Village communities stinted the number of animals on the common pasture (Säihke, 1963; Jutikkala, 1963).

3. Crop rotation was organized on a village basis. As mentioned earlier, in south-western Finland, the dominant field system was two-course crop rotation. In Pohjanmaa there were also some hamlets where a three-course rotation was applied.
4. The cropping, grazing, and other facets of farm management were regulated in village meetings and in district courts (*häradsrätter*). Swedish medieval legislation includes various regulations concerning fencing and the timing of seasonal works. Villagers also made by-laws but these were rare until the 1750s. Regulation, however, was not as institutionalized as it was in central Europe, for instance, where villages appointed special officials responsible for supervising their farming activities (Blum, 1971).

The following excerpt from the district court of Siuntio parish (in the province of Uusimaa) illuminates the way in which farming was regulated during the eighteenth century. In 1737, four peasants from Strörsby hamlet requested the *häradsrätt* to confirm an arrangement by which they agreed that: (1) everybody must take care of fences around the meadows; (2) nobody was to pasture on meadows in spring time, because the grass was too wet; (3) nobody should let their horses, oxen, calves or other animals graze on the meadows during the growing season; (4) everyone was to take care of gates and water fences after the harvest, in order to prevent the animals from entering the meadow fold before the grass has sprouted. In autumn, it was permissible to pasture calves and draught animals in the fold because there was no other way to feed them, but all the neighbours had to agree to it; 5) the common arable was reserved for calves and sheep immediately after the harvest, because there was not enough pasture in the hamlet (Talvitie, 2013: 49–50). It has been argued that a shortage of pasture induced farmers to let their animals graze on the stubbles after the harvest, and on the fallows, and this example seems to strengthen that argument (Hoffman, 1975; Dahlman, 1980). Furthermore, it indicates that common regulations were necessary even in small hamlets such as Störsby, although the level of regulation was certainly much greater in larger settlements.

There were major differences between the Finnish open-field regions and the continental and English rural landscapes, especially the so-called 'champion' landscapes. First of all, until the late eighteenth century only a small percentage of the land area in western Finland was occupied by arable. Open fields were, in essence, small islands in a wider sea of forests. During the mid-seventeenth century, in the southern parts of Finland (the province of Uusimaa), only 2 per cent of the land was cropped. In Varsinais-Suomi, the cultivation rate was about 12 to 14 per cent in the 1550s (Talvitie, 2013: 192–93). The rest of the land consisted of meadows, pastures, forests, marshes and unproductive areas such as exposed bedrock. In eastern-central Sweden, which belonged to the same *tegskifte* area as south-western Finland, the cultivation rate was similar or slightly higher, about 10 to 30 per cent, during the mid-seventeenth century (Hannerberg, 1971: 22–23; Olai, 1983: 151). Such low cultivation rates were in large part a consequence of the fact that in boreal northern Europe, farmers needed a significant amount of winter fodder for their cattle, and as a consequence, large areas of village lands were devoted to grass.

Another feature which clearly distinguished south-western Finland from European 'champion' landscapes was the character of the settlement pattern: villages or hamlets comprised relatively small clusters of farms. This was also the case in many other parts of Scandinavia (Österberg, 1981: 66–69). In south-western Finland, an average hamlet consisted of three to ten farmsteads, and villages with over twenty farms were rare (Saarenheimo, 1983). Single farms were also common. In the coastal parishes of western Uusimaa, about 20 to 30 per cent of holdings were cultivated from isolated farmsteads during the early modern period (Talvitie, 2013). Open fields and enclosed fields existed side by side in these regions. On the other hand, in the Kokemäenjoki river valley (the province of Satakunta), where fields were located next to each other by the riverside, several hamlets combined their common fields into single rotation system (Jokipii, 1974: 41–43), while in Varsinais-Suomi and Uusimaa, the farmers typically had shares in the fields and meadows of neighbouring hamlets. In other words, open fields were often intercommoned by several hamlets.

A third and final characteristic of Finnish open fields was the regularity of their layout. In south-western Finland, the common fields and meadows were divided according to the principles of *tegskifte* (literally 'strip-division'), which was a highly regular land holding pattern typical of the Scandinavian rural landscape. In Finland, the system was often called *svenskt skifte* (Swedish land division), because *tegskifte* spread to Finland from Sweden during the Middle Ages. A regular *tegskifte* had two basic characteristics. Firstly, the width of the strips varied according to the assessment of individual households in terms of units called *byamål*. Secondly, the strips followed the same sequence in each furlong. In highly regulated villages, the size of the tofts was also proportional to the *byamål*. The idea behind this system was to divide village resources according to the fiscal burden of an individual household. That is to say, the more a household paid in tax, the more the members of that household were allowed to use village resources.

Historical geographers often use the term *solskifte* (sun-division) as equivalent to *tegskifte*, but according to Alf Ericsson the two concepts should be kept separate. *Solskifte* refers to the principle by which the allotments of land were arranged in sunwise sequence. The first principle of strip allocation, that is, proportional widths, was probably introduced earlier than sunwise distribution (Göransson, 1976; Ericsson, 2012: 346). Göransson argues that the *solskifte* principle was derived from England during the period of close political, ecclesiastical and cultural contact with England in the tenth to twelfth centuries (Göransson, 1961; Britnell, 2004: 30–32).

In practice, the fields were divided into strips using a rod (*stång*) or a rope. The length of the rod varied regionally, but the most common length in Finland was six ells (*c.* 3.5 meters). The parcelling progressed mechanically – one strip after another – without taking into consideration the contours of the earth or the location of the roads and ditches. If the farmers did not receive equal numbers of strips at the end of the process, the edges of fields were divided crosswise. A result of the mechanical system of land division, the furlongs were in general unimportant in the Finnish rural landscape. The most important components were fields (*gärde*) and individual strips, which were the actual farming units (Jutikkala, 1953; Jutikkala, 1963).

In several south-western Finnish hamlets, the principles of *tegskifte* were applied imperfectly. The sequence of strips varied, and the widths of the strips were not always proportional to the assessment units of farms. The regulated tofts were common only in regions near the city of Turku and around Häme Castle, situated in the province of Häme. According to Roeck Hansen, this was due to the fact that the Finnish rural landscape was later in date than the central Swedish *tegskifte* regions, where regulated hamlets were more prevalent (Roeck Hansen, 1992; Roeck Hansen, 1996).

At this point, it should be emphasized that the farmers of south-western Finland also had a varying number of outlying fields, which could be held by one or several farms. These often insufficiently manured private or semi-private enclosures were located on the edges of meadows or on the common waste. In some regions such private plots were reserved for the cultivation of oats, which were commonly used as horse fodder, but other crops might also be sown there (Soininen, 1974). Outlying fields of this kind seem to have been especially common in western Uusimaa. In some parishes, when single farms are excluded from the calculations, approximately 25 to 35 per cent of the arable was located outside the common fields by the seventeenth and eighteenth centuries (Talvitie, 2013: 57). This could partly be explained by the topography: the landscape is rocky and indented; exposed bedrock covers large areas of the ground surface. Extensive fertile plains are less common in western Uusimaa than in many other parts of south-western Finland (Kujansuu et al., 1993). The large number of outlying fields also reflects the importance of secondary occupations, such as fishing and burn-beating, in these regions. Regular open-field farming was highly regulated, because all the farming activities had to be carried out at the same time and according to a fixed sequence. The farming of private holdings was less strict, and as a consequence the inhabitants had more flexibility to organize their time during the short summer. Swidden cultivation, in particular, was considerably more time-consuming than arable farming. According to a calculation made in 1558, it took thirty-five man-days to cultivate a burn-beating field of one half-hectare including all the necessary stages of work from chopping down the trees to harvesting the yield. In contrast, eight man-days were enough to plough, sow and ditch an arable field of the same size according to an estimation made in the eighteenth century (Korhonen, 1985: 361). These estimates are probably skewed, because the Swedish crown actively strove to decrease the importance of swidden cultivation from the seventeenth century onwards, but they nevertheless illuminate the challenges posed by the communal farming of open fields and forests.

V. The emergence of open fields in south-western Finland and Pohjanmaa

Several Finnish historians date the origins of the open-field system in south-western Finland to the thirteenth century, even though the evidence for this remains relatively thin (Kaukiainen, 1980: 73–77; Orrman, 2003). The only systematic fieldwork has been carried out by Birgitta Roeck Hansen in Åland, where the open field system spread from Sweden between the twelfth and fourteenth centuries. At that time, an important re-organization of agrarian landscape took place. Dispersed, single farms and small privately owned fields were replaced by large agglomerations of habitation in hamlets, and open fields (Roeck Hansen, 1991). It is probable that open-field agriculture also became more common in other parts of south-western Finland from the thirteenth century onwards. During that period Finland, in common with the rest of Scandinavia, entered a period of settlement expansion. In southern Finland, settlement spread to previously unoccupied areas with soils which were less suited to contemporary tillage techniques, consisting as they mostly did of heavy clays, with moraines dominating at higher elevations. Additionally, Swedish colonists settled in coastal areas between the twelfth and fourteenth centuries, when the Swedish kingdom expanded eastward (Orrman, 2003).

Settlement expansion is thought to have been associated with the diffusion of more effective plough technology. The iron shares of ards became considerably larger after the eleventh century, and from the thirteenth century onwards a forked plough called the *socha* spread to the Finnish territories from Estonia and Novgorod (Orrman, 2003: 266–67). The earliest evidence for the use of a two-course rotation comes from southern Finland in the same period. To judge from pollen analysis carried out on Lohjansaari Island, located in western Uusimaa, a transition from one-year to two-year crop rotation occurred around the 1250s (Alenius et al., 2014). This dating corresponds to the findings from Östergötland in Sweden, where the transition took place during the same century. By the early thirteenth century, two-year crop rotation was common across large areas of central Sweden (Ericsson, 2012).

Several Finnish historians have suggested that the adoption of two-year crop rotation led to the development of open-field agriculture in Finland (Jutikkala, 1983; Orrman, 2003). According to Jutikkala, regular fallowing forced neighbours to interchange plots because otherwise the land of one farmer could have been left entirely in the fallow field every second year. The same logic applies naturally to the adoption of the three-year crop rotation. The exchange of plots thus created the intermixture of holdings typical of the open field system. In Sweden, this interpretation has been criticized by Mats Widgren, among others, because fields were also subdivided into strips in those regions where the practice of *ensäde* (the sowing of arable land every year) was common (Widgren, 1997). But in Finland, the distribution of regular open fields does seem to overlap with that of areas in which a two-year crop rotation was practised.

The last phase in the development of regular open fields, as these are depicted on seventeenth-century cadastral maps, was the introduction of *tegskifte*. The first reference to the system comes from the parish of Parainen (in Varsinais-Suomi) in 1332 where the value of a farm was expressed in alns (ells), an assessment unit which was a precondition for division by *tegskifte*. In the other south-western provinces, the new system of land division became common after the 1440s, and by the early sixteenth century the *tegskifte* was widespread throughout the region (Voionmaa, 1912: 202–03; Jutikkala, 1963: 86; Orrman, 2015). The diffusion of *tegskifte* was a gradual process, and the more ancient arrangement, by which fields were divided into irregular blocks (*i hopstycken*), survived until the sixteenth and seventeenth centuries in several hamlets. There were even hamlets where the strip-field and block-field patterns existed side by side. According to the bailiff's account book from 1556, the peasants of Satakunta were reluctant to adapt *tegskifte* if the soil texture of the fields was too stony and gravelly (Säihke, 1963: 9; Suvanto, 1973).

The diffusion of *tegskifte* was mainly based on private initiatives. Recent findings indicate that it was first adopted by ecclesiastical houses and noble landowners, but freehold peasants must also have been involved, given that most of the farms in south-western

Figure 5.4 The distribution of open fields in Finland (*c.* 1500 – *c.* 1750)

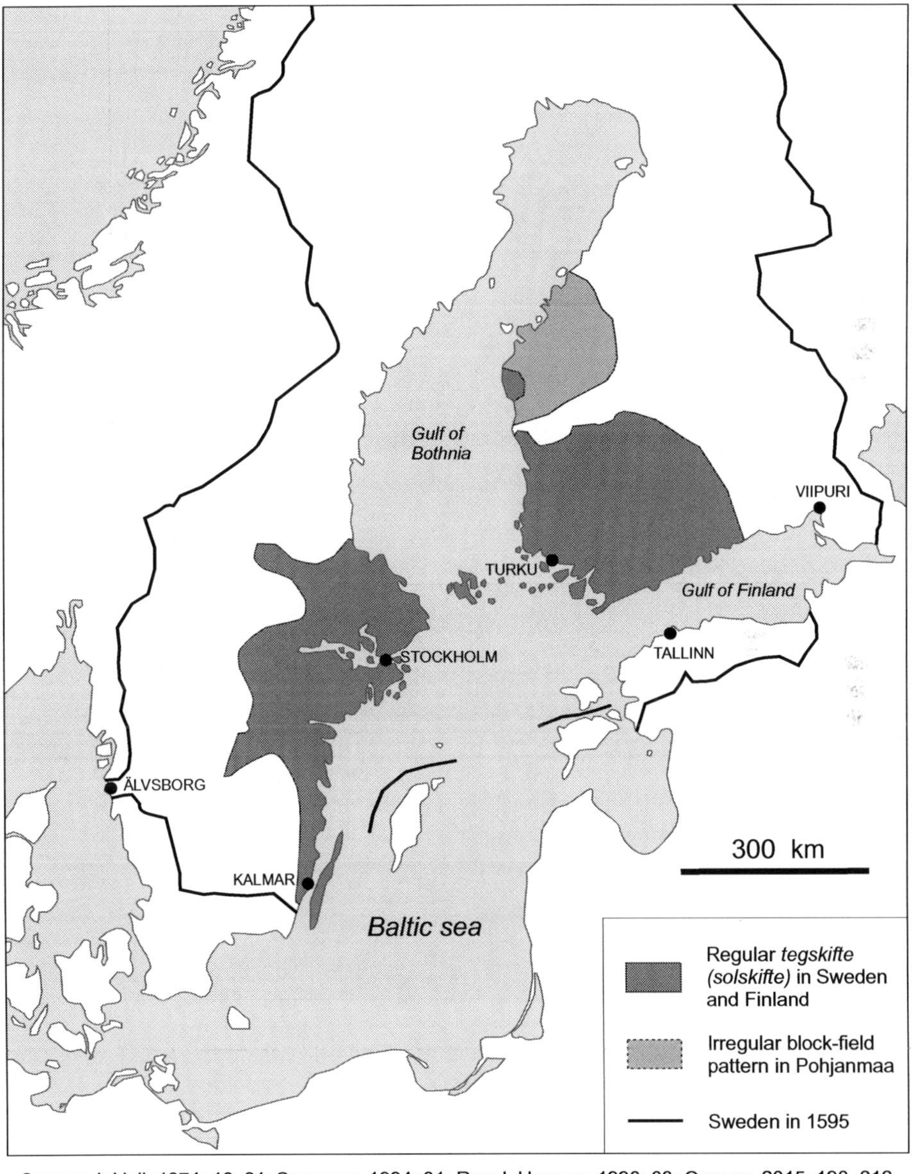

Source: Jokipii, 1974: 18–21; Sporrong, 1994: 31; Roeck Hansen, 1996: 63; Orrman, 2015: 190–212

Finland belonged to such people. The legislation favoured large landowners to some extent. According to King Magnus Eriksson's national law code (*c.* 1350), a landowner was entitled to apply for *tegskifte*, if he owned a minimum of a quarter of the land in a village. This principle was further elaborated in 1442 (King Christopher's code) in such a way that any landowner could demand a new division (Salminen, 2007: 153; Orrman, 2015).

Roeck Hansen has underlined the importance of central government in the spread of the new system. She argues that the regular *tegskifte* divisions in Finland resulted from a deliberate decision by the Swedish administration, aimed at the control of local production and tax collection. According to her, the decision was made at the end of the thirteenth century, when the medieval castle of Häme was built (Roeck Hansen, 1996: 66–70; Roeck Hansen, 1998: 192). The documentary evidence does not support her interpretation, however. Moreover, it is highly improbable that the medieval administration would have been capable of forcing villagers to adopt new forms of land division all over south-western Finland. The highly regulated hamlets found near Häme Castle may have resulted from central planning, but the land divisions with which they are associated were probably carried out after the mid-sixteenth century. During that time the Swedish king Gustav I (1496–1560) gave an order according to which a new, uniform *tegskifte* was to be implemented in south-western Finland, based on a six-ell rod (Orrman, 2015).

In the province of Pohjanmaa, outside the south-western *tegskifte* area, the emergence of open fields followed a different path. In these regions, the most important factor seems to have been partible inheritance. The province was sparsely populated until the latter part of the eighteenth century, which meant there was no shortage of land suitable for clearing. In addition, the amount of cultivable land increased over time due to rapid land uplift and shoreline displacement caused by isostatic recovery following the end of the last Ice Age. As a consequence, local families were able to divide their property equally among all the heirs. This tradition is visible in the field pattern: the plots of the divided holdings are located next to the site of the original farmstead (Roeck Hansen, 1996: 50–51). Over time, and after several subdivisions, the landholding pattern became extremely fragmented. Holdings could consist of 100 or even 150 plots (Jutikkala, 1963). These plots were often named, typically after geographical locations (e.g. Field behind the house; Field in the upper reaches), in order to facilitate land divisions and transactions (Luukko, 1945: 79).

One of the mysteries of Finnish field systems is the absence of *tegskifte* in Pohjanmaa, even though the other aspects of open-field farming were present. Roeck Hansen suggests that this was related to the weakness of feudalism in the locality (Roeck Hansen, 1996), and to some extent, this might be true. Until the seventeenth century, Pohjanmaa was dominated by a freehold peasantry and contained only one royal manor (*kungsgård*). A nobility never became established in the province during the medieval period. On the other hand, 93 per cent of the farms were owned by freehold peasantry all over Finland at the beginning of the sixteenth century; only 6 per cent belonged to the nobility and church (Orrman, 2003). Another reason why *tegskifte* failed to be adopted in the region was the lack of an assessment unit (*byamål*) suitable for regular strip allotment. In comparison to the southern provinces, Pohjanmaa was populated relatively late. The colonization period continued up to the fifteenth century, and the importance of arable farming was weak until that time, with the possible exception of coastal areas colonized by Swedish settlers. The most important sources of livelihood were fishing and hunting. As a result, the taxation system in Pohjanmaa differed from that found in other parts of south-western Finland. Until the early sixteenth century, the collection of annual crown tax was based on contracts between parishes and the crown and not, as it was in the southern provinces, on arable farming and related assessments units (Luukko, 1950; Seppälä, 2009). The situation was the same in Västergötland in Sweden. According to Alf Ericsson, regular *tegskifte* was never implemented in Västergötland as it was in Östergötland, because the province was never included in the military levy (*ledung*) and therefore never underwent a land assessment in *attungar*, a unit typical of Östergötland's farming district (Ericsson, 2012).

During the sixteenth and seventeenth centuries, the Swedish Crown reformed the taxation system several times and finally, in 1608, Pohjanmaa received an assessment unit called the *mantal*. A *mantal* assessment of an individual farm (for example, a third of a *mantal*) was based on its estimated production capacity, taking into account not just arable land but also resources from the forests and fishing. Some of the local peasants began to divide village resources according to the unit during the seventeenth century. Meadows, for instance, were divided into long and narrow strips among the cultivators (Luukko, 1950: 131–34). Arable fields, in contrast, retained their old layout. Roeck Hansen has demonstrated that local peasants began to apply a measurement unit called *revningstunnland* (measuring barrel land), equal to one *öresland* (13,122 square ells ≈ 0.5 ha.), during the same period. It seems that the application of *revningstunnland* did bring some regularity to the arable landscape, but it did not alter the block-field pattern typical of southern Pohjanmaa (Roeck Hansen, 1998: 197).

VI. Open fields and farming methods

According to current knowledge, open-field farming thus spread to Finland during the thirteenth century together with two-year crop rotation. The adoption of more effective tillage technology, especially the adaptation of heavier iron shares, is regularly associated with the same wave of innovations (Orrman, 2003). During the early modern period, arable technology did not come to a standstill as, for example, new plough types capable of turning the earth to some extent, and harrows with iron teeth, spread to Finland from Sweden in the course of the sixteenth and seventeenth centuries. Generally speaking, however, and in contrast to the situation in many western European countries, what were essentially medieval farming methods continued to be employed until the late nineteenth century (Soininen, 1974).

Large manors began to experiment with new rotation systems, such as English-style convertible husbandry and German-style *koppelwirtschaft*, during the early nineteenth century. In the latter part of the century, peasant farmers also began to integrate fodder crops into their rotation systems. The iron plough was adopted in Finland at the same time. The process was slow, however, and in many regions the two-year crop rotation retained its dominant position until the 1870s. Only after the 1870s did agricultural systems began to change rapidly, as new markets opened up for dairy products in Russia and elsewhere, and as an expansion of commercial forestry provided new sources of income for peasants. These changes facilitated investments in new agrarian technology (Soininen, 1974).

Despite the modest technological changes during the early modern period, some noteworthy developments had already taken place before the nineteenth century. The most significant was the gradual decrease of swidden cultivation everywhere in Finland. Burn-beating was a far more extensive form of land use than arable farming, and as the population grew steadily during the early modern period – excluding severe famine years such as at the end of seventeenth century – peasants were constantly forced to clear more arable land in order to feed their families. Open-field farming was not an obstacle in that respect. For instance, in western Uusimaa, where burn-beating was still widespread during the seventeenth century, the cultivated area grew at the same rate before as after *storskifte* reform. In the 1650s, only 2 per cent of the land area was cropped. By the 1760s, when the *storskifte* reform had already been implemented, the figure had increased to nearly 4 per cent. In 1881, about 120 years after the *storkifte*, about 7 per cent of the land was farmed as arable (Talvitie, 2013: 198). Land disputes probably became more common due to active land clearing, since some of the peasants cleared more land than others in terms of the assessment units of their farms (*byamål*), but it was possible to overcome these challenges with renewed *tegskifte* divisions.

Another major change was the introduction of secondary crop rotation (the so-called *toukomaa*) into the two-year crop rotation system, something which occurred during the early modern period, probably by the seventeenth century when rye replaced barley as the dominant crop in the region (Korhonen, 2003: 409–10). Parts of the fallow field were used to grow barley, together with leguminous plants and root vegetables. The crops were sown during spring (*toukomaa* means 'May field') in an enclosure that occupied, at most, one third of the fallow area. The location of the enclosure was changed each time the field lay fallow. The importance of secondary crop rotation was highlighted during the bad years, when the wintering of rye failed. In those days, peasants might sow the whole fallow field with barley. In other words, the secondary rotation system allowed the peasants to minimize environmental risks.

VII. The dissolution of open fields

The dissolution of open fields took place in two phases. The majority of the subdivided fields were consolidated during the eighteenth and nineteenth centuries due to state-led reforms called *storskifte* (with the first regulations enacted in 1749) and *nyskifte* (1848). The purpose of the latter reform was to complement the achievements of the *storskifte*. In south-western Finland, however, the importance of open fields had already begun to decline during the seventeenth century due to the large scale desertion of farms. This intensified in the period between *c.* 1570 and 1630, when the kingdom of Sweden was constantly at war and crop failures were common due to extremely cold winters (this was the period of the Little Ice Age). The extent of desertion varied, but in some regions as many as 30 or even 50 per cent of farms were marked as *öde* (deserted) in the cadasters surveyed around the 1620s. It should be noted at this point that *öde* refers to the inability to pay taxes. Most of the deserted farms were still in fact inhabited (Jutikkala, 1963).

Over time, the occupants of these farms began to pay taxes again, but a large number of holdings came to be attached to neighbouring peasant farms or to cavalry farms (*rusthåll*) owned by wealthy peasants. To some extent, peasant farms also became attached to tax-exempt manors owned by nobility (*säteri*). In the process, entire hamlets disappeared. In the province of Uusimaa, for instance, hamlets that consisted of six or seven farms in the 1550s had turned into single large estates by the end of the seventeenth century, although these cases were exceptional (Talvitie, 2013). In general, the number of farms in south-western Finland decreased by between 20 and 30 per cent between the 1570s and the 1630s. According to Jutikkala, the decline was related to structural change taking place in the Finnish countryside. The better-off peasants concentrated more and more on arable farming in permanent fields because wilderness areas used for burn-beating and hunting were now populated. Estates were enlarged at the expense of impoverished neighbours, however, and the number of landless people rose significantly (Jutikkala, 1980: 162).

During the same period, members of the rural upper class began to disentangle themselves from the village communities. There are several examples of clergymen and noble officers moving their homes away from the toft areas and creating larger cultivation units by interchanging strips with peasants living in the same village. Broadly speaking, however, the extent of these changes was relatively modest. Large numbers of manors continued to have their holdings intermingled with those of peasants during the eighteenth century, probably because their lands were mainly cultivated by tenants. Owners seem to have been more interested in rent than in agriculture. Furthermore, manors were comparatively small in Finland. Half of the noble manors (*säteri*) had only 10 ha. of arable land, and large estates with more than 40 ha. were rare, indicating that both manors and peasant farms were cultivated with the same traditional methods. The issue, however, has not been properly investigated (Jutikkala, 1932; Oksanen, 1981).

Storskifte reform began in south-western Finland during the 1750s, immediately after the 1749 Act. The reform had two objectives: to reduce the fragmentation of fields by consolidating the strips into larger farming units; and to privatize the village commons. The overall goal was to clear the way for the implementation of new farming techniques and hence more productive agriculture. In many regions, however, the results were meagre. In Häme, for instance, the land surveyors managed to reduce the fragmentation so little that there was no reason for peasants to abandon open-field farming. Two-year crop rotation prevailed, and the arable was thrown open for common pasturing after harvest. It took another hundred years to accomplish the reform (Saarenheimo, 1983). The results were better in those regions where villages were small, because it was relatively easy to reduce fragmentation in a hamlet of three to five farms, although peasants still continued to graze their animals on common waste after the *storskifte* was carried out and the commons had been privatized (Talvitie, 2013).

The final stage in the modernization of the Finnish rural landscape began after 1848 when the so-called *nyskifte* (new land division) legislation was enacted. Within the next 50 years, the last remnants of open-field farming disappeared throughout nearly all of south-western Finland.

VIII. Conclusions

Open fields were an essential part of the Finnish rural landscape from the late Middle Ages up to the nineteenth century. Such fields usually formed small cultivated areas surrounded by large tracts of forested land. Yet farming was organized along the same principles as in nearly all other European common-field areas. The emergence of open fields in Finland coincided with several important societal and economic changes. To begin with, the territories of present-day Finland were gradually attached to the expanding kingdom of Sweden during the period of the Crusades, beginning in the mid-twelfth century and continuing until the early fifteenth. During this period, Swedish government established itself by, for instance, constructing several fortifications (including, as already noted, Häme Castle) and by creating systems of regional administration. The earliest evidence concerning the collection of annual taxes for the Swedish king dates from the mid-thirteenth century (Kaukiainen, 1980).

The settled population expanded considerably during the same period, as was the case throughout Scandinavia. Settlement became denser in core regions and previously unoccupied areas were colonized. Moreover, Swedish colonists arrived from central Sweden (e.g. Uppland and Södermanland) and settled in the coastal areas around the Gulf of Finland (Haggrén, 2011). Settlement expansion was primarily conditioned by a number of developments which facilitated grain cultivation, the most important of which were new agricultural innovations such as better tillage technology, as well as by environmental change (Orrman, 2003; Solantie, 2012). Agricultural expansion began during the Medieval Warm Period (*c.* 980–1250), when the July mean temperature was *c.* 0.8°C warmer in Finland than it is today and winter snow cover was probably negligible (Tiljander et al., 2003; Alenius et al., 2014).

The emergence of open fields was not a single event, with a single cause. Instead, it was the consequence of a number of factors: the need to fence common infields, the intermingling of holdings, the transition from one-year to two-year crop rotation, the development of nucleated settlements, and the introduction of common grazing, together with the emergence of regulatory institutions such as legislation and village meetings, and finally the introduction of *tegskifte*, which was probably the final stage in their development. For the present, it is impossible to date the origins of all of these elements, but it is conceivable that the open field-system spread to the south-western parts of Finland during the thirteenth century. The first evidence of two-year crop rotation dates from the 1250s, and the first documentary evidence concerning the implementation of *tegskifte* dates from 1332.

By this time, the open-field system was already highly developed in Sweden, and it is probable that it spread into Finland from east-central Sweden together with two-course rotation (Ericsson, 2012). Factors supporting its adoption were naturally the same as elsewhere. The need to keep animals out of the arable fields and meadows was certainly one of the major motives behind the common fencing practices, and conceivably, the lack of pasture was one of the main reasons why fields were thrown open for common grazing (Myrdal, 2011).

As for the subdivision of fields, it seems to have been caused by both internal and external influences. In the southern and central parts of Pohjanmaa, where there was no shortage of cultivable land during the late medieval period, the fragmentation of holdings resulted from piecemeal colonization of land combined with partible inheritance, meaning that the intermixed field pattern was not a goal in itself but a result of long-term development. The fragmentation might also have resulted from biennial and triennial crop rotations, but their role is debatable. External influences were more pronounced in south-western Finland because of *tegskifte*, promoted by the Swedish crown in order to facilitate the collection of annual taxes.

The open-field system never spread to all parts of Finland. It was restricted to the south-west, an area which thus came to represent the extreme north-eastern corner of the European arable farming region. In the eastern and northern parts of Finland, dominated by burn-beating and animal husbandry, fields were held privately and the settlement pattern was dispersed. In other words, in regions where arable farming was not the dominant source of livelihood, the open-field system with its strips, common rights over the arable and meadow, and regulated farming activities never took root. In the northern agricultural district, summer was too short for intensive field cultivation, whereas in the eastern district, the combination of burn-beating and open-field farming would have been too demanding, given the arduous and labour intensive nature of swidden cultivation.

Bibliography

Alenius, T. et al. (2014), 'The history of settlement on the coastal mainland in Southern Finland: palaeoecological, archaeological, and etymological evidence from Lohjansaari Island, Western Uusimaa, Finland', *Journal of Archaeological Science*, 47, pp. 99–112.

Blum, J. (1971), 'The internal structure and polity of the European village community from the fifteenth to the nineteenth century', *The Journal of Modern History*, 43, 4, pp. 541–76.

Britnell, R. (2004), 'Fields, farms and sun-division in a moorland region, 1100–1400', *Agricultural History Review*, 52, 1, pp. 20–37.

Dahlman, C. J. (1980), *The open field system and beyond. A property rights analysis of an economic institution*, Cambridge.

Ericsson, A. (2012), *Terra mediaevalis. Jordvärderingssystem i medeltidens Sverige*, Uppsala.

Göransson, S. (1961), 'Regular open-field pattern in England and Scandinavian solskifte', *Geografiska Annaler*, 43, 1, pp. 80–104.

Göransson, S. (1976), 'Solskifte: the definition of a confused concept' in: R. A. Buchanan, R. A. Butlin and D. McCourt ed., *Fields, farms and settlement in Europe. Papers presented at a symposium, Belfast, July 12–15, 1971*, Holywood, pp. 22–37.

Haggrén, G. (2011), 'Colonization, desertion and entrenchment of settlements in western Nyland ca. 1300–1635 AD' in: G. Haggrén and M. Lavento ed., *Maritime landscape in change: archaeological, historical, palaeoecological and geological studies on Western Uusimaa*, Helsinki, pp. 152–79.

Hannerberg, D. (1971), *Svenskt agrarsamhälle under 1200 år. Gård och åker. Skörd och boskap*, Stockholm.

Hoffman, R. C. (1975), 'Medieval origins of the common fields' in: W. N. Parker and E. L. Jones ed., *European peasants and their markets. Essays in agrarian economic history*, Princeton, pp. 23–71.

Jokipii, M. (1974), *Satakunnan historia IV. Satakunnan talouselämä uuden ajan alusta isoonvihaan*, Pori.

Jutikkala, E. (1932), *Läntisen Suomen kartanolaitos Ruotsin vallan viimeisenä aikana 1–2*, Helsinki.

Jutikkala, E. (1950), 'Tegskifte in Finland. Några tillägg i diskussionen om den gamla skiftesformens uppkomst', *Fortig og Nutid*, 18, pp. 151–60.

Jutikkala, E. (1953), 'How the open fields came to be divided into numerous selions', *Sitzungsberichte der Finnischen Akademie der Wissenschaften*, pp. 117–42.

Jutikkala, E. (1963), *Bonden i Finland genom tiderna*, Helsinki.

Jutikkala, E. (1980), 'Asutus ja väestö' in: E. Jutikkala, Y. Kaukiainen and S-E. Åström ed., *Suomen taloushistoria 1. Agraarinen Suomi*, Helsinki, pp. 149–70.

Jutikkala, E. (1983), 'Tilusjaot ja tilusmittaukset ennen isojakoa', *Maanmittaus Suomessa 1633–1983*, Helsinki, pp. 7–19.

Kaukiainen, Y. (1980), 'Suomen asuttaminen' in: Jutikkala, Y. Kaukiainen and S-E. Åström ed. *Suomen taloushistoria 1. Agraarinen Suomi*, Helsinki, pp. 11–145.

Kilpelä, V. (2006), 'Jälkiä jo varhain verotusta varten tehdyistä maanmittauksista', *Pirkan maan alta. Arkeologisia tutkimuksia*, 7, pp. 42–51.

Korhonen, T. (1985), 'Kansankulttuurin juuret' in: Y. Blomstedt ed, *Suomen historia 2*, Helsinki, pp. 351–414.

Korhonen, T. (2003), 'Perinne hallitsee maataloustekniikkaa' in: V. Rasila, E. Jutikkala and A. Mäkelä-Alitalo eds, *Suomen maatalouden historia I. Perinteisen maatalouden aika. Esihistoriasta 1870-luvulle*, Helsinki, pp. 405–32.

Kujansuu, R. et al. (1993), *Tammisaaren kartta-alueen maaperä. Maaperäkarttojen selitykset, Lehti 2014*, Espoo.

Luukko, A. (1945), *Etelä-Pohjanmaan historia 3. Nuijasodasta isoonvihaan*, Vaasa.

Luukko, A. (1950), *Etelä-Pohjanmaan historia 2. Keskiaika ja 1500-luku*, Helsinki.

Myrdal, J. (2011), 'Farming and feudalism, 1100–1700' in: J. Myrdal and M. Morell ed., *The agrarian history of Sweden: From 4000 BC to AD 2000*, Lund, pp. 72–117.

Oksanen, E-L. (1981), *Anjalan historia*, Anjalankoski.

Olai, B. (1983), *Storskiftet i Ekebyborna. Svensk jordbruksutveckling avspeglad i en östgötasocken*, Uppsala.

Orrman, E. (2003), 'Rural conditions' in: K. Helle ed., *The Cambridge History of Scandinavia. Volume 1. Prehistory to 1520*, Cambridge, pp. 250–311.

Orrman, E. (2015), 'Iakttagelser om solskiftets utbredning och kronologi i Finland under medeltiden med särskild hänsyn till Nyland', *Historisk Tidskrift för Finland*, 100, 2, pp. 186–216.

Österberg, E. (1981), 'Methods, hypotheses and study areas' in: S. Gissel ed., *Desertion and land colonization in the Nordic countries c. 1300–1600. Comparative report from the Scandinavian research project on deserted farms and villages*, Stockholm.

Pirinen, K. (1982), *Savon historia 2:1, Rajamaakunta asutusliikkeen aikakautena 1534–1617*, Kuopio.

Renes, H. (2010), 'Grainlands. The landscape of open fields in a European perspective', *Landscape History*, 31, 2, pp. 37–70.

Roeck Hansen, B. (1991), *Township and territory. A study of rural land-use and settlement patterns in Åland c. AD 500–1550*, Stockholm.

Roeck Hansen, B. (1992), 'Solskifte i Nyland?', *Västnyländsk årsbok*, 15, pp. 32–47.

Roeck Hansen, B. (1996), *The agrarian landscape in Finland circa 1700. With special reference to Southwest Finland and Ostrobothnia*, Stockholm.

Roeck Hansen, B. (1998), 'Early agrarian landscape in Finland', *Geografiska Annaler*, 80 B, 4, pp. 187–201.

Saarenheimo, J. (1983), 'Isojako ja isojaonjärjestelyt', *Maanmittaus Suomessa 1633–1983*, Helsinki, pp. 20–62.

Salminen, T. (2007), *Joki ja sen väki. Kokemäen ja Harjavallan historia jääkaudesta 1860-luvulle*, Jyväskylä.

Seppälä, S. (2009), *Viljana, Nahkoina, Kapakalana. Talonpoikien maksamat kruununverot Suomessa vuosina 1539–1609*, Helsinki.

Soininen, A. M. (1974), *Vanha Maataloutemme. Maatalous ja maatalousväestö Suomessa perinnäisen maatalouden loppukaudella 1720-luvulta 1870-luvulle*, Helsinki.

Solantie, R. (1988), 'Climatic conditions for the cultivation of rye with reference to the history of Finland', *Fennoscandia Archaeologica*, 5, pp. 3–22.

Solantie, R. (2012), *Ilmasto ja sen määräämät luonnonolot Suomen asutuksen ja maatalouden historiassa*, Jyväskylä.

Sporrong, U. (1986), 'Agrarian settlement and cultivation from a historical geographic perspective: examples of the Mälar district in central Sweden', *Geografiska Annaler*, 68 B, 2, pp. 69–94.

Sporrong, U. (1994), 'Det äldre agrarlandskapet före 1750' in: S. Helmfrid ed., *Sveriges nationalatlas. Kulturlandskapet och bebyggelsen*, Stockholm, pp. 30–35.

Suvanto, S. (1973), *Satakunnan historia III. Keskiaika*, Pori.

Säihke, I. (1963), 'Varsinais-Suomen maanviljelys ja karjanhoito 1500-luvulla' in: E. W. Juva et al. ed., *Varsinais-Suomen historia 5*, Turku.

Talvitie, P. (2013), *Kyläosuudesta yksityiseen maanomistukseen. Isojako Länsi-Uudellamaalla 1700-luvulla*, Helsinki.

Thirsk J. (1964), 'The common fields', *Past and Present*, 29, pp. 3–25.

Tiljander, M. et al., (2003), 'A 3000-year palaeoenvironmental record from annually laminated sediment of Lake Korttajärvi, central Finland', *Boreas*, 32, 4, pp. 566–77.

Virtanen, E. A. (1934), 'Kyläyhteiskunta ennen isojakoa' in: G. Suolahti et al. eds, *Suomen kulttuurihistoria II. Sääty-yhteiskunnan aika 1*, Jyväskylä, pp. 417–77.

Voionmaa, V. (1912), *Suomalaisia keskiajan tutkimuksia. Veroja, laitoksia, virkamiehiä*, Porvoo.

Widgren, M. (1997), *Bysamfällighet och tegskifte i Bohuslän 1300–1750*, Stockholm.

6 *[...] this made the countrie to remayne champion, and without enclosures or hedging*[*]

Open-field landscapes and research in the Netherlands and in Europe

Hans Renes, Utrecht University and Vrije Universiteit Amsterdam, Netherlands

For many centuries, the open field landscapes were at the core of European agriculture. They have a place in public memory, particularly because so many people have worked on these fields in the period before mechanisation (Lemaire, 2013; Fig. 6.1). At first sight, an open field landscape looks uniform, large-scale and sometimes rather boring. On closer inspection this landscape shows a variety of small differences, landscape features and methods of use. This combination of uniformity and variety is also characteristic of the wider picture of open fields in Europe: on the one hand a landscape of grain fields that seems rather similar, on the other hand an enormous variation in detailed forms, in ownership patterns, in management and in history.

Scientific research into the open field landscapes in Europe started in the nineteenth century and goes on until the present day. However, international overviews are rare, particularly because different research traditions and terminologies make the results of studies in different countries difficult to compare (for the most recent exception, see Moorhouse and Bond, 2016). In this chapter, I try to show the variety in open field landscapes and to draw some lines through the different research traditions. After a terminological introduction, I present some data on the Netherlands to show the variety even within a very small country. Next I extend the argument to the European level and look at international developments in open field landscapes. In the following part, these different landscapes are connected to research traditions. In the final part, the present state of knowledge is extended to a series of research questions.[1]

I. Definitions

Many discussions on open field agriculture end in a Babylonian confusion because of inconsistent terminology and different research traditions. Therefore it is necessary to start with some terminological considerations, and with some attention to the individual components of the system.

[*] Gray, 1915, p. 186. The quote comes originally from George Owen's *Description of Pembrokeshire* (1603).

[1] The arguments given in this paper have been built up over the years and have partly been presented in earlier publications (Renes, 1988, 2007, 2010, 2012, 2016). The author thanks Prof. C. Dyer for his very helpful suggestions on an earlier version of this paper.

Peasants and their fields. The rationale of open-field agriculture, ed. by Christopher Dyer, Erik Thoen and Tom Williamson, Turnhout, 2017 (*CORN* Publication Series, 16), pp. 121-161

BREPOLS PUBLISHERS DOI: 10.1484/M.CORN-EB.5.114271

Figure 6.1 People working on an open field

Source: Educational picture 'A day on the field in July', by Cornelis Jetses, 1905

The simplest definition of open fields is: 'Open field agriculture was the means by which land was cultivated by the inhabitants of a township who worked their holdings in unenclosed parcels' (Taylor, 1981: 13). The term 'open field' in fact expresses the same idea twice: the word *field* (Old English: *feld*; Dutch: *veld*) already means open land. The open fields were particularly used for grain growing and in fact grain-growing could be included in the definition. The French equivalent is 'champ' (hence: Champagne), that also occurs in the English 'champion', a term that was already in use around 1600 (see quote in the title of this paper; Gray, 1915, p. 186; Williamson et al., 2013).

The essential characteristic of open fields was the absence of hedges, fences or walls marking the boundaries between the individual owned or tenanted parcels. In open fields that were ploughed in ridge and furrow (see below) it was easy to recognize the individual parcels. Elsewhere, this must have been difficult, especially when large open fields were ploughed or sown at the same time. In many cases boundary stones or wooden stakes have been placed at the corners of the strips (Rackham, 1986: 173–74). Such boundary markers are still in use and corner stones have been found in archaeological research (Verspay, 2016). There has been some discussion on the question whether grass

Figure 6.2 An aerial photograph from 1959 shows the village of Hart (48°22'17" N / 8°46'49" O) in Southwest-Germany

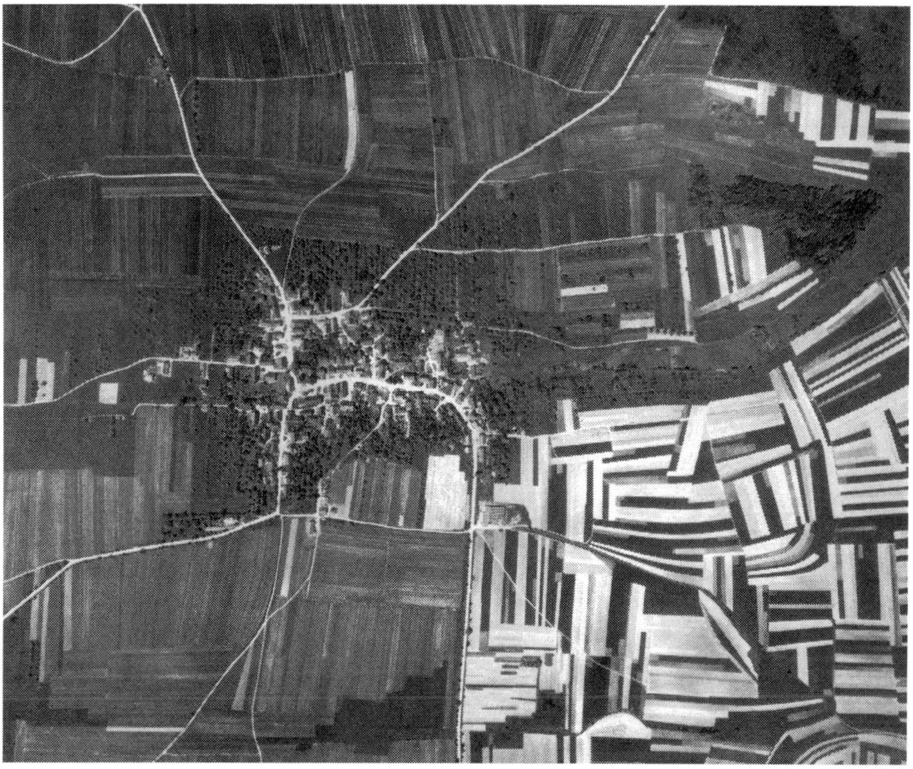

Note: Here a 'regular' three field system still seems to be in active use

strips ('balks' or 'baulks') have been used as strip boundaries. That this happened has been often suggested, for example by McCloskey (1975: 80), who saw this as one of the inefficiencies of the open field system, but evidence is rare. Beecham (1956) found no evidence in written sources, but Roberts (1973: 198) gives some reliable examples. More recently, Hall showed a photograph of the use of balks in an open field in Cornwall (Hall, 2014, Plate 15). He also published a map, dated 1595, of the village of Strixton in the central English open field region, that shows a number of balks, although they seem exceptional and were certainly not used as markers between individual strips throughout the field (Hall, 2014, Plate 5).

'Open field' is an umbrella-term for all arable fields without clearly visible boundaries between the plots of different owners. Often a distinction is made between 'regular' and 'irregular' open fields (see for example Bloch, 1931/1966: 49), in which the first is characterized by a strictly organized management.

Table 6.1 Definitions of open field, common field and Midland system (X = necessary component; 0 = possible component).

Aspect	Open field	Common field	Midland system
Main land use is arable	X	X	X
Individual parcels not surrounded by hedges or drystone walls	X	X	X
Strip-field pattern	0	X	X
Used as pasture after harvest and in fallow periods	0	X	X
Existence of common pastures and waste lands	0	X	X
Management of the open field is regulated by an assembly of cultivators	0	X	X
Two or three field system on a village base, with enforced crop rotations *(Flurzwang)*	0	0	X

Source: Mainly based on Thirsk, 1964.

The best known regular type are the so-called *common fields*, defined by Thirsk (1964: 3) as having the following characteristics: [1] the arable and meadow is divided into strips among the cultivators, each of whom may occupy a number of strips scattered about the fields; [2] both arable and meadow are thrown open for common pasturing by the stock of all the commoners after harvest and in fallow seasons. During the fallow periods, the arable becomes part of the common grazings; [3] there is common pasturage and waste, where the cultivators of strips enjoy the right to graze stock and gather wood, peat and other commodities, and [4] the ordering of these activities is regulated by an assembly of cultivators (the manorial court or a village meeting). This aspect suggests common regulations of the fields.

The ultimate development goes even further and divides the whole arable of a village in two or three large fields and in fact manages the whole village as one large productive unit. In this situation, all farmers have to follow the same crop rotation. This system is described by Gray as the Midland system (Fig. 6.2).

The development of 'regular' open fields is explained in different ways. The two most popular explanations are technical / organizational and risk management, both explanations were originally proposed more than a century ago. The technical and organizational line of thought – the necessity of villagers to cooperate – can be found by authors such as the Orwins (1938/1954). The explanation of the fragmentation of fields by risk management has been followed by many authors, McCloskey (1976) being probably the best-known. The argument of risk management seems logical, but is difficult to prove. A particular problem is, that it rather explains patterns, but not the processes behind it (Baker, 1979). Another problem is, that some of the most extensive open fields with fragmented ownership seem to lie in rather uniform landscapes, in which the risks seem equally spread over the whole village territory. But in regions with more differences in soil and hydrology, risk management can certainly explain part of the dispersion of lands belonging to an individual farmer.

Figure 6.3 Regular open fields after Hopcroft (1999), with additions

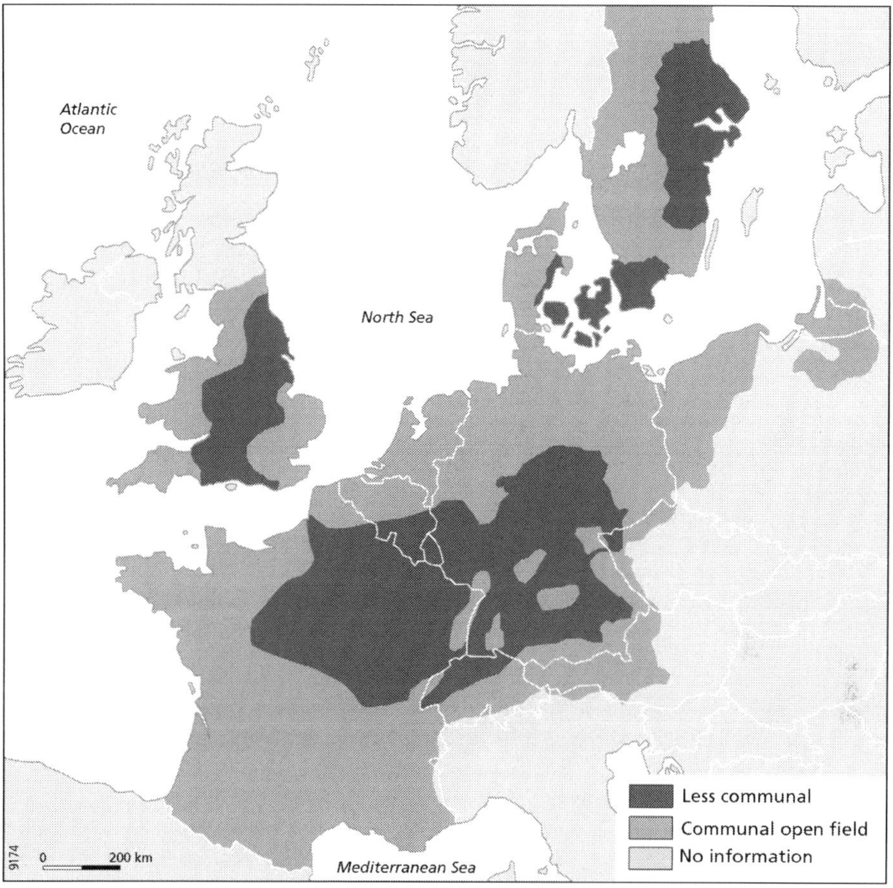

Table 6.1 shows the main characteristics of the open fields in general and of the two types of 'regular' systems just mentioned. The dispersal of the regular systems is still not yet very clear. The best, but still unsatisfying, map is made by Hopcroft (Fig. 6.3). But this is only one aspect of an enormous variety.

During the high Middle Ages, a three-field rotation system seems to have been in general use in most of the open fields of north-west and Central Europe. In this case, summer grain during the first year was followed by winter grain during the second year and fallow during the third year. An indication of a three-yearly rotation is the existence of many lease contracts that ran for a succession of three-year cycles. However, other systems, such as a two-field system, have existed during the Middle Ages and the early modern period; in the Mediterranean this system was even dominant.

In 'open-field landscapes' the arable land was so extensive that little room was left for pastures and meadows. It was necessary to use the grazing after the harvest and when it lay fallow in order to accommodate the animals that had to pull the plough and to provide the manure. In the so-called regular systems, this was organized on a village basis, but elsewhere it seems to have been implemented by individuals. Many open fields had temporary fenced areas (Bloch 1931/1966, p. 36; Spek 2006, p. 230), and sometimes might even have contained small parcels of completely individual pastures, with tethered animals.

II. Variety in open fields in the Netherlands

For a better insight into variations, we now turn to the Netherlands, a country where much research has taken place, but only a relatively small amount has been published in English. Figure 6.4. shows the main landscape types in the Netherlands. The main distinction is between 'pluriform' and 'uniform' landscapes. The first group is based on different land use areas, each with their own field-patterns, within the same village territory. In the second group a unified field-pattern, covering (almost) the whole settlement area, dominates the landscape. Here, different types of land-use found their place within this uniform field-pattern. Most of these landscapes are a result of planned settlement. An example is shown in Figure 6.5. Here, a medieval settlement (Sonnega, province of Friesland) was divided into large strip fields, each farm consisting of a single strip. In this strongly individualized landscape, the heathlands, although privately owned, were used in common. Also on the arable, the individual fields had no physical boundaries and had the characteristics of an open field.

On the basis of physical geography and cultural landscapes, a division into three main zones is often made, which is relevant for the present paper.

The extensive open fields in south Limburg

South Limburg is the only region in the Netherlands with extensive open fields. It lies in the extreme south-eastern part of the country (Fig. 6.6). The region is part of the large Central European zone of extensive open fields. The South Limburg landscape consists of loess-covered plateaus, dissected by deep stream valleys. On the plateaus, uninterrupted open fields developed already during the High Middle Ages.

Historic landscape research started relatively late in this region. A landmark study was the analysis of the medieval land-ownership in the southwestern corner of the region by Hartmann (1986). The history of settlement in this region shows long-term continuity in the valleys. The loess-covered plateaus, which had been settled during the Iron Age and the Roman period, seem to have been almost completely deserted in the following centuries of population decline (third to sixth centuries AD). The plateaus were not resettled until the period around 1100 AD. Around 1300 most of the plateaus were reclaimed and were probably turned into large grain fields (Hartmann, 1986). In a regional survey, we could confirm and extend the results for a much larger region (Renes, 1988).

The sandy regions

The second zone, including almost half of the Netherlands, is dominated by poor sandy soils. It is part of a zone that runs from Belgium to the Danish Jutland peninsula (fig. 6.7). The historic landscape, as it is depicted on nineteenth-century topographical maps,

Figure 6.4 Landscape types in the Netherlands

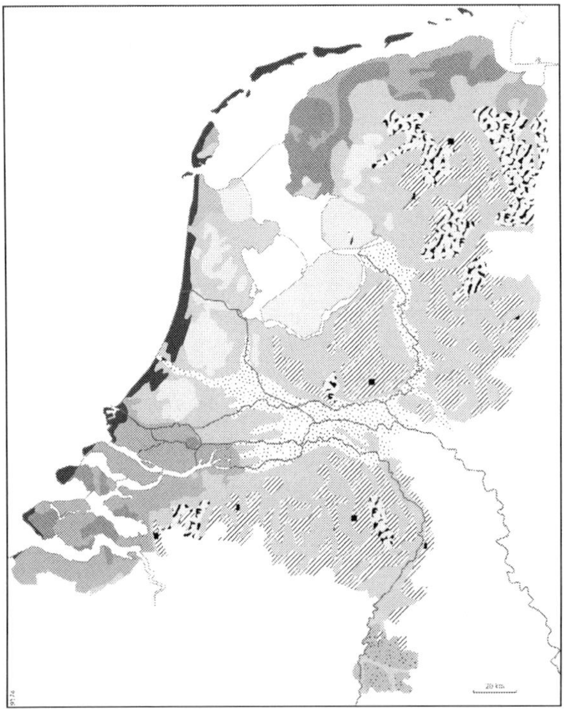

Physical geography	Man-made landscape (situation 19th century)	Field-pattern	
		Pluriform	Uniform
Higher parts of the Netherlands			
Plateaus South-Limburg	Open fields		
Sand, boulder-clay	Enclosures and small open fields <1850		
	Common heathland, forests and wetlands 1850		
Fluvial landscape	Enclosures and small open fields		
Coastal dunes	Enclosures and small open fields		
Coastal lowlands			
(Former) peat-and fenlands	Fenland reclamations (strip-fields)		
	Peat-colonies		
	Drained lakes (all >1500)		
Coastal marshes	Old landscapes		
	Reclaimed land (> Late Middle Ages)		

Figure 6.5 Sonnega (province of Friesland, NL), situation after the cadastre of 1832

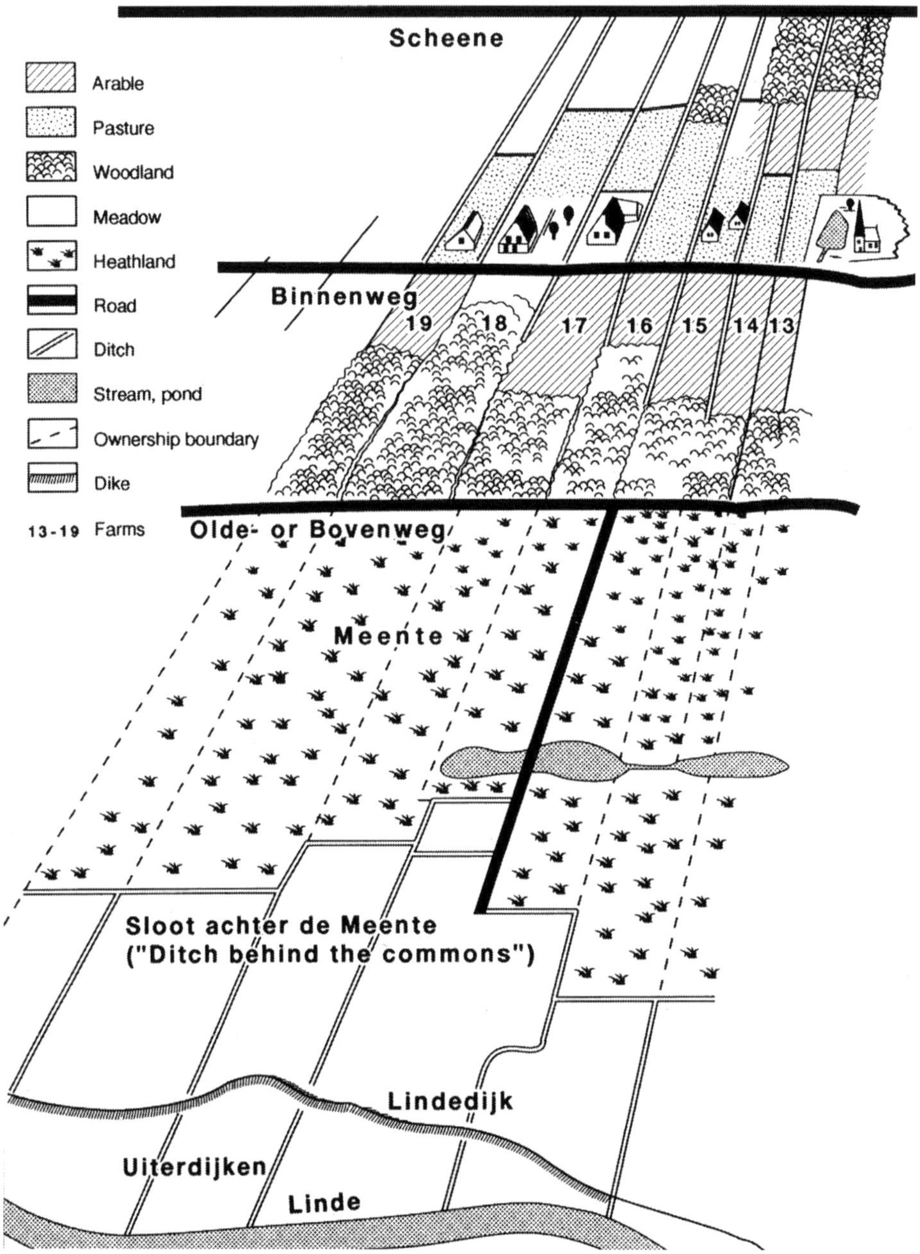

Source: after Bouwer, 1970, figure 4

Figure 6.6 Open fields in the southern part of the province of Limburg

The region is characterized by extensive open fields and by villages

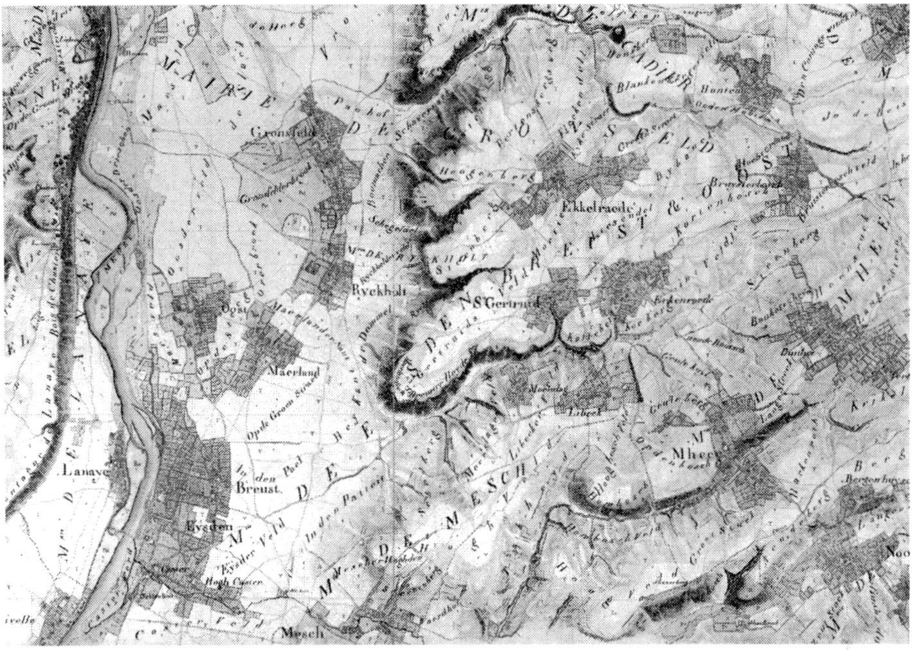

Source: map by Tranchot, 1806

consisted of limited amounts of arable land and extensive common forests and heath. The arable was situated on the best lands, which were mainly defined by geomorphology and hydrology: usually the highest parts of the landscape were used as arable. During the early modern period the cultivators in these landscapes practised an integrated system of mixed farming, in which the fertility of the arable was sustained by manure from herds of cows and sheep that roamed the pastures and the heathlands. The manure was combined with sods from the heathland and the resulting mix of organic and inorganic materials was brought on the arable. In the course of time the arable was covered by a dark, man-made plaggen soil (Spek, 1992; Van Smeerdijk et al., 1995).[2]

At second glance, these landscapes reveal much variety. Three late nineteenth-century topographical maps show settlements in different regions in the sandy landscapes. The first one (Fig. 6.8) depicts part of the province of Drenthe in the north-eastern Netherlands, a region with huge common heathlands. The second map (Fig. 6.9) depicts a sandy

[2] It is remarkable that plaggen soils seem not to have existed on the sandy soils of Breckland (East-England), although they are known in other parts of the British Isles (Scotland; Grime and Guttmann-Bond, 2011).

Figure 6.7 The 'Geest' (sandy landscapes) in the Low Countries and North-western Germany

Source: after W. Müller-Wille (Haartsen et al., 1989, p. 98)

region in the central Netherlands. In this region, the settlement pattern was dominated by individual farms, each with its own arable field. This resulted in an enclosed, small-scale landscape. Each farm appearing on the map is named. The third map (Fig. 6.10) depicts a landscape in the southern part of the Netherlands. The contrast with Drenthe is remarkable: the landscape is densely settled, even crowded and the common heathlands were smaller.

There is a long tradition of research into these landscapes, reflecting a Dutch tradition in historical geography which was closely connected with physical geography. This resulted partly from the early post-war development of historical geography as a discipline that was much stimulated by questions arising from the national systematic soil survey that had started in the late 1940s, and from water management (Renes, 1999). But it is also difficult to deny the strong influence of the physical environment on the early modern agrarian landscape in these regions.

Figure 6.8 Topographical map of the village of Borger (Drenthe, the Netherlands)

The map is showing a village with an open field, surrounded by common heathlands

Sources: Topographical map nr 188 (1898), © Topografische Dienst Kadaster, Apeldoorn and Hisgis.nl.

Note: The small map shows the ownership pattern on the open field of Borger.

Figure 6.9 Topographical map of the landscape around the villages of Achterveld and Barneveld in the central Netherlands

The map is showing dispersed settlements and enclosures. The agrarian system was comparable with that in Borger, but on the scale of individual farms.

Source: Topographical map n 429 (1870), © Topografische Dienst Kadaster, Apeldoorn

During the 1970s, most authors agreed on a number of issues. The individual enclosures and dispersed settlement were connected to a fragmented physical landscape. The higher ground was used for arable, as pasture and meadow lay mainly in the stream valleys, and the rest of the land formed the common pasture. The poor sandy soils could only be used for arable when large enough pastures and commons delivered enough manure. These landscapes were often described in terms of almost completely self-supporting, closed systems. The difference between landscapes with open fields and landscapes with individual enclosures was explained by the scale of the physical landscape, with the region of figure 6.9 containing small sand ridges only large enough for the arable of one or two farms. The system was seen as very stable, although in the course of time small intensifications and reclamations took place.

Difficult to explain were the differences in population density between the regions, with the regions of Drenthe and Brabant (Fig. 6.8 and 6.10) as extremes. The Groningen geographer H. J. Keuning (1838, 1964) saw the settlements of Drenthe as a basic type that had changed little since the (early) Middle Ages. The denser settlement pattern in the southern part of the country was explained by the light hand of management on the commons: whereas in Drenthe the commons were completely controlled by the large farmers

Figure 6.10 Topographical map of a landscape near Eindhoven (Noord-Brabant, the Netherlands)
This map is showing again a comparable landscape, but much denser populated, with more arable land and less common heathlands.

Source: Topographical map n 670 (1900), © Topografische Dienst Kadaster, Apeldoorn

in the village, who held their properties together by a system of undivided inheritance, in the southern provinces the number of farmers was made relatively large by practising a system of partible inheritance. Here, all villagers shared in the use of the commons. In this last situation, reclamations were less strictly opposed, leading to the gradual disappearance of large parts of the commons.

Interpretations of the medieval history of the sandy landscapes have changed remarkably during the last decades, although the regional distribution of research is quite unbalanced. Drenthe has a long history of research by archaeologists, historians and historical geographers. During the 1980s, the agrarian historian Jan Bieleman (1987) demonstrated the dynamism of the region's early-modern history. On settlement history and field-patterns there is the landmark study by Spek (2004), whose case studies show that the existing villages could be traced back to smaller settlements and that particularly in periods of high prices (1200–1350, 1570–1650) farms were divided. At the same time the case studies made clear that each village had its own, unique development. The nineteenth-century landscape of Drenthe was not the result of long-term stability but had in fact a complex history. Instead of a closed, self-sufficient landscape, this region had transformed from a mixed economy towards an emphasis on the production of rye on the open fields.

In the central and eastern parts of the country the amount of in-depth research is still limited, although there are recent exceptions (Spek et al., 2010). In the southern provinces much research has been done by medieval archaeologists, but there are still few detailed studies of landscape and settlement history. As in other regions, the initial construction by archaeologists of a landscape transformation (in this case in the middle of the thirteenth century) has recently given way for a more gradual and regionally differentiated development (Vangheluwe and Spek, 2008).

The recognition that these landscapes were in fact open economies, made it possible to devise new interpretations of the differences. The variations in population density are nowadays explained by the distance from the main urbanized regions: the closer to the towns of Flanders (late Middle Ages), Brabant (sixteenth century) or Holland (seventeenth and eighteenth centuries), the greater the rural population density was and, hence, the higher land prices and the more intensive agriculture. Brabant had a very much higher population density and percentage of arable than Drenthe. In Brabant, the open fields had grown over a very fragmented geomorphological landscape, that was partly levelled when plaggen layers were spread out over the land. In this context, the different regimes for the commons must be seen as an effect, rather than as a cause for the different intensities in the use of the landscape.

The sandy landscapes were transformed since the late nineteenth century. First, the crisis of the 1880s brought a new emphasis on animal husbandry and a shift in the use of the open fields from bread grain (rye) to animal fodder (nowadays the former open fields are planted mainly with maize). During the 1890s, when agriculture was recovering from the crisis, a combination of artificial fertilizers, improved education and new credit facilities (the farmers' loan banks) made it possible to reclaim the former heathlands, that mostly disappeared in the following four decades.

The delta of Rhine and Meuse

In the delta, also the subtle differences in geomorphology were reflected in the structure of the medieval agrarian landscape. On the higher natural levees that had been built up by sea or rivers – lands that were fertile and well-drained – permanent arable was possible and small open fields could develop. Figure 6.11 shows a late nineteenth-century topographical map of part of the fluvial landscape, with villages and open fields. In this region, there is still a need for detailed settlement studies.

Names

Not just the open fields themselves but also their names show a large variety. Figure 6.12 shows the regional names for open fields in the Netherlands. These names can be distinguished into three groups:

[1] *Names for open land*. The main example of these is 'veld', related to the German *Feld* and the English *field,* that originally means 'open' land (Philippa et al., 2009; Fowler, 2002: 127; Hooke, 1988: 51). The word 'open' could probably mean visually large-scale as well as 'accessible for everyone', but both meanings could be used for arable fields (that were open for grazing after the harvest) as well as for the common heathlands. In some regions, in the eastern Netherlands as well as in Northern Germany, *veld/Feld* meant the

Figure 6.11 Topographical map of part of the Dutch fluvial landscape, with villages and arable on the fertile natural levees

The lower lands were enclosed and surrounded by water-filled ditches during the High Middle Ages, but still drainage was so problematic that they could only be used as meadows and extensive pastures.

Source: Topographical map n° 508 (1870), © Topografische Dienst Kadaster, Apeldoorn

common pastures: the heathlands of the sandy landscapes and the common grasslands in the riverine region, whereas in the southeastern Netherlands and in most of Germany it became the common name for the visually open arable lands (Bach, 1953: 384; Schönfeld, 1950). This last meaning is also known in England, for example in the contrast between the neighbouring Warwickshire regions of Feldon (dominated by open arable fields) and the small-scale 'forest' landscape of Arden (Hooke, 1981: 53; this book p. 00).

[2] *Names for arable*. The most typical of these is 'akker', in Friesland 'ikker' (Wiersma, 2015), that is known in most of the Low Countries and originally means arable. The word *akker* is mentioned in Flanders in the early ninth century (Philippa et al., 2009). A derived meaning is 'a piece of (arable) land of a certain size', as in English 'acre'. A *kouter* is typical of Flanders. According to Belgian historians, an earlier pattern of dispersed settlement and small enclosures developed during the high Middle Ages by new reclamations into a landscape of villages with open fields. These fields were called *kouter*, from Old French *colture* (Spek 2004: 634; Thoen 1990). The only known example in the Netherlands is a small open field in the village of Rijsbergen, close to the Belgian border. Here, the *kouter* is mentioned in 1376 and the name was possibly brought here by the St Baafs Abbey

Figure 6.12 Regional names for open fields in the Netherlands

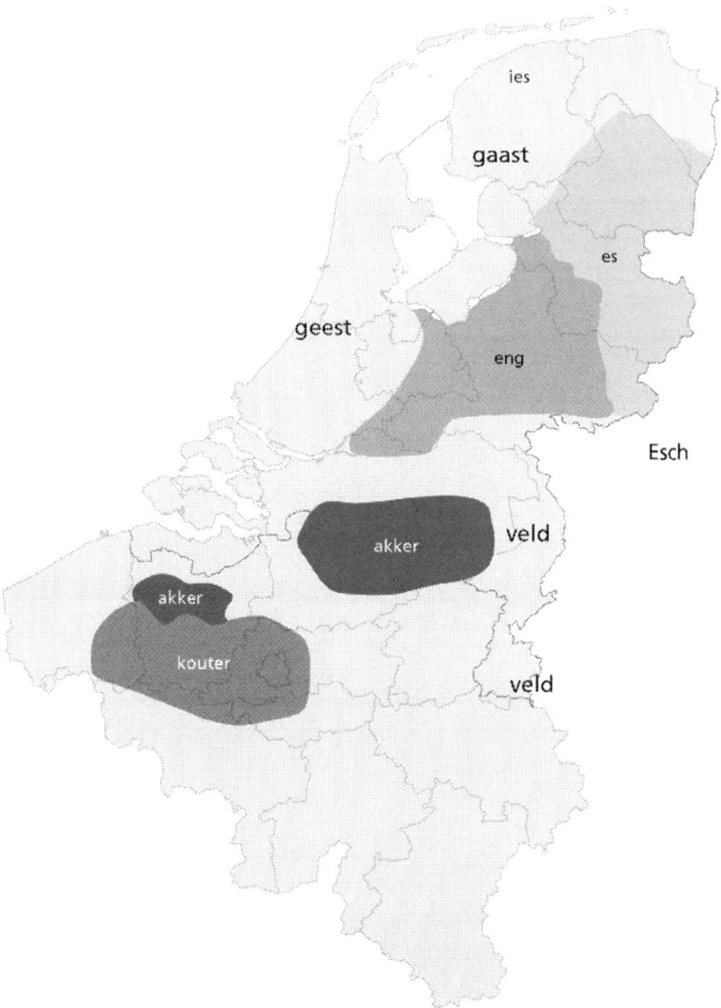

Source: after Blok (Verhulst and Blok, 1981), with additions

from Ghent, that owned lands in this village (Buiks, 1997: 108; Leenders, 1996: 458; 2000). The word 'es', the common term in the north-eastern part of the Netherlands and in neighbouring Germany (as 'Esch') is explained by Bach (1953: 384) as 'the arable lands of a settlement'. Yet another term within this category is 'valge', that is known in parts of the north-eastern Netherlands as a name for arable (Wiersma, 2015) and that seems originally to refer to the ploughing of land (Van Veen and Van der Sijs, 1997). In English the same word developed into the term 'fallow', referring to a stage in the crop rotation system in which the land was not sown, but was ploughed several times for weed control.

[3] The third group of names is derived from *soil and geomorphology*, and mainly consists of names for high and dry sandy ridges in regions dominated by wetlands. As in such regions the possibility for arable were limited to the higher grounds, these became almost synonymous with arable. The words 'eng' and 'enk' originally must have meant 'higher land' (Blok, 1958). Another term is 'geest' (German 'Geest'; Friesian 'gaast'), that originally meant dry, infertile land and developed into a name for the sandy ridges in the coastal regions of the Netherlands and Northwest-Germany (see above fig. 6.7). In the Dutch regions, the word developed further and is now in the coastal zone almost synonymous with the small open fields (Baas et al., 2014).

Land use and field-patterns

So the variety is enormous. These differences extend also to field-patterns. Whereas the open fields in Drenthe and in South-Limburg more or less fit into the standard pattern of subdivided strip fields, most open fields in Brabant are characterized by block-fields.

Small open fields are not by definition simple in field-pattern and in organization. Many of the relatively small open fields of Drenthe have a complex structure. Also, the only known written sources that mention forced cropping regulations (German: *Flurzwang*) in the Netherlands come from this region (Bieleman, 1987: 577–80). Perhaps the best source comes from a court-roll from the village of Gees that, as late as 1772, mentions a bye-law saying that 'first one, then the other part of the 'es' was sown with summer and winter rye' (Bieleman, 1986: 577).

Such cropping regulations on a village scale have never been discovered elsewhere in the Netherlands, not even in the open-field landscapes of southern Limburg. Some individual large landowners seem to have used cropping regulations on their own lands, that were dispersed over the village fields and were mostly rented out. For an example of such a 'dispersed demesne' (Hall, 2014: 96) see the example of the village of Vlijtingen, just across the Dutch-Belgian border, where the largest landowner, the Maastricht abbey of Saint Servatius, seems to have run its own three-field system with forced crop rotation on its estate (Hackeng, 2006; Fig. 6.13). This suggests that the other farmers in the village used a more individual agriculture.

III. From the Dutch to the European picture

The distinction between regions that specialized in grain production and were dominated by open fields on the one hand, and regions with small open fields within systems of mixed farming on the other, can be extended to the European level. The distinction is a basic ingredient in literature on historic landscapes in England ('champion' and 'woodland') and France ('champs' and 'bocage', see Bloch, 1931/1966; Smith, 1967). It goes back to the High Middle Ages. The earliest open fields, known from the Early Middle Ages, were small and surrounded by pastures, meadows, forests and wilderness. From the tenth century onwards, these landscapes started to develop in the two different directions just mentioned, leading to a distinction between 'open field landscapes' and 'landscapes of mixed farming'. Figure 6.14 gives an impression of the European distribution of both types, based on the well-known map of European settlement types by René Lebeau, translated by Hugh Clout (Lebeau, 1969; Clout, 1988). The map needs

Figure 6.13 The village of Vlijtingen

Source: Hackeng (2006)

Figure 6.14 Landscapes in Europe

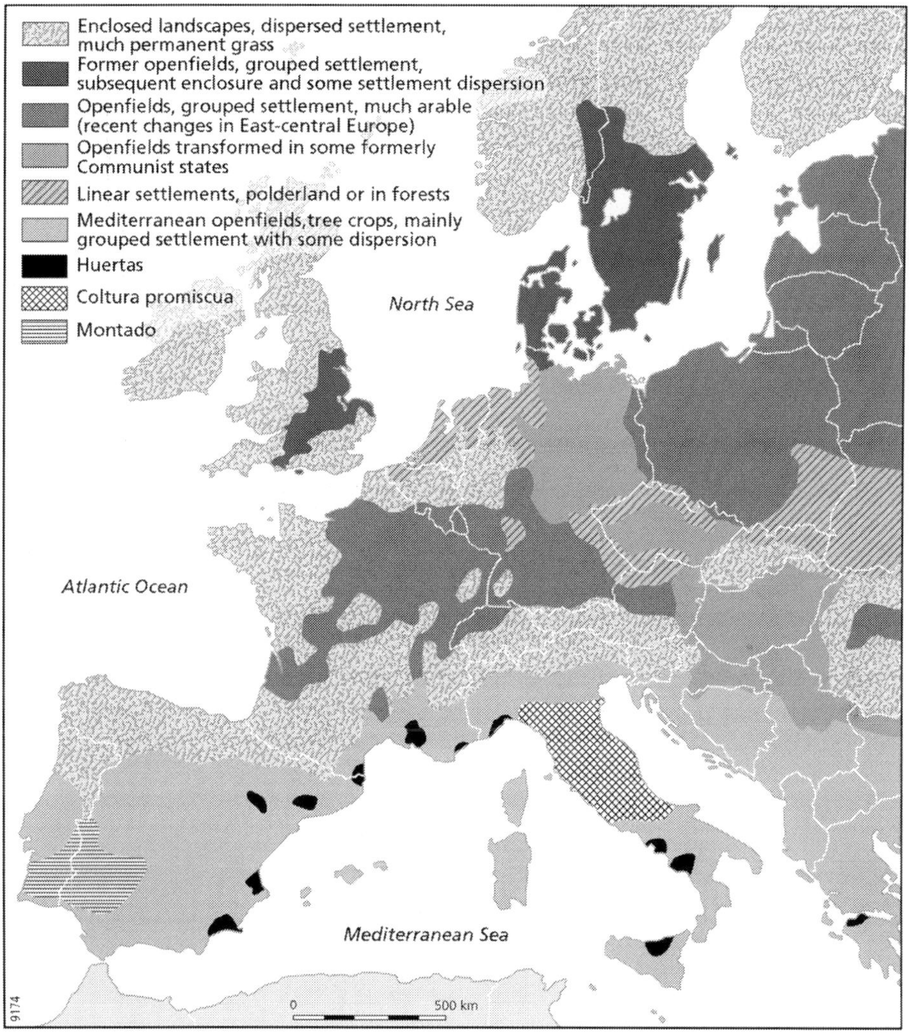

Source: after Lebeau (1969) and Clout (1998), with minor additions

revision, and for some regions (for example for southeast-England) enough data are available to do so. For many other regions, however, the necessary data are lacking.

Open-field landscapes

A number of regions started to specialize in grain production. In these regions, the open fields grew into large complexes, often taking in most of a village territory. These visually open landscapes were known under terms such as Champagne (French) or 'champion'; I will use the term *'open-field landscapes'*. The open field systems overran older agrarian structures, which were partly erased and partly incorporated in the new system. In English historiography there has been a prevalent notion of a replanning of the landscape around the ninth and tenth centuries (Williamson, 2013: 13; Brown and Foard 1998), mainly based on archaeological evidence. However, more archaeological research made clear that the picture is varied and that on a regional scale developments were more gradual (Williamson, 2016).

Many older, sometimes even prehistoric, landscape features have been discovered in open-field systems (for some interesting examples: Chadwick, 2013), but they provide few clues about the origins of the open fields. The open-field landscapes are often situated on the best soils, such as the large east-west zone of loess-soils that runs through Western and Central-Europe. However, the location in relation to markets, and particularly the distance to major urbanized regions, was also relevant.

In some regions the expansion of the open fields must have reached its limits already during the tenth or eleventh centuries, when the arable of some villages in the English Midlands had reached the edges of village territories (Dyer, 2003: 15).

Landscapes of mixed farming

Other regions were dominated by mixed farming with a large component of animal husbandry. These regions were characterized by enclosed landscapes. Small open fields existed in most of these regions, but grain was mainly for subsistence. These regions were known as 'bocage' or 'woodland'. The English landscape historian Oliver Rackham uses the terms 'ancient countryside', but in fact these landscape usually have a long history of gradual reclamation. Therefore, I prefer the term *'mixed-farming landscapes'*. In general, these landscapes can be found in regions with poor soils, in which large common grazings were necessary to obtain enough manure, but also and in low-lying, wetland regions in which only the higher ground was suitable for arable. But in general, we may not see these landscapes as marginal, peripheral or backward: they were a component of the same economic system that fed the growing population of medieval Europe. Hooke (2004) cited English figures for the eleventh century that suggest that the open-field regions were substantially more densely populated than the less specialized enclosed landscapes. Williamson (2003: 21, 35), however, showed that this was certainly not everywhere the case.

Mixed-faming landscapes with small open fields could be found in many parts of Europe. We already mentioned the belt of sandy landscapes that runs from Flanders through the Netherlands and north-western Germany to western Denmark. But these landscapes share many characteristics with the landscapes of western France. The typical bocage country of Brittany, for example, consisted in fact of complex landscapes with individual enclosures as well as small open fields (Astill and Davies, 1997, figure on p. 13; Uhlig, 1961). These landscapes have a long history of gradual reclamation and were even during the nineteenth century still characterized by extensive common grazings ('landes'). The same stories apply to the British Isles outside the Central Province of England. Examples come from the Scottish Highlands, where most hamlets had small open fields ('clachans') for their grain supply. Complex mixed-farming landscapes with small open fields are also known from the Mediterranean, for example in Catalonia (Bolòs, 1999) and Galicia (Calvo-Iglesias et al., 2009).

In the post-medieval period, many of these regions were ever more dominated by enclosed landscapes, as the small open fields were gradually enclosed. In England outside the 'Central Province', this process was almost complete by the late sixteenth century (Williamson et al., 2013: 9). In Western France, the process took longer, but here also most open fields had disappeared by the twentieth century. Typical for these piecemeal enclosures is that some of the medieval arable strips were surrounded by hedges but retained their distinctive shapes. Modern air photographs of Normandy, for example, show many old open fields that developed in this way.

The changing distribution of open fields in Europe

In my opinion, the origin of open fields is rather simple. When different fields are used year after year for growing grain, hedges and field walls are not only unnecessary but also lead to loss of land and, by the shadow they give, to lower harvests. This means that the growth of open fields is connected to the expansion of grain production, a process known as 'cerealisation' (Bartlett, 1994; Hoffmann, 2014). This again makes it probable that population pressure is behind the emergence of ever larger open fields from the ninth and tenth century. The heydays of open field agriculture were the early decades of the fourteenth century.

We may expect a relation between the location of the main open-field landscapes and the major urban landscapes in high medieval Europe (Fig. 6.15). We can think of Von Thünen-like configurations, with dairy farming and market gardening close to the main towns. Further away, the large open fields supplied the grain for bread, porridge and beer. London and Paris were surrounded by grain fields, the Flemish towns were supplied from southwest Flanders and north-western France. Outside these arable zones, extensive animal husbandry took place; at the end of their lives, animals could walk over long distances to the urban slaughter-houses.

The distribution of open fields changed after the early fourteenth century. During the late medieval economic and demographic crisis grain production was concentrated in the best regions; elsewhere former grain lands were converted to pasture or even gave way to forest.

During the early modern period, these changes continued. The open fields gradually disappeared from the British Isles, although a few examples of functioning open fields survived (and even exist nowadays; Hooke, 2010). Much of the huge open fields of central England gave way to sheep pasture, that preserved many traces of the older landscape, such as ridge and furrow. But as the open fields disappeared from the British Isles new open fields were laid out in the eastern Baltic, showing a movement of commercial grain production towards the Baltic region. At the same time, the use of the remaining open fields in Central and Western continental Europe became more diverse and individualized, making the 'regular' open fields ever more difficult to organize. Most of the common regulations gave way to individual management. Within the present article, these changes cannot be dealt with in detail (more on these changes can be seen in Renes, 2010).

Figure 6.15 Towns and open fields in late medieval Europe

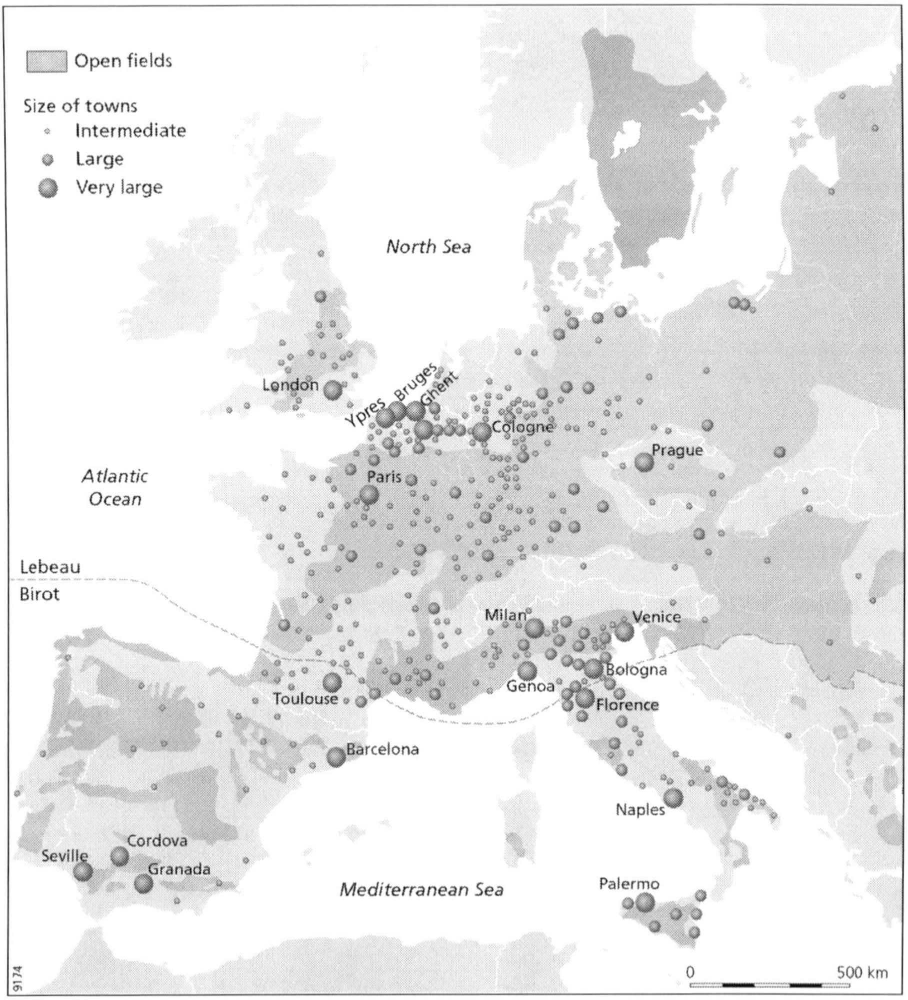

Sources: towns after Pounds, open fields after Lebeau (1969), both with additions.

IV. Open field research in different landscapes

British and Continental traditions

On a European scale, two prominent research traditions can be distinguished: the German geographic tradition of morphogenetic settlement studies and the British interdisciplinary landscape history and archaeology. These different traditions are highly connected to the different histories of the landscape.

On the British Isles, nineteenth-century historians described open fields mainly as a historic system. The first earthworks of former villages that had been associated with open fields were already recorded around 1900 (Hurst, 1986: 198). In the course of the twentieth century, the growing availability of air photographs made historians aware of the remains of open fields, particularly in ridge and furrow in pastures that had remained unploughed for centuries. From there, landscape archaeologists started mapping these remains. This method resulted in a strong bias towards the large open field landscapes that were characterized firstly by ridge and furrow and secondly by the early disappearance of open field agriculture. Moreover, the evidence of ridge and furrow led to a focus on agrarian technology, even when the patterns were sometimes combined with the ownership patterns as shown on estate maps.

In Germany (and in the Netherlands), historic landscape research was mainly in the hands of geographers, who used contemporary landscapes and cadastral sources as starting points for their research. The main line of research was in settlement morphology, and the research tradition that developed was known as 'genetische Siedlungsforschung' (genetic or morphogenetic settlement studies; see Burggraaff, 1988; Nitz, 1974).

This tradition started in the late nineteenth century with August Meitzen, who was involved in the introduction of the Prussian land tax, giving him access to large numbers of cadastral maps that showed land ownership patterns (Harnisch, 1975: 99). Having studied law, economics and history, Meitzen was a pioneer in using the cadastral maps as a historical source (Aldenhoff, 1990). In 1863 he published a book on villages in Silesia, combining cadastral maps with the study of medieval charters. During the following years he was busy collecting agrarian statistics for the Prussian government. In 1875 Meitzen received a professorship in Berlin (Harnisch, 1975: 102). Then, in 1895, he published the four volumes on settlement and agriculture (Meitzen, 1895; see Fig. 6.16), that are generally seen as the start of morphogenetic studies of settlement types and field-systems.

Meitzen distinguished different types of field-patterns and suggested an ethnic origin of these differences. He connected the complex strip-field patterns of many open fields with German populations. Meitzen's book reached the English-speaking world (for example: Ashley, 1898) and was adapted to the British Isles by Howard Levi Gray in 1915. Gray suggested an introduction of the Midland system by the invading Anglo-Saxons, so from Germany (Gray, 1915).

Later generations of geographers used the same nineteenth-century cadastral maps as a starting point for historic research and for theories on the origin of settlements and fields. Some geographers focused on the small open fields in the sandy regions: Wilhelm Müller-Wille (1944), Georg Niemeier (1938, 1944), Herman Hambloch (1960, 1962). Niemeier and, later, Hambloch, saw a single group of long and narrow strips (a *Langstreifenflur*) as the earliest core of the open fields of a village. Others took up the task to bring structure in the extremely complex open field landscapes of Central and South-Germany. A major breakthrough was the work of Krenzlin and Reusch (1961) in Lower-Franconia. They followed the changes in ownership back into the late Middle Ages, a method that was known as 'Rückschreibung' ('writing back').

The German morphogenetic tradition reached its summit around 1960. The highlight was an international conference in Vadstena, the second meeting of what came to be

Figure 6.16 The village of Maden as published by Meitzen (1895)

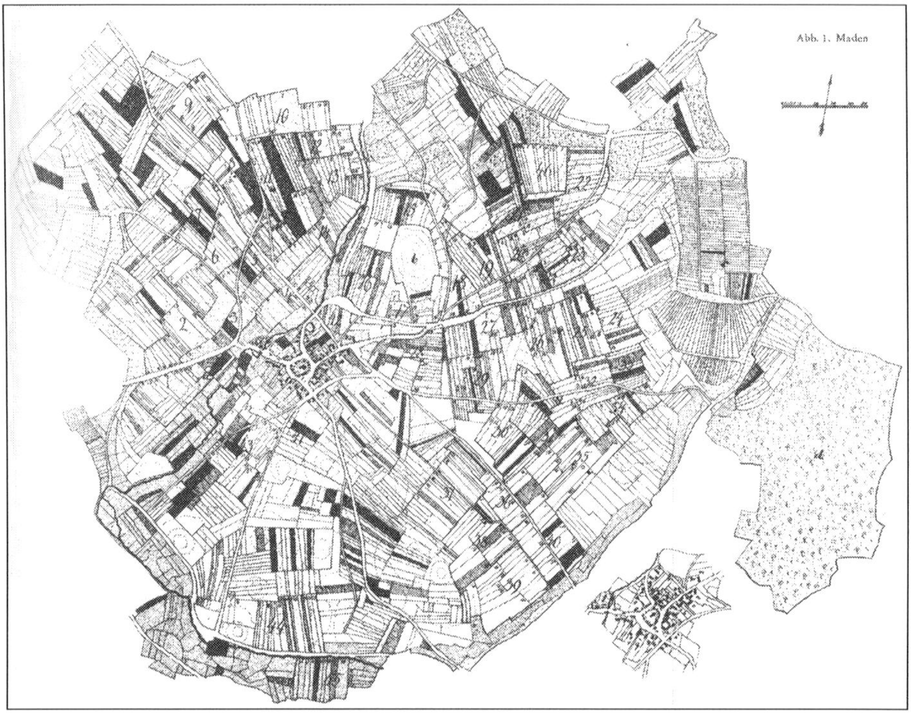

known as the Permanent European Conference for the Study of the Rural Landscape, a series of meetings that continues to the present day. The Vadstena meeting was dominated by German geographers and their international offshoots.

Another branch of historical geography involved research into deserted settlements. This was a theme with a long history. During the nineteenth century, local historians found large numbers of traces of 'lost' settlements in archives and during the first half of the twentieth century academic geographers such as Otto Schlüter connected those settlements with a medieval crisis period. In 1935, the agrarian historian Wilhelm Abel summarized the results in a model of economic fluctuations between the thirteenth and nineteenth centuries (Abel, 1935). But only during the 1950s, geographers started a fieldwork campaign to map deserted settlements and particularly their field-systems, using methods of field survey, collection of potsherds and sometimes small excavations, that are nowadays mainly used by landscape archaeologists (Simms, 1976). In this way, the study of deserted settlements had a strong link with morphogenetic studies, getting better insights into the medieval settlement forms by examining sites which were abandoned in the medieval period. The deserted field systems were described using the terminology of settlement studies and were connected to the archival studies. As many of these continental deserted settlements are now under forest, air photographs are difficult to use. Only recently, with LiDAR-techniques, the traces of former arable have become available for large-scale systematic mapping (see for example Bofinger and Hesse, 2011: 168; Ewald and Klaus, 2010: 87).

Combination of the results of Rückschreibung and field-mapping was not easy and provoked much debate. The meeting at Vadstena was followed by a *'Kolloquium über Fragen der Flurgenese'* ('colloquium on questions of the origins of field-patterns') in Göttingen in 1961. The proceedings (Mortensen and Jäger, 1962) still make good reading because of the sometimes robust language of those who were in disagreement.

So an English tradition based on air photographs and fieldwork contrasts with a continental tradition of archival study. The results are not always easy to compare. The continental tradition focuses on land ownership, whereas the English tradition gives insight into agrarian techniques. Characteristically, the German literature rarely mentions the term open field, but instead uses terms for different types of field-patterns.

But there is another difference. The British research has a strong bias towards the open field landscapes of Central England and in particular the regions where traces of ridge and furrow were still recognizable on mid-twentieth-century air photographs. Although the interest in the smaller open fields in the mixed-farming landscapes is growing (for example Herring, 2006), they are rarely the theme of intensive historical research. In Germany, on the other hand, the morphogenetic studies of open fields developed in the open field landscapes as well as in the mixed farming landscapes (Table 2).

It was in particular the research tradition in the mixed farming region of the sandy landscapes of North-western Germany that appealed to Dutch geographers. From the 1930s to the 1970s, in a period in which Dutch mainstream human geography was under the influence of French regional geographers, Dutch historical geography was completely oriented towards German approaches (see for example Keuning, 1938).

The influence of German morphogenetic historical geography was prominent in Eastern Europe too. The influence went partly through local geographers who took part of their education in Germany, such as the Polish geographer Halina Szulc (https://www.daad.de/alumni/netzwerke/vip-galerie/mitteleuropa/12740.en.html [2–3-2015]) and the Jugoslavian (Slovenian) geographer Svetozar Ilešič (Ilešič, 1959), partly by German scholars who studied other countries. An important but often overlooked aspect of

Table 6.2 Regional differences in the research about the history of open fields

	Open-field landscapes	Mixed farming landscapes
British Isles	Much research, but focused on regions with ridge and furrow	Little research (exception: Uhlig, 1961)
Continental Europe	Long tradition of research in Germany (Krenzlin and Reusch, 1961)	Long tradition of research in Germany (Müller-Wille, 1944; Niemeier, 1938, 1944; Hambloch, 1960, 1962; Glässer, 1967)

German geography was the double specialization of most scholars, who were obliged to write their 'habilitation' on a theme that was not the same as their PhD. Geographers who wrote about settlement geography in Germany also used their skills in settlement analysis in very different regions: Hermann Hambloch on the Low Countries, Harald Uhlig on the Atlantic fringe of Europe and on the Himalaya, Ewald Glässer on Norway, Georg Niemeier on Spain, Hans-Jürgen Nitz on India, etc. Some of their publications introduced the field of settlement studies into their second country of study and are still relevant today.

In the course of the 1960s, the interest of human geographers moved away from settlement studies. There were several reasons for this. One was, that geography as a discipline moved in other directions. From the 1960s onwards, a career in geography could better be built on urban, economic or social geography, and on the use of quantitative data and statistical methods. The few who continued the old morphogenetic studies started to present themselves as historical geographers. In fact, in contradiction of the common opinion, the numbers of historical geographers in Germany (and elsewhere in North-Western Europe) have not dwindled, but the occupiers of chairs in mainstream geography that did genetic settlement studies have all but disappeared.

A second reason may have been, that the research methods of historical-geographical settlement research had reached their limits. With Rückschreibung-techniques, one could follow changes in field-patterns and settlement morphology back to the late Middle Ages, but not much further. Renewal had to come from archaeological data or from new research questions. But archaeologists and historians (the 'spatial turn' in history) only became interested in medieval settlements during the 1970s. Nowadays, these disciplines are covering some of the subjects that geographers pursued half a century ago, albeit often without much knowledge of that tradition.

Continental historical geographers have tried to keep the old tradition alive in three ways. Firstly, by sharpening their terminology. An early adaptation to the new, quantitative orientation in geography led to the huge project of the development of a terminological framework (Uhlig and Lienau, 1967). However, this has turned out to be a dead end, that did not inspire new research nor lead to interesting new insights. Secondly, scholars sought to renew the subject by developing new research questions, such as the changing geography of open fields in the European economic system. This theme was particularly taken up by the German historical geographer Hans-Jürgen Nitz who, belonging to the post-war generation that had been educated in the old morphological tradition, during the 1980s moved to research in which he connected changes in land-use to international economic developments (Nitz, 1987). The main inspiration were the world-system theories of Immanuel Wallerstein and Fernand Braudel (Wallerstein, 1974, 1980; Braudel, 1985). Nitz succeeded in connecting these theories with concrete regional developments and landscapes (Nitz, 1989; 1993). Thirdly knowledge of historic landscapes could be applied to planning, as planners show an increasing interest in landscape as heritage. This planning-oriented study of historic landscapes started during the 1970s in the Netherlands (Renes, 1999) and in Germany (Burggraaff, 1997), in both countries connected with the rapid transformations of the agrarian landscapes. The national programmes of land consolidation (German: *Flurbereinigung*, Dutch:

ruilverkaveling), in which historic field patterns and individual landscape features disappeared fast, met with growing opposition. The open fields, with their sometimes extreme fragmentation, were among the first regions that were consolidated and, in the process, mapped by historical geographers.

In the British Isles, on the other hand, the historians, geographers and archaeologists who started research into historic settlements and field-patterns developed the still flourishing interdisciplinary field of 'landscape history and archaeology'. Here, the stream of books on field systems in general and open fields in particular never stopped (Baker and Butlin, 1973; Rowley, 1981) and even showed a renewed growth in recent years (Williamson et al., 2013; Hall, 2014).

The limited amount of interaction between the English and continental traditions is remarkable. Even the Vadstena-conference had little noticeable influence on British studies. The effect came only a few years later, when Joan Thirsk (1964, 1966; see also Williamson, 2016) translated the German ideas to the English context.

Land ownership and field-patterns

So on the Continent the morphogenetic study of field-patterns had a long tradition. This resulted in a rather broad spectrum of theories, which were generated by local examples and were then applied to explain general patterns. Figure 6.17 shows a number of developments that could lead to a small-scale pattern of subdivided fields. Most of these developments have been described in research in historical geography, showing that different processes can lead to the same results.

The fragmentation process is usually explained by population growth within societies that knew partible inheritance, sometimes combined with McCloskey's risk management. An additional factor could be the development of a Midland system, that meant that the lands of individual farmers had to be divided equally over the two or three fields (Fox, 1981: 66). Figure 6.18 shows a very recent example of a medieval farm (the Thuserhof, near Roermond Netherlands) that was still a single block of land in 1833. After the death of the private owner in that year, the farm was divided between her seven children. Three of them lived on the site of the original farmstead. In 1900, after further subdivisions, the farm had grown into a hamlet of ten houses and the two original enclosed fields had grown into a subdivided open field (data: Ruiten, 1982).

All these processes are linear, but in reality the history of field-patterns can have been even more complex. The small open fields on the sandy ridges in the Netherlands have for a long time escaped archaeological fieldwork by the development of plaggen layers, mainly during the early modern period, that cover older remains. When archaeological excavations started during the 1980s, early medieval settlements were discovered on the ridges, suggesting a replanning of at least the village site.

Agrarian technique and field-patterns

Another way of tackling the development of open fields starts with agrarian techniques. Particularly the patterns of ridge and furrow that are studied from air photographs or fieldwork, stimulated research in the individual field strips. These can, for example, be

Figure 6.17 Development of field-patterns after Vervloet (1984)

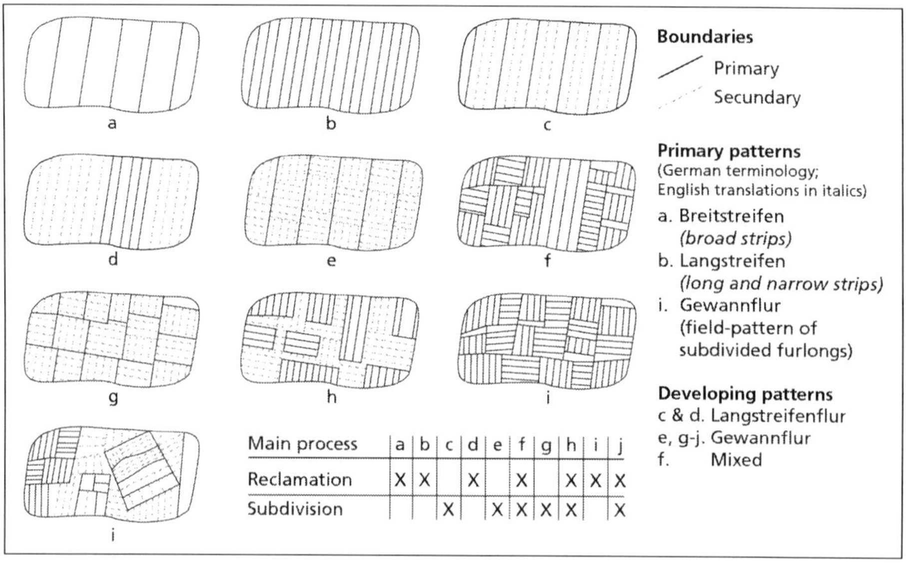

[a] shows a pattern of broad strips that, together with large block fields, is often described as a possible original layout.

[b] and [d] show small strips as an original form. This type of layouts is described for the small open fields of north-western Germany. An original pattern of long and narrow strips ('*Langstreifenflur*') is described by Niemeier (1944) in his so-called *Eschkern*-theory ('core field'-theory). Type [d] is suggested by Hambloch (1960, 1962) in his study of the village of Ostquenhorn. In this case it is likely that the open field itself is a secondary development, with older farms and arable enclosures situated around the open field.

[i] There are more examples of field-patterns that were small-scale from the start. Examples are where open fields were extended with new reclamations in a joint operation of the farmers of a village: as every farmer got his share, a primary pattern of narrow strips was laid out. Such examples are known from the Early Modern period. But already during the High Middle Ages planned subdivided fields are described (Fehn, 1975).

[e, g] Highly subdivided fields could originate from earlier patterns of block fields or broad strips, as was described by Krenzlin for villages in Lower Franconia (Krenzlin and Reusch, 1961). In Central England (Hall, 1981, 1982; Harvey, 1982; Williamson et al., 2013, p. 7) as well as on the Continent (Matzat, 1988) very long (sometimes more than a kilometre) strip fields have been discovered that had been broken down to shorter strips or to patterns of furlongs with their strips running into different directions [e].

explained from the use of the heavy plough with its limited manoeuvrability (Møller, 2015). This is also the usual explanation for the reversed s-shape of many strips (Eyre, 1955, followed by Scharlau, 1956). Ridge and furrow (German: *Wölbäcker*) results from ploughing along the same lines, turning the furrow inwards, for many years. The ridges must have been regarded by the farmers as having some utility: with the same ploughing equipment, flat ploughing was possible cutting one season's furrow through the previous season's ridge (Astill, 1988: 70; O'Keeffe, 2000: 64).

We have to realize that ridge and furrow only partly coincides with open fields. On the one hand, it also occurred in other landscape types, for example on enclosed land. The cultivators of crofts and closes used the same ploughing techniques and in some regions saw the practical benefits of ridged fields. On the other hand, many open fields have never known ridge and furrow, which seems to have been particularly useful on heavy soils that were difficult to drain. However, the medieval distribution is still unclear, although they are seen as typical for northern Europe (Møller, 2015). In many regions that remained under arable during the Early-Modern period and in the nineteenth century, ridge and furrow disappeared by agricultural modernization that included cross-ploughing and underdrainage after enclosure (Liddiard, 1999). The former headlands (German: *Ackerbergen*) on the boundaries of furlongs may survive much longer (Taylor, 1975: 84).

Not only does the distribution of ridge and furrow not coincide with that of open fields, it also does not coincide with landownership (see, for example, Mead, 1954 and Hooke, 1991). A strip belonging to a single owner often contained two or more ridges. A ridge must have been the minimal size of exploitation, but even this 'building block' could be jointly owned by two or even more owners. Therefore, the complex patterns of ridge and furrow grouped in furlongs that are often described in English literature, cannot be equated to the ownership-based *Gewannflur* of Continental literature.

Moreover, in the whole discussion it is often overlooked that not all open fields consist of strip fields. In the Netherlands and in Flanders, open field with block-field patterns exist (Fig. 6.19). This is not always explicable by block-shaped ownership fields that are technically subdivided into strips. In the examples shown here, the fields have obviously been used by the farmers in blocks. Possibly, these fields were worked with spades and small ploughs rather than by heavy ploughs. Late medieval Flemish agriculture was characterised by an extreme labour intensity.

Figure 6.18 The Thuserhof

The farm belonged to the Carthusian monastery at Roermond between 1460 and the end of the eighteenth century. After the death of the private owner in 1833, it was divided between her seven children. Three of them lived on the original farmyard. In 1900, after further subdivisions, the farm had grown into a hamlet of ten houses and the two original enclosed fields had grown into a subdivided open field. Data: Ruiten, 1982

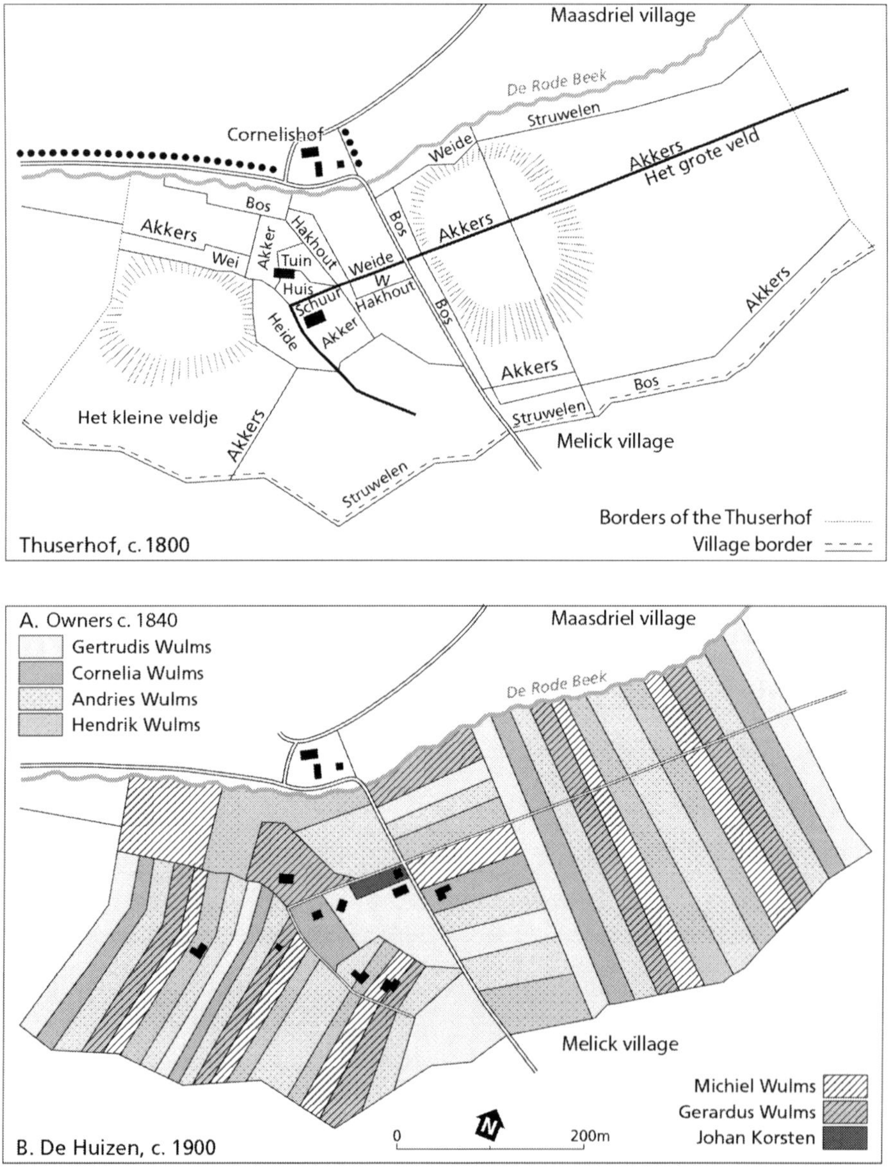

Figure 6.19 An open field with a block-field pattern in the sandy landscapes of the Netherlands

The field of Tiggelt (southwest of Breda, in the south-western Netherlands) shows a block-field pattern not only in ownership, but also in terrain models. Probably the differences in height are caused by spitting instead of ploughing.

Sources: left: Vervloet, 2010, p. 159

V. State of knowledge and research questions

After more than a century of research, a number of questions are still unsolved. Concluding this paper, I will summarise the state of knowledge and some interesting research themes.

Land use and field-patterns: combining British and continental traditions

The very different research traditions on both sides of the North Sea have delayed international cooperation. Such cooperation, however, seems necessary, as both traditions have their limitations.

In my opinion, particularly the research on the British Isles, that tends to be rather insular, could profit from a more international perspective. A good example to illustrate the problems is the list of factors behind the origins of villages and common fields, that is provided by Rippon (2008). Of the factors he mentions, he rightly dismisses the old ethnic theories of Meitzen (1895) and Gray (1915). Another group of factors, the fragmentation of 'federative' estates and antecedent landscapes, refer to processes in the transition phase and explain the existence of older landscape structures in the 'common fields', but they do not explain the emergence of common fields.

More helpful is the argument of land use, in particular the balance of arable, pasture, and woodland. This would demand efficient use of resources (but McCloskey sees common fields as inefficient). Indeed, a lack of pasture may cause a more intensive use, including cooperation, in the exploitation of open fields as pasture after the harvest. But this type of arguments could certainly profit from a larger and wider series of case studies from different regions. The same is true for the influences of local or regional lords. This question requires a wider scope of case studies then a single country might be able to provide.

Another factor is the influence of soils in relation to agricultural practices. The example of the Netherlands shows that open (and common) fields existed on many different soil types. In the Netherlands open fields could be found on sand, marine clay, fluvial clay, boulder clay and, most extensive, on loess. Soil is certainly an argument (and has rightfully been emphasized recently by Williamson, 2003), but mainly as a limitation rather than as an explanation.

A final argument is the process of emulation, that refers to the question of innovation-diffusion as against the independent development of an innovation. It seems logical that the inhabitants of villages kept an eye on practices in neighbouring villages and imitated what they saw as useful. But on a larger scale, this is more doubtful. Meitzen and Gray wrote in a period in which diffusionist perspectives were dominant among archaeologists as well as geographers and historians.

Present ideas adopt a more nuanced approach. On the one hand, open field systems or elements of them seem to have popped up independently in different regions in Western and Mediterranean Europe and even elsewhere (for example: Twidale, 1972; Zimmerer, 2002). But there are also examples of diffusion. In medieval Ireland, open fields are only known from Norman estates suggesting an introduction by the twelfth-century Norman invaders (O'Keeffe, 2000: 62). In parts of Central Europe, open fields are probably, together with the heavy plough (Bartlett, 1993: 148–49), introduced by colonists from the German lands within the so-called Ostkolonisation.

For the three-field system the question of diffusion is also unclear. Around the eighth century it is known from monastic estates in Northern France and the regions around the southern Rhine Valley (Hildebrandt, 1988: 275–79), but these are almost the only written sources available, so the system may have been more widespread. Much later, during the sixteenth century, open fields with three-field systems were founded in what was then called Lithuania, and some of the local terminology suggests a German origin.

In general, we might expect that the more complex systems are, the greater the probability of diffusion. Very special types, such as sun division, are unlikely to have developed independently. The Swedish geographer Göransson (1961) has suggested that it was exported from England to Scandinavia, although in fact most evidence occurs relatively late (Roberts, 2008: 84).

However, common field systems are known from the British Isles, different parts of central Europe and even some regions in the Mediterranean (The *vidazzoni*, that once occupied a large part of Sardinia, suit all criteria for a common field system; Delano Smith 1979, pp. 35–37). When these have developed independently, the reasons behind them must be functional rather than political.

Fields and villages

An intriguing question concerns the often suggested relationship between open fields and settlement nucleation. The connection is not as simple as has often been suggested in the past (Rippon, 2008: 13; Oosthuizen, 2013). In the first place, open fields occur in combination with very different settlement types. In many regions, small — and sometimes also larger — open fields are combined with hamlets and dispersed farms (see for example Flatrès, 1957, p. 420 and Roberts and Wrathmell, 2002. The last authors also show the nuances and complexities).

Still in the core regions of the medieval open-field landscapes, the large common fields are almost everywhere combined with nucleated villages. One reason for this must have been functional: those who lived in scattered farms surrounded by their own fields found it difficult as their lands fragmented to maintain their separate existence. The village, with its central position in the pattern of roads leading out of the fields, became the best place to build a farm. This must be one of the explanations for the change from a pattern of dispersed settlement into one of concentrated villages and the desertion of older dispersed hamlets and farms. Such developments have been described in Britain as well as on the Continent (Williamson, 2003: 13–14; Lewis et al., 1997; Schreg, 2006: 153, 158; see also Fig. 6.17).

In tracing the early stages of villages and open fields, there are still questions of chronology. The dating of the process of settlement concentration is still not completely clear; opinions vary between the eighth and twelfth centuries and Brown and Foard even concluded that the concentration of villages preceded the development of the common fields, making the connection between the two still more complex (Brown and Foard, 1998; Higham, 2010: 11). Moreover, the relationship between settlements, their fields, and the agrarian system is not always obvious.

The medieval nucleated villages must have had a relatively open structure, often with farms situated around a village green. The very large, densely-built settlements that are known in German literature as *Haufendörfer* mainly date from early-modern population growth, when the houses of cottagers and labourers filled the gaps between the old-established farms as well as, in many cases, the village green (for examples: Vits, 1999: 104 and Williamson et al., 2013, Plate 27).

Open fields in the European economic system

Analyses of the European distribution of open fields have not made much progress since the maps by Lebeau (1969) and others. The making of such a map could now be much easier than in the past, because of the large number of regional publications (although syntheses are rare), the growing availability of scans of old topographical maps and, for the present situation, by the availability of systematic series of satellite and air photographs, particularly Google Earth. At the same time, distribution maps give a static picture and are therefore less popular than half a century ago, and research into the changing distribution is scarce.

Some of the more promising approaches start from Von Thünen's model of the most ideal location of different types of land use around a central place, mainly based on the influence of transport costs. This was one of the main ideas behind a large research project on the grain supply of medieval London (Campbell et al., 1993). It was also one of the building

Table 6.3 Terminology

	English	German	French
Field systems			
	Open field		Champ ouvert
	Three field system	Dreifelderwirtschaft	Assolement triennal
	Common field	Gewannflursystem	
	Midland system	Zelgensystem	
Field patterns			
	Furlong	Gewann	
		Gewannflur	
	Strip fields	Streifenflur Langstreifen Breitstreifen	Lanière
Landscape features			
	Ridge-and-furrow	Hochäcker, Wölbäcker	
	Headland	Anwand, Ackerberg	Tournière, fourrière
Landscape change			
	Land consolidation, enclosure	Flurbereinigung	Remembrement

Sources: Egli, 1985; Uhlig and Lienau, 1967.

stones for Nitz's work on the changing geography of European agriculture during the early modern period, when regional grain markets gave way to a European market for grain.

Of course, the picture is much more complex, as individual regions used their unique qualities, limitations and opportunities to find their own place in a changing world. These regional characteristics include the quality of infrastructure (such as the availability of natural or man-made navigable waters), legal and governmental institutions (for example the distribution of power and the organization of common lands), education and the availability of capital. Also the physical landscape, particularly climate and soil conditions, influenced a region's suitability for large-scale arable farming.

Such systematic approaches still have potential for future research. Many aspects of the early-modern spatial transformation of the European landscape, in which the open fields disappeared in substantial parts of the Atlantic fringe of Europe and new open fields appeared in the Eastern Baltic, are still insufficiently researched and explained.

Heritage: traces in England; small-scale structures in Eastern-Europe

The complex histories of open fields have led to regional differences in survival and in the visibility of physical remains. In continental Europe, large areas of open fields have survived and are still functioning as large-scale arable lands. However, the subdivided strip-field patterns that were characteristic for open fields, have mostly been subject to land consolidation measures (including the post-war collectivization in parts of Eastern Europe) and have become extremely rare. Only in a few parts of Eastern Europe that escaped collectivization, as well as in a few parts of Central Europe where little or no enclosure or other types of land consolidation took place, fragmented field patterns still survive. The best examples can probably be found in parts of Austria, Poland and former Yugoslavia (Rugg 1985: 57; fig. 15).

Elsewhere, remains of field systems can be seen in regions where the open fields have been enclosed earlier. In some regions, individual strips have been enclosed and surrounded with hedges. Such fossilized strips are easily recognizable on modern air photographs. Other traces consist of ridge and furrow, that has been preserved in former arable fields that have been under pasture (particularly in Central England) or forest (in deserted settlements in Central European mountains) during recent centuries. When such lands come under the plough again, traces erode quickly. These remaining vestiges are sources for scientific research and provide a point of contact with the medieval agrarian past.

VI. Discussion

This short review made clear that open fields have functioned in many regions and under many different physical as well as human geographical circumstances. Many decades of research have made clear that landscapes of open fields often have a long and complex history. There are still many questions. To solve these, not only local and regional in-depth studies are necessary, but also international comparative research.

Bibliography

Abel, W. (1935), *Agrarkrisen und Agrarkonjuktur in Mitteleuropa vom 13. bis zum 19. Jahrhundert.* Parey, Berlin.

Aldenhoff, R. (1990), 'Meitzen, August', in: *Neue Deutsche Biographie* 16, p. 734. http://www.deutsche-biographie.de/ppn116871423.html [12–3-2015]

Ashley, W. J. (1898), 'Meitzen's Siedelung und Agrarwesen der Germanen', *Political Science Quarterly* 13 (1), pp. 143–55.

Astill, G. and Davies, W. (1997), *A Breton landscape*, London/Bristol, PA.

Baas, H., Snippenburg, B. van, Renes, H. and Rentenaar, R. (2014), 'De Geesten van Holland; stand van kennis over de oude akkercomplexen op de strandwallen van Holland', *Historisch-Geografisch Tijdschrift*, 32, pp. 131–50.

Bach, A. (1953), *Deutsche Namenkunde. Band II, Die deutschen Ortsnamen*, Winter.

Baker, A. R. H. and Butlin, R. A. ed. (1973), *Studies of field systems in the British Isles*, Cambridge.

Beecham, H. A. (1956), 'A review of balks as strip boundaries in the open fields', *Agricultural History Review*, 4, pp. 22–44.

Bieleman, J. (1987), *Boeren op het Drentse zand 1600–1910; een nieuwe visie op de 'oude' landbouw,* Wageningen (A. A. G. Bijdragen 29).

Bloch, M. (1966), *French rural history; an essay on its basic characteristics*, Berkeley/Los Angeles (Original ed. 1931).

Blok, D. P. (1958), 'De enken'. *Driemaandelijkse Bladen, New Series*, 10, pp. 1–16.

Bofinger, J. and Hesse, R. (2011), 'As far as the laser can reach[...] Laminar analysis of LiDAR detected structures as a powerful instrument for archaeological heritage management in Baden-Württemberg, Germany' in: D. C. Cowley ed., *Remote sensing for archaeological heritage management*, Brussels (EAC Occasional Paper, 5), pp. 161–71.

Bolòs, J. (1999), 'Landscape and society in the county of Besalú (Catalonia) in the Middle Ages', *Landscape History*, 21, pp. 19–30.

Bouwer, K. (1970), *Cultuurlandschapsvormen aan de westzijde van het Drents Plateau*, unpublished PhD thesis Groningen State University.

Braudel, F. (1985), *Civilization and capitalism 15th-18th century*, London (3 vols.).

Buiks, C. (1997), *Laatmiddeleeuws landschap en veldnamen in de Baronie van Breda*, Assen.

Burggraaff, P. ed. (1988), *Genetische Siedlungsforschung in Mitteleuropa und seinen Nachbarräumen*, Bonn.

Burggraaff, P. (1997), 'Angewandte Historische Geographie im ländlichen Raum. Erfahrungsbericht und Perspektiven'in: K.-D. Kleefeld and P. Burggraaff ed., *Perspektiven der Historischen Geographie. Siedlung – Kulturlandschaft – Umwelt in Mitteleuropa*. Bonn, pp. 235–43.

Calvo-Iglesias, M. S., Fra-Paleo, U. and Diaz-Varela, R. A. (2009), 'Changes in farming systems and population as drivers of land cover and landscape dynamics: the case of enclosed and semi-openfield systems in Northern Galicia (Spain)', *Landscape and Urban Planning*, 90, pp. 168–77.

Campbell, B. M. S., Galloway, J. A., Keene, D. and Murphy, M. (1993), *A medieval capital and its grain supply: agrarian production and distribution in the London region c. 1300*, London.

Chadwick, A. (2013), 'Some fishy things about scales. Macro–and micro-approaches to later prehistoric and Romano-British field systems', *Landscapes*, 14, pp. 13–32.

Delano Smith, C. (1979), *Western Mediterranean Europe; a historical geography of Italy, Spain and Southern France since the Neolithic*, London.

Dyer, C. (2003), *Making a Living in the Middle Ages: the people of Britain 850–1520*, London.

Egli, H.-R. (1985), 'Die Rückschreibung zur Rekonstruktion der Gewannflurgenese im bernischen Seeland', *Geographica Helvetica,* 40, pp. 19–24.

Ewald, K. C. and Klaus, G. (2010), *Die ausgewechselte Landschaft; vom Umgang der Schweiz mit ihrer wichtigsten natürlichen Ressource*, Bern/Stuttgart/Wien (3rd ed.; 1st ed. 2009).

Eyre, S. R. (1955), 'The curving plough-strip and its historical implications', *Agricultural History Review*, 3, pp. 80–93.

Fehn, K. (1975), 'Angerdörfer und Plangewannfluren in Mittelschwaben und auf der Fränkischen Alb', *Berichte zur deutschen Landeskunde*, 49, pp. 87–97.

Fox, H. S. A. (1981), 'Approaches to the adoption of the Midland system' in: T. Rowley ed., *The origins of open-field agriculture,* London/Totawa, pp. 64–111.

Glässer, E. (1967), 'Zur Entstehungsgeschichte der bäuerlichen Kulturlandschaft', *Geographische Rundschau,* 19, pp. 383–89.

Göransson, S. (1961), 'Regular open-field patterns is England and Scandinavian solskifte', *Geografiska Annaler*, 43, pp. 80–104.

Haartsen, A. J., Klerk, A. P. de and Vervloet, J. A. J. (1989), *Levend verleden; een verkenning van de cultuurhistorische betekenis van het Nederlandse landschap*, 's-Gravenhage (Achtergrondreeks Natuurbeleidsplan 3).

Hackeng, R. (2006), *Het middeleeuwse grondbezit van het Sint-Servaaskapittel te Maastricht in de regio Maas-Rijn*, Maastricht.

Hall, D. (1981), 'The origins of open-field agriculture – the archaeological fieldwork evidence' in: T. Rowley ed., *The origins of open-field agriculture*, London/Totawa, pp. 22–38.

Hall, D. (1982), *Medieval fields*, Aylesbury (Shire Archaeology 28).

Hall, D. (2014), *The open fields of England*, Oxford.

Hambloch, H. (1960), *Einödgruppe und Drubbel. Ein Beitrag zu der Frage nach den Urhöfen und Altfluren einer bäuerlichen Siedlung*, Münster.

Hambloch, H. (1962), 'Langstreifenfluren im nordwestlichen Alt-Niederdeutschland', *Geographische Rundschau*, 14, pp. 346–57.

Harnisch, H. (1975), 'August Meitzen und seine Bedeutung für die Agrar–und Siedlungsgeschichte', *Jahrbuch für Wirtschaftsgeschichte 1975*, pp. 97–119.

Hartmann, J. L. H. (1986), *De reconstructie van een middeleeuws landschap; nederzettingsgeschiedenis en instellingen van de heerlijkheden Breust en Eijsden bij Maastricht (10e–19e eeuw),* Assen/Maastricht (Maaslandse Monografieën 44).

Harvey, M. (1982), 'Regular open-field systems on the Yorkshire Wolds', *Landscape History*, 4, pp. 29–39.

Herring, P. (2006), 'Cornish strip fields', in: S. Turner ed., *Medieval Devon and Cornwall: shaping an ancient countryside*, Macclesfield, pp. 44–77.

Hooke, D. (1988), 'Early forms of open-field agriculture in England', *Geografiska Annaler*, 70 B, pp. 123–31.

Hooke, D. (2010), 'The past in the present; remnant open field patterns in England', *Hungarian Journal Landscape Ecology (Special issue),* pp. 43–51.

Hopcroft, Rosemary L. (1999). Regions, institutions, and agrarian change in European history. University of Michigan Press.

Hurst, J. G. (1986), 'The medieval countryside' in: I. Longworth and J. Cherry ed., *Archaeology in Britain since 1945; new directions*, London, pp. 196–236.

Ilešic, S. (1959), *Die Flurformen Sloweniens im Lichte der europäischen Flurforschung*, Kallmünz/Regensburg (Münchner Geographische Hefte, 16 / Materialien zur Agrargeographie, 5).

Keuning, H. J. (1938), 'Eschsiedlungen in den östlichen Niederlanden', *Westfälische Forschungen; Mitteilungen des Provinzialinstituts für Westfälische Landes–und Volkskunde,* 1, pp. 143–57.

Krenzlin, A. and Reusch, L. (1961), *Die Entstehung der Gewannflur nach Untersuchungen im nördlichen Unterfranken*, Frankfurt/Main (Frankfurter Geographische Hefte 35–1).

Leenders, K. A. H. W. (1996), *Van Turnhoutervoorde tot Strienemonde; ontginnings–en nederzettingsgeschiedenis van het noordwesten van het Maas-Schelde-Demergebied (400–1350)*, Zutphen.

Leenders, K. A. H. W. (2000) *Oorkondenlijst voor Rijsbergen.* http://users.bart.nl/~leenders/oork/rijsberg.html [2-1-2015].

Lemaire, T. (2013), 'Korenvelden' in: T. Lemaire, *Verre velden; essays en excursies 1995–2012*, Amsterdam, pp. 123–64.

Lewis, C., P. Michell-Fox and C. Dyer (1997). *Village, hamlet and field. Changing medieval settlements in central England.* Manchester.

McCloskey, D. N. (1975), 'The persistence of English common fields' in: W. N. Parker and E. L. Jones ed., *European peasants and their markets; essays in agrarian economic history,* Princeton, pp. 73–119.

McCloskey, D. (1976), 'English open fields as behaviour towards risk' in: P. Uselding ed., *Research in economic history, 1,* Greenwich (Conn.), pp. 124–70.

Matzat, W. (1988), 'Long strip field layouts and their later subdivisions; a comparison of English and German cases', *Geografiska Annaler*, 70B, pp. 133–47.

Meitzen, A. (1895), *Siedelung und Agrarwesen der Westgermanen und Ostgermanen, der Kelten, Römer, Finnen und Slawen*, Berlin (3 Volumes and Atlas).

Moorhouse, S. and Bond, J. (2016), 'An approach to understanding medieval field systems across Europe: the structure and anatomy of township field systems' in: J. Klápště ed., *Agrarian technology in the medieval landscape*, Turnhout (Ruralia, 10), pp. 1–48.

Mortensen, H. and Jäger, H. (1962), 'Kolloquium über Fragen der Flurgenese am 24.-26. Oktober 1961 in Göttingen', *Berichte zur deutschen Landeskunde,* 29–2.

Müller-Wille, W. (1944), 'Langstreifenflur und Drubbel; ein Beitrag zur Siedlungsgeographie Westgermaniens', *Deutsches Archiv für Landes–und Volksforschung*, 8 (1), pp. 9–44.

Niemeier, G. (1938), 'Fragen der Flur–und Siedlungsformenforschung im Westmünsterland', *Westfälische Forschungen; Mitteilungen des Provinzialinstituts für Westfälische Landes–und Volkskunde,* 1, pp. 124–42.

Niemeier, G. (1944), 'Gewannfluren; ihre Gliederung und die Eschkerntheorie', *Petermanns Geographische Mitteilungen*, 90, pp. 67–74.

Nitz, H.-J. (1974), *Historisch-genetische Siedlungsforschung. Genese und Typen ländlicher Siedlungen und Flurformen*, Darmstadt (Wege der Forschung 300).

Nitz, H.-J. ed. (1987), *The medieval and early-modern rural landscape of Europe under the impact of the commercial economy; papers presented at the meeting of the Permanent European Conference for the study of the Rural Landscape, held at Rastede and Hagen, Federal Republic of Germany, 2–9 September 1985*, Göttingen.

Nitz, H.-J. (1989), 'Transformation of old and formation of new structures in the rural landscape of northern central Europe during the 16th to 18th centuries under the impact of the early modern commercial economy', *Tijdschrift van de Belgische Vereniging voor Aardrijkskundige Studies*, 58, pp. 267–90.

Nitz, H.-J. (1992), 'The temporal and spatial pattern of field reorganization in Europe (18th and 19th centuries); a comparative overview' in: A. Verhoeve and J. A. J. Vervloet ed., *The transition of the European rural landscape: methodological issues and agrarian change 1770–1914,* Brussels, pp. 146–58.

Nitz, H.-J. (1993), 'The European world-system: a Von Thünen interpretation of its eastern continental sector' in: H.-J. Nitz ed., *The early-modern world-system in geographical perspective*, Stuttgart (Erdkundliches Wissen, 110), pp. 62–83.

Orwin, C. S. and Orwin, C. S. (1954), *The open fields*, Oxford (2nd ed.; original ed. 1938).

Philippa, M., Debrabandere, F., Quak, A., Schoonheim, T. and Sijs, N. van der ed. (2009), *Etymologisch woordenboek van het Nederlands*, Amsterdam. http://www.etymologie.nl/ [2–1-2015].

Renes, J. (1988), 'Some aspects of open fields in the southern part of the province of Limburg (the Netherlands)', *Geografiska Annaler*, 70 B, pp. 161–67.

Renes, J. (1998), 'Dutch commons: variety and change' in: P. Sereno and M. L. Sturani ed., *Rural landscape between state and local communities in Europe, past and present; Proceedings of the 16th Session of the Standing European Conference for the Study of the Rural Landscape (Torino, 12–16 September 1994),* Alessandria, pp. 85–93.

Renes, J. (1999), *Landschappen van Maas en Peel: een toegepast historisch-geografisch onderzoek in het streekplangebied Noord–en Midden-Limburg*, Leeuwarden.

Renes, J. (2007), 'De veengebieden rond de Noordzee; landschapsgeschiedenis en landschapsonderzoek' in: J. J. J. M. Beenakker, F. H. Horsten, A. M. J. de Kraker and J. Renes ed., *Landschap in ruimte en tijd*, Amsterdam, pp. 287–300.

Renes, J. (2010), 'Grainlands; the landscape of open fields in a European perspective', *Landscape History*, 31 (2), pp. 37–70.

Renes, J. (2012), 'Historische transformatieprocessen van Europese cultuurlandschappen', *Historisch-Geografisch Tijdschrift*, 30, pp. 195–210.

Renes, J. (2016), 'Landscape history and archaeology of open field landscapes in Europe'in: J. Klápště ed., *Agrarian technology in the medieval landscape*, Turnhout (Ruralia 10), pp. 255–65.

Roberts, B. K. (1973), 'Field systems of the West Midlands' in: A. R. H. Baker and R. A. Butlin ed., *Studies of Field Systems in the British Isles*, Cambridge, pp. 188–231.

Roberts, B. K.(2008), *Landscapes, documents and maps. Villages in Northern England and beyond, AD 900–1250*, Oxford.

Rowley, T. ed. (1981), *The origins of open-field agriculture*, London/Totawa, NJ.

Ruiten, J. (1982), 'De Thuyserhoff te Maasniel', *Roerstreek 1982, Jaarboek Heemkundevereniging Roerstreek 14*, pp. 13–24.

Scharlau, K. (1956), 'S-Formen und umgekehrte S-Formen unter den deutschen und englischen Langstreifenfluren', *Zeitschrift für Agrargeschichte und Agrarsoziologie*, 4, pp. 19–29.

Schönfeld, M. (1950), *Veldnamen in Nederland*, Amsterdam (Mededelingen K. N. A. W., Afd. Letterkunde, New Series, 12–1).

Schreg, R. (2006), 'Die Archäologie des mittelalterlichen Dorfes in Süddeutschland; Probleme – Paradigmen – Desiderate', *Siedlungsforschung: Archäologie-Geschichte-Geographie,* 24, pp. 141-62.

Simms, A. (1976), 'Deserted medieval villages and fields in Germany; a survey of the literature with a select bibliography', *Journal of Historical Geography*, 2, pp. 223–38.

Smith, C. T. (1967), *An historical geography of Western Europe before 1800*, London/Harlow (Geographies for Advanced Study).

Spek, T. (1992), 'The age of plaggen soils; an evaluation of dating methods for plaggen soils in the Netherlands and Northern Germany' in: A. Verhoeve and J. A. J. Vervloet ed., *The transformation of the European rural landscape: methodological issues and agrarian change 1770–1914,* Brussels, pp. 72–91.

Spek, T. (2004), *Het Drentse esdorpenlandschap; een historisch-geografische studie*, Utrecht.

Spek, T. (2006), 'Entstehung unt Entwicklung historischer Ackerkomplexe und Plaggenböden in den Eschlandschaften der nordöstlichen Niederlande (Provinz Drenthe)', *Siedlungsforschung; Archäologie-Geschichte-Geographie* 24, pp. 219–50.

Spek, T., H. van der Velde, H. Hannink & B. Terlouw (2010), *Mens en land in het hart van Salland; bewonings–en landschapsgeschiedenis van het kerspel Raalte*, Utrecht.

Taylor, C. C. (1975), *Fields in the English landscape*, London.

Taylor, C. C. (1981), 'Archaeology and the origins of open-field agriculture' in: T. Rowley ed., *The origins of open-field agriculture*, London/Totawa, pp. 13–21.

Thirsk, J. (1964), 'The common fields', *Past and Present*, 29, pp. 3–25.

Thirsk, J. (1966), 'The origin of the common fields', *Past and Present*, 33, pp. 142–47.

Twidale, C. R. (1972), "Lands' or relict strip fields in South Australia', *Agricultural History Review,* 20, pp. 46–60.

Uhlig, H. (1961), 'Old hamlets with infield and outfield systems in Western and Central Europe', *Geografiska Annaler,* 43, pp. 285–312.

Uhlig, H. and Lienau, C. (Internationale Arbeitsgruppe für die geographische Terminologie der Agrarlandschaft) (1967), *Flur und Flurformen*, Giessen.

Vangheluwe, D. and T. Spek (2008), 'De laatmiddeleeuwse transitie van landbouw en landschap in de Noord-Brabantse Kempen', *Historisch-Geografisch Tijdschrift*, 26, pp. 1–23.

Van Smeerdijk, D. G., T. Spek and M. J. Kooistra (1995), 'Anthropogenic soil formation and agricultural history of the open fields of Valthe (Drenthe, the Netherlands) in mediaeval and early modern times', *Mededelingen Rijks Geologische Dienst*, 52, pp. 451–80.

Van Veen, P. A. F. and N. van der Sijs (1997), *Van Dale Etymologisch woordenboek,* Utrecht.

Verhulst, A. and Blok, D. P. (1981), 'De agrarische nederzettingen' in: D. P. Blok and A. Verhulst ed., *Algemene geschiedenis der Nederlanden*, Vol. 1, Haarlem, pp. 153–64.

Verspay, J. P. W. (2016), 'Structuring landscape, shaping community' in: J. Klápště ed., *Agrarian technology in the medieval landscape*, Turnhout (Ruralia 10), pp. 247–54.

Vervloet, J. A. J. (2010), 'Zandlandschap' in: S. Barends, H. G. Baas, M. J. de Harde, J. Renes, R. Rutte, T. Stol, J. C. van Triest, R. J. de Vries and F. J. van Woudenberg ed., *Het Nederlandse landschap: een historisch-geografische benadering*, Utrecht (10th edn), pp. 132–61.

Wallerstein, I. (1974), *The modern world-system. Capitalist agriculture and the origins of the European world-economy in the sixteenth century*, London (Studies in Social Discontinuity).

Wallerstein, I. (1980), *The modern world-system 2. Mercantilism and the consolidation of the European world-economy 1600–1750*, London (Studies in Social Discontinuity).

Wiersma, J. (2015), 'Essen, iezen, ikkers en gaasten. Ruimtelijke variatie van historische open akkercomplexen in de provincie Friesland', *Historisch-Geografisch Tijdschrift*, 33, pp. 38–47.

Williamson, T. (2003), *Shaping medieval landscapes. Settlement, society, environment,* Macclesfield.

Williamson, T. (2016), 'Joan Thirsk and the "common fields"' in: R. Jones and C. Dyer ed., *Farmers, consumers, innovators. The world of Joan Thirsk*, Hatfield (Explorations in Local and Regional History 8), pp. 35–48.

Williamson, T., Liddiard, R. and Partida, T. (2013), *Champion. The making and unmaking of the English Midland landscape,* Liverpool.

Zimmerer, K. S. (2002), 'Common field agriculture as a cultural landscape of Latin America. Development and history in the geographical customs of resource use', *Journal of Cultural Geography* 19 (2), pp. 37–63.

7 Open fields, capital and labour in medieval and early modern Flanders

Erik THOEN, Ghent University, Belgium

Up until now, research into open fields in Flanders has been extremely restricted in its focus. Apart from some unreliable 'studies' from amateur historians, most of the valuable literature has concentrated on the origins of 'micro-open fields' or on their geography and/or toponymy, neglecting to a large extent their social contexts. In this chapter I will analyse the meaning and use of open fields in a broad socio-economic context, in order to explain more fully their character and distribution within Flanders. While most past research has been restricted to the former County of Flanders, I will deal with the larger Flemish-speaking part of Belgium, which covers more or less half the surface of that country, as presently constituted.

I will use the term 'open field' to describe a cluster of intensively used plots of arable land, cultivated according to communally applied rules or customs. Intermingled parcels, held or owned by different cultivators, were farmed in such a way that the same types of crop were cultivated together in the same 'field' in the same seasons; and the same periods of fallow were respected by all cultivators. To facilitate such common practices, open fields were 'open landscapes', in the sense that individual plots of land were not surrounded or separated by physical barriers such as hedges, fences or ditches.

Only a small amount of evidence is available to study open-field systems in Flanders.[1] Where relevant documents exist, they mainly date to the period after 1300; even for this later period, they are rather limited in number. Studies based on such scarce contemporary data must of necessity be accompanied by retrogressive and comparative methodologies. As a consequence, some of the results of such an enquiry – especially relating to the origins of the system(s) – will remain, to a significant extent, hypothetical.

I. Open fields, social agro-systems and the environment

To explain the agricultural use of open fields in the past, it is necessary to understand that they took many different forms, even within a relatively small area such as Flanders. Moreover, to obtain a full understanding of the way such fields were used, it is essential to examine them within their social-economic context: that is, we need to view them in terms of how people in the past ensured their long-term survival. Many new insights in rural economic history have emerged over the last fifteen years that can help us to explain open-field agriculture with more confidence: indeed, it is arguable that they have been examined for too long from a predominantly geographical perspective, to the neglect of social and economic considerations. Historians have shown relatively little interest in what

[1] For a state of the art analysis of the open fields in the south of Belgium, see the chapter by Nicolas Schroeder in this volume.

Peasants and their fields. The rationale of open-field agriculture, ed. by Christopher DYER, Erik THOEN and Tom WILLIAMSON, Turnhout, 2017 (*CORN* Publication Series, 16), pp. 163–182

BREPOLS PUBLISHERS DOI: 10.1484/M.CORN-EB.5.114272

Figure 7.1 The regional social agro-systems in late medieval Flanders with references to the types of open fields mentioned in the text

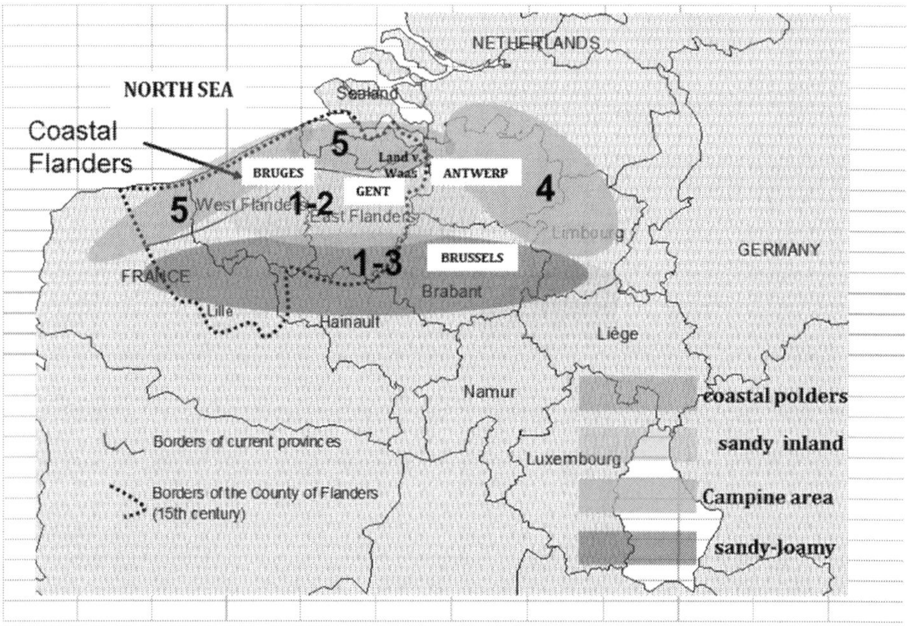

has wrongly been considered an essentially geographical problem, leaving the study of Flemish field systems, to a large extent, to amateur historians.

Production for survival was the most common aim of cultivators in the pre-industrial period. People collaborated (within the family as well as with other peasants and land owners and with the ruling strata of society) to organize production, within the limits set by natural and feudal frameworks. This collaboration, over space and time, took a divergent range of forms – 'regionally collaborative' systems which are best considered as 'social agro-systems', that is, rural production systems based on geographically-specific social relations involved in economic reproduction. Especially in the period before the intense effects of globalization in the twentieth century, the organization of production displayed forms which varied greatly from region to region (Thoen and Soens, 2015). This regional variety was the consequence of different kinds of feudal structure, different traditions (in terms of labour, commodity and land) and variations in market structures, and was significantly influenced by path-dependency (Thoen, 2004). Such structures mostly changed only slowly over time, in response to both internal developments and external influences (such as ecological changes or urbanization) (Figure 7.1).

Within this complex pattern of geographical and temporal variety the physical environment played a significant role, structuring the boundaries of different agro-social systems. Across an area in which social structures were broadly similar, that is, diverse environmental conditions could ensure that they were expressed, on the ground, in a range of material ways. Even within the comparatively small area as Flanders, soils, relief and other natural factors displayed significant differences from one sub-area to another, ensuring in particular significant variations in settlement patterns and settlement intensity. The result was that even within an area with a very similar or identical social organization, landscapes could still be organized differently. The occurrence of different types of open fields was thus linked to both regional social agro-systems and to specific geographical-soil conditions.

II. A typology of open fields in Flanders and their links with the divergent social agro-systems and soil conditions

It is important to remember that, so far as current research goes, no 'champion' landscapes – comprising regularly structured and extensive open-fields, farmed on a village basis, of the kind familiar from the Midlands of England, or parts of central and southern France) – have ever existed in Flanders. Instead, open fields in this region took a number of other forms in the pre-industrial period and, depending to some extent on when we look, and in which geographical and social-agrarian context, we can (to an extent subjectively) discern five main variants or types, although it easier to describe them than to define the periods during which each existed. This typology can certainly be improved and fine-tuned with more research, but it forms a useful basis for discussion, as long as we remember that it is also partly chronological in character, in the sense that not all the defined types necessarily co-existed at the same time. I will, in the following sections, briefly set out the principal characteristics of each type in turn (Figure 7.2).

Type 1: the early 'cultura' or 'kouter' type

Although this can be considered as the oldest type of open field in Flanders, it is to date also the most intensively researched. 'Culturae', as they were called in Latin since the early Middle Ages, or 'Kouters', as they were named since the twelfth century in vernacular Dutch, were relatively small micro-open fields, each usually covering an area of between 10 and 60 ha.[2] They continued to exist in the early modern period, sometimes as relic features within *bocage* landscapes (type two in our classification, see below), sometimes occupying the higher areas of ground within more open environments (the 'velden' type, see type three in our classification). Most villages or demesnes only ever had a few fields of this kind within their boundaries.[3] For many decades, historians have debated their origins, use and organization. However, new comparative and interdisciplinary research, combined with the investigation of new sources, is now bringing this debate to a decisive phase.

Kouters and landscapes: two subtypes

There is general agreement in the published literature that we can usefully subdivide 'kouters' into two sub-types (Verhulst, 2001, Thoen, 2011). The *hofkouters* (Engl.: 'demesne

[2] The – *kouter* names are compounded with proper names, points of the compass, geographical terms, and with terms reflecting the character of the local environment.

[3] Verhulst, 1995 and 1991.

Figure 7.2 Scheme of the five types of open fields in Flanders in the preindustrial period discussed in the text

Type 1: The early *kouter* type

Type 2: The *kouter in bocage* type

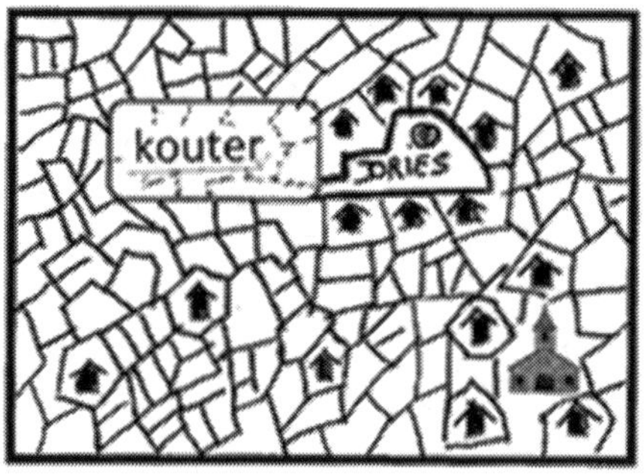

Type 3: The *velden* type

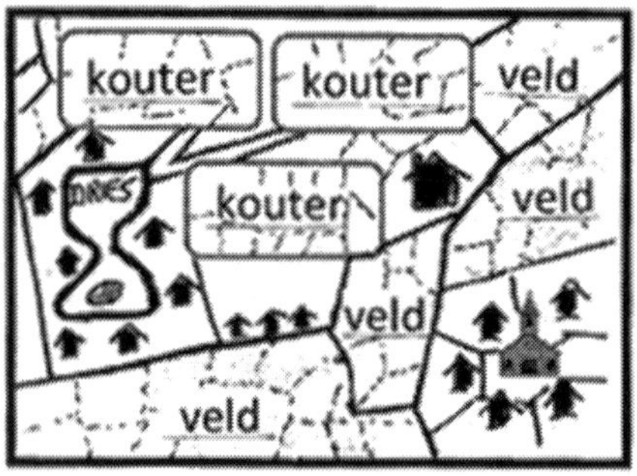

Type 4: The *dorpsakker* type

Type 5: The *polder* type

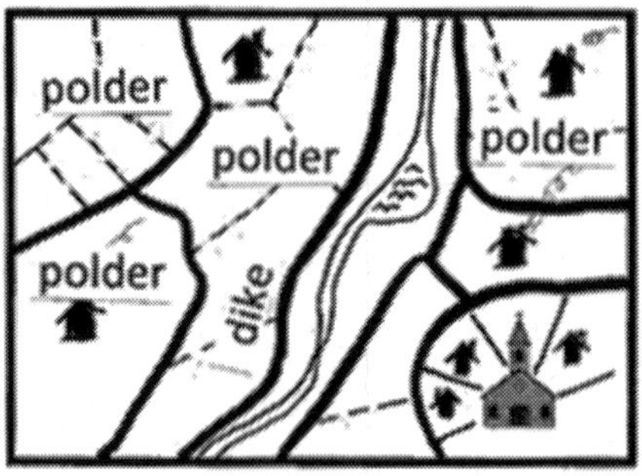

kouters') had their origins as early as the Carolingian period (the ninth century), although many examples came into existence much later, continuing to develop into the twelfth or thirteenth centuries. They were the infields of large demesnes and were usually divided into only a few extensive plots, owned by one large landowner. Fields of the *Dorpskouter* type ('village kouter'), on the other hand, seem to date to the period 1050–1300, and were made up of a larger number of small plots owned, and worked, by different peasants. Many examples of both types of field were, from an early period, divided into three more-or-less equal parts. Both types could be found all over the sand and sandy-loam districts of Flanders, although *dorpskouters* seem to have been more common on the lighter soils of the inland areas. Often both *hofkouter* and *dorpskouter* could be found on the lands of the same village community. There also existed some 'hybrid' examples, where the core of a micro-open field could be designated as 'hofkouter' and the periphery as 'dorpskouter'.

Kouters and social agro-systems

In the past, the interpretation of the many kouter names that can be found in the sources, and on early maps, gave rise to the so called 'kouter debate'. This was largely concerned with the antiquity of such fields, with amateur historians often advancing unscientific arguments for their Merovingian or even Roman origins, as well as using them as reliable indicators of the oldest reclaimed lands of a settlement. This is in spite of the fact that academic research has established beyond reasonable doubt that they only developed, as noted, after the ninth century, and that they do not necessarily represent the oldest 'reclaimed' lands, whatever we mean by this term.[4] A related debate has concerned the origins of the term 'kouter' in field-names. Important specialists in onomastics, such as J. Lindemans and H. Draye, and more recently L. Van Durme (1988) and M. De Vos (1991), agree that the 'Germanic' (Dutch) term began to be used only from *c.* 1100, but disagree over whether it was directly derived from the Latin precursor 'cultura' or via the Romanic language term 'coltura/coultre'. Either way, they have shown little interest in how such agrarian units actually functioned in the landscape.

Amongst professional historians, it was in particular A. Verhulst who undertook the most extensive research into both the occurrence and the function of these micro-open fields. He – together with a number of other specialists in the early medieval period – is clearly correct in arguing that 'culturae' or 'hofkouters' were large open plots used for grain cultivation on the typical early medieval bipartite demesnes (Verhulst, 2001), some of which continued to exist, even after the disintegration of the 'bipartite system', as the core of large demesnes, rented out from the thirteenth century. The emergence of the second type of kouter – the dorpskouters – in the eleventh and twelfth centuries he associates with the emergence of true nucleated villages. The key features which distinguished these kouters from the rest of the village fields were, according to him, that they were worked under a collective regime for rotation and fallowing, and that they represented 'the most important fields of the village community', often the first lands to have been reclaimed and cultivated. Such a characterization of kouters is rather unsatisfactory as it fails to explain fully why, and how, kouters became the most important village fields. This can only be

[4] Land reclamation during the Middle Ages in Europe is now seen as a slow process, leading from a more extensive use towards a more intensive use of the land. Even in periods of low population density most land was 'used' in one way or another.

achieved by considering them within a wider socio-economic context, and against the background of settlement history. Without this, it remains unclear why some, at least, of these fields conserved their open character until the end of the Ancien Régime; and when, how and especially why they continued to provide the space for communal practices.

Only a few early texts allow us to see how the kouter functioned. One particularly useful example dates from the thirteenth century, and provides a description of the *curtis* Desselgem,[5] part of the huge estates of St Peter's Abbey in Ghent, and situated next to the Lys river.[6] In this text, the lands belonging to the demesne are divided into two groups. The first are described as *'ista terra semper potest coli temporibus ad hoc aptis'*, a phrase which probably means 'land which was put under the plough each year', given that the other group was described as '*terrae que possunt aliquando coli [...]*', lands whch were not permanently under the plough. In other texts, land of the latter kind is often described as *prata arabilia* (Lat.) or *prairies arables* (Fr.). These fields were temporarily rather than permanently cultivated. They functioned as part of a primitive up-and-down husbandry system, and for much of the time were used as temporary grazing. Such land, sometimes called *pré arable* in French or in Dutch *dries*, *tries*, *groenzwaarde* or *wede*, was always, and explicitly, situated outside the permanently-cultivated *culturae*.

This form of agriculture, in which periods of cultivation were interspersed with long periods in which the land lay under grass, seems to have been very important in the twelfth and thirteenth centuries, and continued to be a feature of Flemish agriculture into the early modern period. It represented an intermediary stage between temporary and more permanent cultivation of arable land. Such 'up and down' husbandry (German *Koppelwirtschaft*) was markedly different from the situation on the kouters, which were intensively used for the cultivation of crops. This contrast, between permanent and temporary cultivation, was thus the principal difference between the 'kouterlanden' and the lands outside the kouters, more important than the practice of compulsory crop rotation, seen by Verhulst as the major distinguishing characteristic of the former. The more intensive use of kouters for crop cultivation[7] is confirmed by another feature: the fact that they were more intensively manured than the land lying outside them. In the pre-industrial period, manure was the 'gold' of the farmer, and a detailed study of short-term leasing contracts from the fourteenth, fifteenth and sixteenth centuries reveals that while on the kouters the land had to be manured at least once every three years, this was not the case outside them: here the land was manured only once in every five, six or seven years (Thoen, 1988: II, 792 ff.). In the eleventh and twelfth centuries especially, when much of the 'Flur' of each village remained unintensively farmed, or still survived as unenclosed common land used for grazing, wood gathering and the like, the kouterlands must have stood out starkly in the landscape.

Kouters, exclusively used for intensive crop (and especially grain) cultivation, evidently represent a form of *infield*, broadly similar to those found in many areas of Europe, from

[5] Desselgem is a village between Courtrai and Ghent. For a detailed analysis, see Thoen 1993.

[6] For more details, ibidem.

[7] From at least the thirteenth century the lands on the kouters were cultivated with a full-year fallow every three years. Contrary to what is sometimes thought, during the fallow period, the kouterland was cultivated very intensively and ploughed between four and six times: during the most important of these operations, the manure was ploughed into the soil.

Spain to Scandinavia, in the pre-industrial period. The temporary cultivated lands outside were forms of *outfield*. A similar contrast, beteen intensively-cultivated arable cores, and a more extensive, less intensively farmed penumbra, has been described in the sandy areas of the northern Netherlands, in regions such as Drenthe (Spek, 2004): in these regions the infields were termed *essen* or *enken*. Here, however, such a system appears to have developed at a rather later date – between the fourteenth and the eighteenth or even the nineteenth centuries. It emerged as a consequence of an intensification of cultivation on the *essen* employing sod-manuring, a practice for which there is, so far, no evidence in Flanders. The resemblances between kouters and essen/engen are nevertheless remarkable, even more so given the differences in chronology, with the system reaching its zenith in Flanders in the twelfth and thirteenth centuries, and then evolving on rather different and more intensive lines due to the scale of population pressure.

What I have not yet explained is why dorpskouters were collectively-managed, and what explains their concentration at certain places within the wider landscape of the village. In some areas in the east of Ghent, Adriaan Verhulst was able to demonstrate that dorpskouters were created out of a collection of smaller (open) fields, each covering from *c.* 0.5 to 3ha., which are referred to as *accar* or *akker* in texts from the seventh, eighth and ninth centuries.[8] It seems likely that these plots were the earliest fields, and were under permanent cultivation, or so the early texts seem to suggest. They contained a single plot, or only a small number, and were scattered across the landscape, still separated from each other by woods and areas of outfield. A number lying in reasonable proximity were, probably in the ninth century, extended in area and integrated into a kouter system while most of the others disappeared, as the kind of infield-outfield system just described was introduced – a change which was probably accompanied by the concentration of settlement in hamlets or villages focused on small commons or 'greens'.[9] This appears to be reflected in the close connections evident between settlement structures and field systems, for settlements of this type – generally with a place-name featuring the element – *dries*, and covering an area of between 0.5 to 10 ha – are closely associated with early kouters. The greens were not only serving as social meeting points, but also more importantly as areas in which livestock (sheep and cows)[10] could be protected and where, especially in the earliest period, manure accumulated prior to being put – collectively – on the infield-kouters (Thoen, 1993). Some of these greens disappeared in the later Middle Ages but others survived until they were enclosed in the seventeenth or eighteenth centuries. In most villages, the greens were situated more or less between the infield and the outfield, and it seems likely that they came into existence at the same time as the infield-outfield system itself.[11]

[8] This could at least be reconstructed for the sandy area east of Ghent with research based of the s.c. *Liber Traditionum* which is an inventory of the estates for the Saint Peter's abbey of Ghent and one of the most valuable sources for rural history of early medieval NW Europe. However it is likely that the same process occurred elsewhere in inland Flanders as well: some '-akker' place-names that were not integrated in *kouters* survived there as well while a lot early medieval – akker- place names disappeared. Of course we cannot develop this here due to lack of space (Verhulst, 2001 and Thoen, 1993 and 2011).

[9] Further archeological research is here necessary.

[10] All had a natural or manmade pond for watering stock that is often conserved until today.

[11] Of course these village structure had other functions as well in relation to the social agro-system based on intensive collaboration of the peasants. There is no room to develop this in this context. See the literature mentioned.

It is clear that the kouters and their associated settlement patterns emerged as a consequence of the development of a social agro-system which had the collective management of fields at its core (Thoen and Soens, 2015). This pattern of social organization was based on the survival of the family, facilitated by collaboration between peasant households. Indeed, the evidence suggests that most peasants were not interested in enlarging their holdings, or even in maintaining their size, given that division of holdings through inheritance was the accepted norm. This does not mean that larger holdings, mostly rented out by 'absentee' land owners, were absent. Since the early Middle Ages, and increasingly during the high Middle Ages, religious institutions, the nobility and (from the late thirteenth century) rich townsmen, often maintained larger farms. These, however, seldom cultivated more than half the available acreage and generally enjoyed a kind of economic coexistence with the majority of smaller holdings. Indeed, in most areas of 'inland Flanders' compulsory labour services had become rather scarce by the eleventh or twelfth centuries, if they had ever existed. An economic relationship thus replaced, at an early date, a 'judicial' one: smaller landholders and their children could hire themselves out as cheap temporary labour during the 'high season', such work forming part of a portfolio of employments which might include proto-industrial textile production (wool, linen). In this crowded countryside, they might also find intermittent work selling products such as dyes made from plants, or from peat digging. Indeed, peasants were gradually forced more and more towards commercialization, especially with the development of the towns from about 1050. In other words, commercialization became an extra 'tool' for survival, together with purely agricultural activities. This development reached its zenith during the thirteenth century: and it was at this time, and in relation to increasing population pressure, that the landscape described above evolved further in two different directions, diverging mainly in response to different soil conditions.

Type 2: 'kouters in bocage'

Due to the huge population pressure in Flanders the infield-outfield system did not last long. In those parts of the region where lighter soils were abundant, the area outside the kouters, which constituted for a long time most of the land area, evolved towards an enclosed landscape made up of numerous small fields surrounded by hedges and/or ditches. The open fields that survived within such a landscape I refer to as 'kouters in bocage'. On the heavier and more loamy soils evolution took another direction (below: type 3): here a large amount of new micro-open fields developed next to the existing kouters in the village, giving birth to a complicated patchwork of open fields.

Kouters in a 'bocage' landscape

The development towards what we might call, on analogy with the situation in France, 'bocage' landscapes occurred across large areas on the lighter sandy and sandy-loam soils of west-inland Flanders. In these areas most fields were enclosed with hedges (often very wide, and frequently containing standard trees); or with lines of isolated trees, often pollarded. This development started in the Middle Ages but, as in France, accelerated during the early modern period (see Annie Antoine's study of French Britanny: Antoine, 2002). The Flemish version of 'bocage' may have been even more minutely subdivided, and more densely hedged, than was the case in France, or in the 'enclosed' parts of

England or Ireland. As population increased and more land was reclaimed a huge part of the Flemish landscapes came to display this kind of 'honeycomb' structure, reducing the extent of panoramic views in a rather flat landscape.[12] These hedges or tree-lines, often accompanied by ditches next to the hedges, were multifunctional: they served to keep livestock from roaming, where fields under temporary or permanent pasture, and fields cultivated as arable, lay intermingled;[13] they assisted in drainage; and above all they provided the teeming peasant population characteristic of inland Flanders with wood and timber, compensating for the almost complete absence of forested land.[14]

What is typical of Flanders, and particularly important in the context of this chapter, is the fact that this dense 'honeycomb' of hedges and tree lines was interrupted by only limited areas of more open land (Figure 7.3). One kind was to be found beside the dense network of streams and rivers, where wide alluvial plains and wet meadows occurred. More important were the 'kouters' or micro-open fields (from Type 1, above) which often survived and retained their character as open fields within these largely enclosed landscapes. Since they were often situated on well-drained, slightly sloping areas well suited to cereal cultivation, it is likely that most continued to be farmed on 'traditional' lines well into the early modern period, although evidence to support this has so far only been recovered in relation to the 'hofkouters', and not yet for the 'dorpskouters'. The former, mostly covering an area of between *c.* 15 and *c.* 50 ha., had once been the most productive areas of the village, but now become rather traditional, even backward elements in the landscape, compared to the enclosed fields around them, since it was hard to adopt new crops or cultivation techniques (for which Flanders was gradually becoming famous) within them (Thoen, 1988; Verhulst, 1991).

The new settlement patterns which developed in this landscape during the later Middle Ages have not yet been studied in any detail. It seems likely that irregular nucleated settlements gradually emerged, accompanied by increasing numbers of small–and medium-sized farmsteads scattered across the landscape. The greens of the dries-villages, which had originated in the earlier period, now lost their value and were systematically subdivided and privatized from the later Middle Ages onwards.

'Kouters in bocage' and social agro-systems

In this northern part of inland Flanders, with its light soils, ploughing was to a significant extent gradually replaced by spade cultivation, especially after the thirteenth century, and as a result individualism was stimulated. Where soils were too heavy for spade cultivation, in contrast, as was the case with loamy and polder soils, plough teams with oxen or horses were necessary, enforcing a degree of collaboration between peasants (as was the case in

---

[12] A study of the evolution and disappearance of the 'bocage' landscapes in Flanders has not so far been undertaken. I am currently preparing an article on this phenomenon. The bocage landscape is extremely well mapped on the military maps of the eighteenth century (see especially the so called Ferraris map dating back to the 1770s).

[13] A strip a few metres wide running beside the hedges, where shadow hindered grain growing, was often reserved for grazing; the inner part of the field was used for cultivation.

[14] That is, hedges supplied both firewood and also timber, from standard trees, which could be used for building. Timber, especially from oak, was steadily increasing in importance during the early modern period (unpublished data).

Figure 7.3 Type 2 'Kouter in bocage' within a 'honeycomb' of bocage (Ferraris map *c.* 1777)

the 'velden' areas; type 3 below). By the eighteenth century around a third of Flanders was cultivated by spade, and there is little reason to doubt that this form of agriculture had been equally important for many centuries, especially on the light soils around Ghent. This is one of the explanations for why, on these lighter soils, a bocage landscape could develop, while in the areas with heavier soils, more open landscapes emerged. Moreover, bocage landscapes held other advantages: hedges not only prevented soil erosion but also, as noted, provided a source of fuel in a region otherwise deficient in firewood. In general, these were areas which did not experience a large amount of social polarization, with only a limited number of larger holdings which needed, and could afford, to use a plough to cultivate their enclosed lands. A range of interconnected factors thus ensured that bocage landscapes emerged in north Flanders, and not in regions of heavier, more loamy soils.

Type 3: the 'velden' type and the 'kouters in velden' type

Landscape evolution and the 'kouters in velden'

The third type of open-field system found in Flanders is what we might call the 'Velden' type. Like the bocage landscapes just described, this represents a secondary development from the type 1 systems discussed earlier. In the areas where this type occurred, most of the land came to be occupied by open fields: village territories consisted of a 'patchwork' of more-or-less distinct micro-open fields which covered the majority of the cultivated

Figure 7.4 Type 3 Patchwork of open fields in the neighborhood of Brussels on the map of Ferraris (*c.* 1777)

acreage (Fig. 7.4). Most of these micro-fields consisted of a range of small plots, not usually laid out with any degree of regularity, and they were separated from each other by natural or manmade elements such as roads, settlements, ditches or watercourses. Landscapes of this type emerged, in particular, on the heavier sandy-loamy and loamy soils of southern Flanders. They appear, from the documentary record, to have developed both here, and in neighbouring regions of southern Brabant and northern France, during the late Middle Ages (Derville, 1988 and Thoen, 1988 and 2011). We sometimes find an intermediate phase in this process documented in the eleventh, twelfth or thirteenth centuries, in which three 'kouters' first developed (associated with a three-course rotation system); to these, further micro-open fields were gradually added which were not referred to in the documents as 'kouters', and which were often situated in low lying areas prone to be wet and and less suitable for growing grain. In the first stage of their existence, these new lands were not as intensively cultivated or manured as the original 'kouters', and may not initially have been under permanent cultivation: that is, there are indications that, beginning as pasture, they were farmed for a period under a convertible or 'up and down' system, before evolving into permanently cultivated fields. These later micro units of open field were often (although not invariably) designated with field names ending in the element – veld, probably a reference to the previous stages of extensive use. Beside them, in certain areas, limited numbers of – kouter names survived, probably indicating the older areas of permanent arable, the original infields. Elsewhere, however, – kouter names were either never employed, or have not survived, so that – velden names are applied even to these more ancient cultivation areas.[15] Whatever their precise origins, a measure of common regulation was perpetuated on these 'patchwork open fields' well into the early modern period, and may help explain why open landscapes survived in many districts up until the time when the first local maps were surveyed.

The settlement structures in this area gradually became less ordered but a certain degree of 'nucleation' appears to have continued. Nucleated villages and the dries-villages in which the smaller peasants lived occurred in the same neighbourhoods, allowing a measure of economic inter-dependence to ensure family sustenance, and gradually expanded in size. In some areas early maps indicate the emergence of linear villages, situated in between the 'patchwork' of velden: but systematic research on these matters has not yet been carried out.

Social agro-systems and the landscapes in the areas with a 'kouters in velden' – type

Although in the 'velden areas' open fields remained much more important within the village-'Flur' than in the type 2, 'kouter in bocage' areas, the social agro-system with which they were associated was rather similar. In both areas, for the large majority of smallholders the survival of the family was the primary goal. Collaboration between the smaller holdings, but also between smaller and larger holdings, was based on the exchange of labour and capital. Population pressure was huge and the legal system allowed a progressive increase of micro-holdings (especially through partible inheritance), so that commerce became an essential element in the survival system. Commercialization was, and increasingly became, a tool for survival and reproduction.

Detailed lists of share cropping, as well as lists of tithe collections, dating back to the fifteenth and sixteenth centuries clearly reveal that by this time open fields had come to occupy most of the cultivated acreage. Only a small amount of bocage survived, or had come into existence, in the immediate neighborhood of farmhouses.[16] Within each patch of open field, owned and cultivated by many different peasants living on different holdings, the majority of plots was cultivated according to the same crop rotation. Adjacent land plots were sown with the same type of cereal, and all plots within a particular field were fallowed together.

It is important in this context to note that the semi-collective character of open-field management did not much hinder the development of more intensive systems of cultivation. It was relatively easy to abandon the fallow in each of these relatively small 'patches', and already by the late fifteenth century, as more detailed documentary evidence becomes available, full-year fallows were becoming rare. Yet on the other hand the system had allowed collective sharing of capital, especially horses, which was necessary given that spade cultivation was difficult on these relatively stiff soils.

[15] This seems to be the idea of De Vos, 1991. According to this specialist in place names, kouter-names were absent in the eastern part of the – velden area, which was more or less the area east of the Dender river in the former Duchy of Brabant. However, the more are analyzing this area in detail, the more 'kouter' names can be traced in the villages of this area (see e.g. our study on the area North of Brussels (Dilbeek), Thoen, 2011.

[16] See Thoen, 1988 II and some tithe registers of villages south of Ghent in the late sixteenth and early seventeenth centuries, conserved in the state Archives Ghent.

Although further study is needed, it is likely that this voluntary 'open-field character' with 'collective cultivation of the micro-open fields' gradually disappeared from the end of the Middle Ages, due to the ever more intensive use of the 'progressive' techniques for which Flanders became famous.[17] But the physically open character of the fields still partly survived at least until the early nineteenth century, and is widely depicted on topographical maps of that date.

Type 4: the 'dorpsakker' type

The landscape evolution of the 'akkers'

The dorpsakker type was similar in many ways to the dorpskouter fields (type 1) described earlier, but it probably only emerged in the later Middle Ages, and survived well into the period of the Ancien Régime. It consisted of a single small 'akker' (normally with a size of about 10–30 ha., and thus smaller than most 'kouter' fields) cultivated by a community which, in medieval times, was mainly involved in extensive agriculture and (sheep-) breeding. It emerged in an area of very light and sandy 'loess' soils, reclaimed at a relatively late date, mostly within what is known as the Campine district (Figure 7.5). This remained an important sheep farming area, and commons and outfields occupied as much as two-thirds of the land area of many villages well into the nineteenth century. The 'akkers', as they were called in field names, often seem to have lost their common-field rules in the course of the early modern period (Van Onacker, 2014).[18] However, their physically open character survived into modern times. In many villages, sod-manuring of the akker, using hay-sods from the outfield, was widely practised (Bastiaens and Verbruggen, 1996). In the adjacent areas of the northern Netherlands a similar kind of micro-open field existed, often termed an '*-es*'.

Social agro-systems in the dorpsakker-region

The landscape in the *dorpsakker* region followed a different path of development to that seen in the *kouter* and *velden* areas. Here, in sharp contrast to the situation elsewhere in the County of Flanders, or in the south and east of Brabant, large amounts of common land continued to exist for a long time, together with a small, central open field. There were few large farmers in the area: the majority of peasants possessed small or medium-sized holdings. Family survival was largely based on the extensive common lands, which were used as a source of firewood and peat, and as a place for hunting and grazing, reducing the need for local peasants to become involved in commerce (see recently for the Campine area in Belgium: Van Onacker, 2014). These various characteristics largely arose from the fact that demographic pressure remained low in this district, towns were few and far between and holdings did not become excessively subdivided: a lower middle class could survive, heavily involved in sheep breeding, and in many places this ensured the survival of both open fields, and commons, until the end of the Ancient Régime.

[17] Such as the cultivation of clover and the gradual disappearance of the long fallow periods stimulating also more individual cultivation (Thoen and Dejongh, 2011).

[18] This type survived also in the Dutch part of the Campine area.

Figure 7.5 Type 5 Relic of the village akker -openfield in the Campine area on the Ferraris map (*c.* 1777)

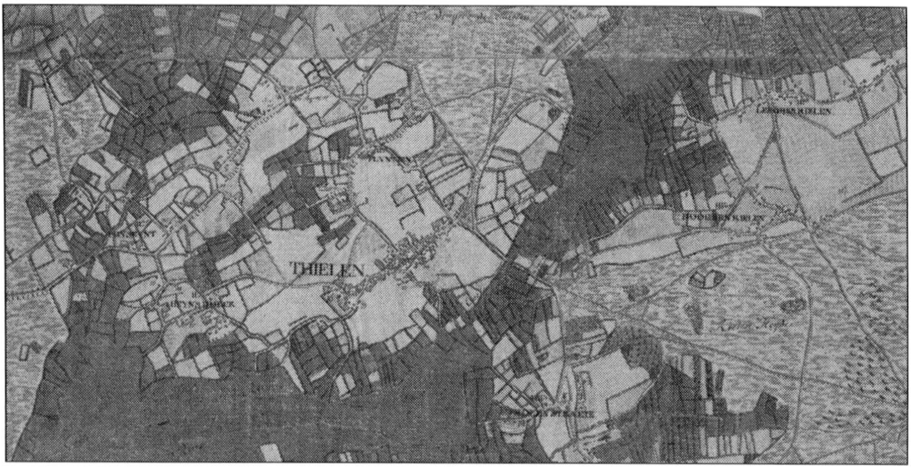

Type 5: the 'polder' type. Open landscapes without open fields

This type is completely different from the systems so far described, and arguably only partly relevant to the subject under discussion. The area in which it was found, the coastal districts of Flanders, developed along similar social lines to the rest of the county in the period between the tenth and thirteenth centuries. But during the thirteenth and fourteenth centuries new and completely different forms of social organization emerged here, in which medium-sized and larger farms largely replaced smaller holdings. Landholdings continued to grow steadily in size, absorbing their neighbours, until the first half of the eighteenth century. This 'new' social agro-system was more commercial and capitalist in character than anything found elsewhere in Flanders (Thoen, 2004, Thoen and Soens, 2015). Within this agro-social context, a landscape emerged which, while essentially open in character, displayed almost no recognizable 'open field' features.

The landscape evolution of the 'polders'

Polders are reclaimed lands situated close to the North Sea and their development was strongly influenced by the sea's continuing presence. From as early as the twelfth century, the landscape consisted of a connected irregular patchwork of micro-open landscapes called *polders*, varied in size but all comprising numerous parcels owned and cultivated by different landowners and peasants. They were surrounded by man-made dykes, constructed as embankments against floods caused by storm surges. From the thirteenth century, these were maintained by special offices led by the landowners (s.c. *wateringues/wateringen/waterschappen*), who also took care of the drainage systems designed to remove excess water from the land. Before large-scale embankment and reclamation in the high Middle Ages, the economy had been based on sheep and cattle breeding, carried out on areas of marsh naturally protected by sand dunes and saltings. Due to rising sea levels and local population growth, the 'natural' ecosystem became unbalanced during the eleventh and twelfth centuries and the need for man-made protection against flooding increased. The area of embanked polders expanded rapidly so that, by 1350, they made up most of the area exploited by coastal villages. Only a minority of the land remained undyked (see Thoen et al. ed., 2015). Following this process, former villages were destroyed or shrank in size, and in their place a pattern of large, isolated farms came into existence.

The polders persist as the archetypal landscape feature of the coastal region of Flanders, although few of the original dykes remain. They have been altered and rebuilt over the centuries following breaks and collapses, the consequence in part of poor maintenance resulting from a decline in the organization of the *wateringues*. Although more research is needed into the issue, it seems likely that polders achieved their open character at an early date: only a minority resembled 'bocage', when we first see them clearly in the documents, although the situation may have been different in earlier periods.[19] There is certainly, so far, no firm evidence that any form of common-field system ever existed in the polders: if it had done, it too must have disappeared at an early stage. Some 'common' rights and common duties did exist, however, which were specific to the area. Landowners had to contribute towards the maintenance of dykes and drainage systems, an obligation later replaced by tax payments made to the *wateringues* who took over this activity (especially Soens, 2006); and, at least in the early modern period, the poor had the right to glean on the grain fields after the harvest, a practice necessary to ensure the continuance of the local social-agro system.

Social agro-systems and the landscape of the polders

The coastal system evolved between the thirteenth and the sixteenth century from a 'commercial survival' system towards a 'commercial capitalist' system: fewer and larger farms emerged which could survive the increasing exploitation costs and tax burdens of the region, and the lands of small peasants were progressively absorbed into larger holdings. This development was exacerbated by environmental challenges: floods became more common during the twelfth and thirteenth centuries, increasing the costs of maintaining dykes. This was partly because over-exploitation of the land caused problems such as soil shrinkage, lowering the land surface and making the polders more vulnerable to storm surges. As a result, huge areas were covered with marine sediments, creating new soil conditions: heavy soils progressively replaced light, peat soils. This made the use of extended (and expensive) horse teams necessary to drain and cultivate the land. Only the wealthiest farmers could survive in these changed and changing environments.

The population was decreasing from the thirteenth century, and much land was taken over by absentee landowners who leased it out to (new) farmers. Large numbers of people must have moved towards inland Flanders where the opposite evolution was taking place: increasing population and ever smaller holdings. To survive, farmers on the polders were obliged to rely more and more on markets, but bankruptcy always lurked around the corner. It is these developments which explain the radical changes in the settlement pattern on

[19] However, some sources seem to indicate that at least lines of standard trees were a more common feature of the polder landscape than they are today. They were important for building the larger farms in the area, as well as for providing material for drainage works. It is noteworthy in this context that many of the relatively few woods that existed in Flanders were situated in between the coastal area and the inland sandy area.

the polders, just noted. New larger farm houses were built outside the former villages: the latter dwindled in size or disappeared altogether. In contrast to the situation in inland Flanders, where a (smaller number of) large farmers could employ local smallholders on a temporary basis when labour was needed, here in the polders full-time resident wage earners were used to cultivate the land, together with poor migrants from inland districts. The result was higher wages, which could only be paid by the largest farmers.

The polder landscape became increasingly open, as land was exchanged to created larger plots around the new, isolated farms. It is probably due to this relatively rapid process of re-allotment and concentration of land into large fields, cultivated by large, more capitalist farmers, that open-field rights and duties, and open fields themselves, did not come into existence here in the way that they did elsewhere in Flanders. A rare exception was the right, already mentioned, of gleaning (collecting leftover grain in) the fields after harvest, enjoyed by the local poor. According to Thijs Lambrecht and Anne Winter, this would have provided a substantial measure of support for the poor since around 10 per cent of the total harvest of grain would have been gathered by gleaners (Lambrecht and Winter, 2013). The fact that this practice was characteristic of these coastal areas (elsewhere in Flanders it was rarely found) is not a coincidence. Social inequality was nowhere so pronounced as it was in this area. Whereas, elsewhere in Flanders, the average size of holdings declined during the early modern period, in the coastal area it increased markedly, and smallholdings almost disappeared. In such a system, support enabled the poor to survive, and therefore to be available fo work, providing the labour supply on which the wealth of the large farmers ultimately depended.

The general absence of 'open-field rules' in this area is a further indication that open-field systems generally developed in environments in which a majority of small and middle-sized subsistence or 'survival' holdings existed, or in areas where a minority of large and a majority of small holdings co-existed. In areas and periods with high levels of individualism and rising capitalism, such as the coastal areas from the fourteenth century, such communal arrangements were absent. By the same reasoning, elements of open-field organization may have existed in the coastal areas in the earliest (medieval) period, when systems of social organization had been different. But if so, no trace of them has survived.

Apart from these coastal regions, there were only a few limited areas within Flanders where forms of open field do not appear to have come into existence. Most were, compared to other Flemish regions, zones of rather late colonization. One example is the 'giant-velden'-region in the neighborhood of Bruges and south-west of Ghent, around Ypres (not to be confused with the 'velden' areas discussed as type 3, above). This was an area of very infertile, light and drifting sands or dune-sands, some of them with an area equivalent to the territory of one village or more (the *Bulskampveld*, *Maldegem Veld*, the *Scheldeveld*, as well as others). True reclamation, to make these areas suitable for more intensive agriculture, only took place in the late early modern period, and even then not always with success (Verhulst, 1995). A relative absence of open fields was also a feature of large parts of the sandy portions of the Land Van Waas and the Land Van Dendermonde. Here extensive tracts continued to be occupied by woodland until the later Middle Ages. Before this, few large landowners had been interested in financing the reclamation of the area; and this, together with its environmental character, dissuaded small and medium-sized peasants from developing holdings there. The area has a 'bocage' landscape, with irregularly-shaped fields resulting from the late reclamation of strips of woodland strips, and from the practice of rather individualized husbandry from the beginning (see Verhulst, 1995; Thoen, Vervaet and Ronsijn, in press).

III. To conclude: micro-open fields, capital and labour relations in Flanders

In this chapter I wanted to stress the role of changing modes of social organization in the development of different forms of open-field agriculture in the north-western ('Flemish') part of Belgium. It has been shown how regionally and temporally divergent forms of social organization adapted an equivalent diversity of environmental possibilities and constraints into a range of specific landscape types, in which 'open-field agriculture' played a more or less important part. Open fields, with their attendant social constraints and duties, were structured by the needs of different social actors to survive: or, to put it another way, the character of open-field agriculture was related to that of regionally varied 'social agro-systems' displaying different forms of collaboration between small peasants and larger farmers. This collaboration was based on specific labour and capital relations, vital for the survival of the individual peasants and the societies of which they formed a part. The particular degree of social collaboration was determined by specific power and property structures, and open-field landscapes and their attributes were as much a tool as an outcome of patterns of social relations. Constraints and possibilities were, however, also determined by the natural environment, and not least by soil conditions.

As I have demonstrated, in Flanders before the end of the Ancien Régime different power and property relations determined different labour relations in a very regional and even local way. This resulted in different 'degrees of individualism', and different forms of 'field' management. But capital relations were also determined by variations in social organization. Indeed, it is likely that the shortage of capital was the most important problem faced by pre-industrial societies. Although we have good evidence only from the fourteenth century, it is clear that, in many areas, a lack of capital was a huge problem for the smaller peasants. While Flanders was known for its high crop yields, and was very intensively cultivated, most peasants were not even able to manure their land every three years, only at intervals of four, five or even six years (Thoen, 1988): depending on time and place, they were often forced to focus their attention on the intensive manuring of only some of their land parcels. The 'kouters' found in the sandier part of the County of Flanders in the twelfth and thirteenth centuries, and the 'akkers' in the Campine area between the fourteenth and the sixteenth centuries, were thus manured every three years, but not the wider 'outfields' in which these were set. Open fields had the advantage that they could be manured easily and relatively intensively at a time when manure was in short supply. Many peasants had only a few cattle and often no horses; horses were (and still are) physiologically very vulnerable draft animals. There is good evidence that in late medieval and early modern Flanders, many poor peasants kept their horse as long as possible, even if it was blind and crippled (Thoen, 1988).

Amongst the smaller peasants, and between this group and the minority of larger farmers, a collaboration developed as a consequence of this lack of capital. In particular, horses and other animals were exchanged for ploughing and transport. This increased the intensity of labour relations between large farmers and small peasants: the former ploughed the land of the latter, and transported commodities for them, such as manure. The smallholders, in turn, paid for this by providing labour, in part unpaid, on the larger farms. The earliest evidence for such a system comes from the fifteenth century (Thoen, 1988) but can be explored in detail for the seventeenth and eighteenth centuries (Lambrecht, 2002 and 2003). It is also possible to discern other forms of cooperative activity amongst the peasants, aimed at increasing social security, which increased in the course of the early modern period (Lambrecht, 2009). In addition to regional variations in the structure of social relations, however, variations in soil conditions were also important in shaping the particular character of landscapes, and the importance of open fields within them. Soil conditions had a determining influence on the demand for labour and capital, having an impact, for example, on whether land could be cultivated using spades, or whether horse-drawn ploughs needed to be employed.

Let us finish by addressing an apparent contradiction. It is well know that Flanders, from the late Middle Ages, had a well-deserved reputation as the most agriculturally advanced region in Europe, and the one in which the highest cereal yields were achieved. Yet it was also one in which, in many districts, open fields played a major role in the farming economy. How are we to explain this? It seems likely that the particular character of the open fields found in Flanders did not hinder the evolution of progressive agricultural techniques. The fields were flexible in their management: the 'patchwork of micro-open fields' ensured that the constraints of the system were very limited, and could combine the advantages of more intensive land use (involving in particular a decrease in the importance of long fallows) with those arising from the more efficient use of labour and capital. They also served to sustain a system of social organization in which the poor – who had a vital function within the social agro-system, as a source of cheap labour – could, to an extent, be sustained. The relatively limited rules and regulations which ordered the open fields were not, moreover, applied to the smaller enclosed plots and gardens associated with farmsteads, and some areas existed on the margins of the fields where new techniques could be tried out. When, eventually, open-field practices lost their rationale – through an increased need for even more intensive and more individualized forms of cultivation, and population pressure – they were progressively abolished. This clearly occurred in the course of the early modern period (between the sixteenth and eighteenth centuries), as some data seem to prove – in a large part of the Belgian Campine area it had disappeared already in the sixteenth century (Van Onacker, 2014)-, but with a chronology and a regional variety which remains to be elucidated.

Bibliography

Antoine, A. (2002), *Le paysage de l'historien. Archéologie des bocages de l'ouest de la France à l'époque moderne*, Rennes.

Bastiaens J. and Verbruggen C. (1996), 'Fysische en socio-economische achtergronden van het plaggenlandbouwsysteem in de Antwerpse Kempen', *Tijdschrift voor Ecologische Geschiedenis,* 1, pp. 26–32.

Derville, A. (1988), 'L'assolement triennal dans la France du Nord au Moyen Âge', *Revue Historique*, 280, pp. 337–76.

Devos, M. (1991), *Bouwlandtermen in de Vlaamse dialecten. Spreidings–en betekenisgeschiedenis*, Brussel / Tongeren.

Gysseling, M. (1950), *Een inleiding tot de toponymie van Oost-Vlaanderen*, Ghent.

Lambrecht, T. (2002), *Een grote hoeve in een klein dorp. Relaties van arbeid en pacht op het Vlaamse platteland tijdens de 18de eeuw*, Ghent.

Lambrecht, T. (2003), 'Reciprocal exchange, credit and cash: agricultural labour markets and local economies in the southern Low Countries during the eighteenth century', *Continuity and Change*, 18, pp. 237–61.

Lambrecht, T. (2009), 'Peasant labour strategies and the logic of family labour in the Southern Low Countries during the eighteenth century' in: S. Cavacciochi ed., *The economic role of the family in the European economy from the 13th to the 18th centuries*, Florence, pp. 637–49.

Lambrecht, T. and Winter, A. (2013), 'Migration, poor relief and local autonomy: settlement policies in England and the Southern Low Countries in the eighteenth century', *Past and Present*, 218, pp. 91–126.

Ronsyn, W. (2014), *Commerce and the countryside: the rural population's involvement in the commodity market in Flanders, 1750–1910*, Ghent.

Soens, T. (2006), 'Explaining deficiencies of water management in the late medieval Flemish coastal plain, 13th-16th centuries,' in: H. Greefs, et al. eds, *Waterbeheer, gemeenschappen en de natuurlijke omgeving. De Lage Landen in comparatief perspectief, c. 1000 – c. 1800. Jaarboek voor Ecologische Geschiedenis*, pp. 35–61.

Spek, T. (2004), *Het Drentse esdorpenlandschap. Een historisch-geografische studie*, Utrecht.

Thoen, E. (1988), *Landbouwekonomie en bevolking in Vlaanderen gedurende de late Middeleeuwen en het begin van de Moderne Tijden. Testregio: de kasselrijen van Oudenaarde en Aalst*, 2 vols., Ghent.

Thoen, E. (1993), 'Dries versus kouter. De wisselbouw in de Vlaamse landbouw van de Middeleeuwen tot de zestiende eeuw. Bijdrage tot de historische landschapsecologie en de geschiedenis van de agrarische techniek', *Heemkring Scheldeveld, Jaarboek*, 22, pp. 71–102.

Thoen, E. and Dejongh, G. (1999), 'Arable productivity in Flanders and the former territory of Belgium in a long-term perspective (from the Middle Ages to the end of the Ancien Régime)' in: B. van Bavel and E. Thoen, eds, *Land Productivity and Agrosystems in the North Sea Area (Middle Ages – 20th Century). Elements for Comparison*, Turnhout, pp. 30–64.

Thoen, E. (2001), 'A "Commercial survival: economy in evolution". The Flemish countryside and the transition to capitalism (Middle Ages – 19th century)' in: P. Hoppenbrouwers, and J. L. van Zanden eds, *Peasants into farmers? The transformation of rural economy and society in the Low Countries (middle ages-19th century) in light of the Brenner debate*, Turnhout, pp. 102–57.

Thoen, E. (2004), '"Social agrosystems" as an economic concept to explain regional differences. An essay taking the former county of Flanders as an example (Middle Ages-19th century)' in: B. van Bavel and P. Hoppenbrouwers eds, *Landholding and land transfer in the North Sea area (late Middle Ages-19th century)*, Turnhout, pp. 47–66.

Thoen, E. (2011), 'Aspecten van de historiek van de regio Dilbeek. Een blik op het landschappelijk verleden met de nadruk op de evolutie van de bewoning' in: P. Uyttenhove, G. Châtel, E. Thoen, V. Van Eetvelde, *Kijken naar het bebouwde landschap: Ontwikkeling en toetsing van een praktijkgerichte methodologie voor de inventarisatie van erfgoedgehelen*, unpublished report for Agentschap Onroerend Erfgoed VIOE, Ghent.

Thoen, E. with the collaboration of Lambrecht, T. (2011), 'Le paysage flamand et les structures de la société médiévale: nouvelles hypothèses sur l'origine et la fonction des 'kouters'' in: J. Yante ed., *Etablissements humains: finages et communautés rurales entre Seine et Rhin (IVe-XIIIe siècles)*, Louvain-la-Neuve, pp. 379–99.

Thoen, E. (2013), 'Clio defeating Neptune: a pyrrhic victory? Men and their influence on the evolution of the coastal landscapes in the North Sea area' in: E. Thoen, G. J. Borger, A. de Kraker, T. Soens, D. Tys, & L. Vervaet eds, *Landscapes or seascapes? The history of coastal environment in the North Sea area reconsidered*, Turnhout, pp. 397–428.

Thoen, E. and Soens, T. (2015), 'The family or the farm: a Sophie's choice? The late medieval crisis in the former county of Flanders' in: J. Drendel ed., *Crisis in the later Middle Ages : beyond the Postan-Duby paradigm*, Turnhout, pp. 195–224.

Thoen, E. and Soens, T. (2015), 'Contextualizing 1,500 years of agricultural productivity and land use in North Sea area. Regionally divergent paths towards the world top' in: E. Thoen and T. Soens et al. eds, *Struggling with the environment. Land use and productivity*, Turnhout, pp. 455–99.

Thoen, E., Vervaet, L. and Ronsijn W. (in press) 'Real landscapes and imaginary landscapes: the Land Van Waes in the Old Regime Landscape History' *Landscape History*.

Thoen, E. (in preparation) *Labour relations in the late medieval Flanders* (14th–early 16th century Flanders).

Van Durme, L. (1988), *Toponymie van Velzeke-Ruddershove en Bochoute*, Brussels.

Van Onacker, E. (2014), *Leaders of the pack? Village elites and social structures in the fifteenth and sixteenth-century Campine area*, unpublished PhD thesis University of Antwerp.

Verhulst A. (1991), *Précis d'histoire rurale de la Belgique*, Editions de l'Université de Bruxelles.

Verhulst A. (1995a), *Landschap en landbouw in middeleeuws Vlaanderen*, Bruxelles, Crédit Communal de Belgique.

Verhulst A. (1995b), *Le Paysage rural: les structures parcellaires de l'Europe du Nord-Ouest*, Turnhout, Brepols, 1995.

Verhulst, A. (2001), 'Oud en nieuw in het 'Kouter'-debat', *Historisch Geografisch Tijdschrift*, 19, pp. 33–41.

Vervaet L. (2009), 'De sociaal-economische organisatie in een Wase plattelandsgemeenschap en de relatie met het landschap: Sinaai in de 17de eeuw', *Annalen van de Koninklijke Oudheidkundige Kring van het Land van Waas*, 112, pp. 19–106.

8 Medieval and modern open fields in southern Belgium: a summary review and new perspectives

Nicolas SCHROEDER, F. R. S.-FNRS, Université Libre de Bruxelles, Belgium

I. Introduction

Although[1] several definitions coexist (see Renes, 2010: 37–41; and the same, this volume), 'open fields' can be described as arable land with a particular *appearance and physical structure*, and also particular associated *social aspects* (see the introduction). Open fields were made up of several parcels of arable land, lying together, which lacked physical boundaries. These parcels were often private property, but they were subject in some degree to collective property rights, such as common grazing. This implies some degree of collective decision-making. In the Middle Ages and the modern period, various types of open fields existed in the southern provinces of modern-day Belgium. A concise, but not exhaustive, overview of literature relating to their geographical distribution and origins is undertaken in the first and second parts of this paper. The models presented in the overview are discussed critically. In the third section, new perspectives for research are suggested, focussing more specifically on the rationale of open fields.

II. The geography and distribution of open-field types

The southern part of modern-day Belgium is divided into several geographical regions (Figure 8.1). The Sambre and Meuse Valley separates Middle Belgium (elevation 100–200 m.), also called the central Belgian Plateau, and Upper Belgium (200–700 m.). The central Belgian Plateau can in turn be divided into the regions of Hesbaye, Brabant and Hainaut, belonging to the 'Western European loess belt' (Bakels, 2009). This is a vast plain with a gently undulating surface which is covered by loess, an aeolian deposit. South of the Sambre and Meuse valleys, the land is more rugged. The Ardennes are a low mountain range – a region of extensive forests, rough terrain, and rolling hills – with soils mainly based on sandstone and schist. The foothills to the north of the Ardennes are geologically and topographically complex. The Famenne and Fagne are traversed by the

[1] I am grateful to the Wiener-Anspach Foundation, the University of Oxford, the F. R. S.-FNRS, the Université Libre de Bruxelles, the Alexander von Humboldt-Stiftung and the Universität Heidelberg, for supporting the postdoctoral fellowships that enabled me to produce the present paper. I thank the participants of the two panels on open fields at the Rural History Conference in Bern (19th-22th of August 2013) and the Symposium in Ghent (12th-13th of June 2014), for their comments and criticism. I am greatly indebted to Jean-Pierre Devroey who commented on various drafts of this paper, and to Justin Demmerle who edited the final version.

Peasants and their fields. The rationale of open-field agriculture, ed. by Christopher DYER, Erik THOEN and Tom WILLIAMSON, Turnhout, 2017 (*CORN* Publication Series, 16), pp. 183-205

BREPOLS PUBLISHERS DOI: 10.1484/M.CORN-EB.5.114273

Figure 8.1 The geographical regions of southern Belgium

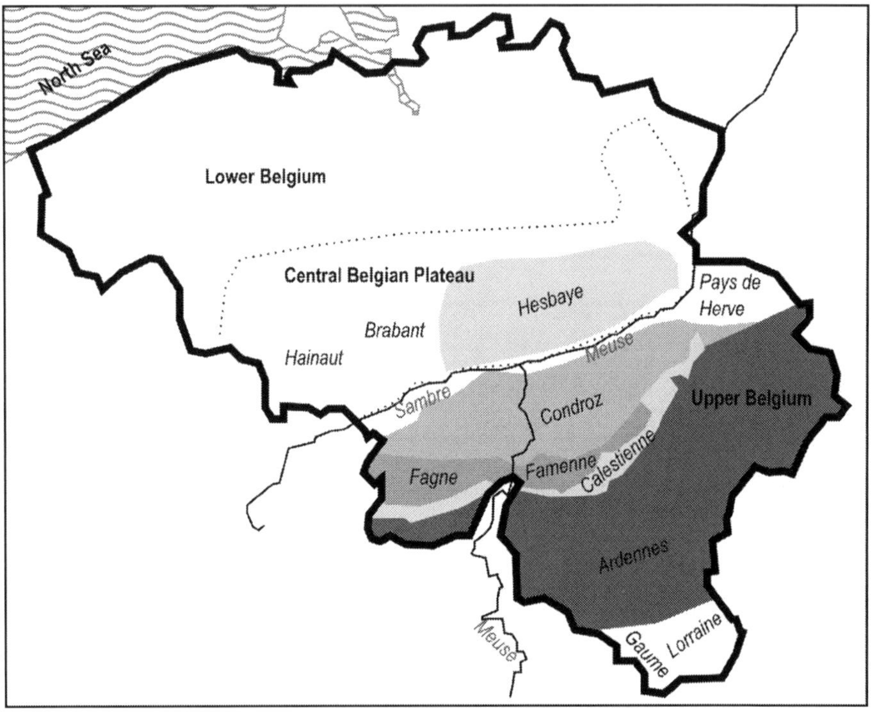

Source: N. Schroeder, based on Genicot et al., 1983, p. 7.

Calestienne, a band of limestone resulting in a depression with calcium-rich soils. In the northern part of the Fagne-Famenne, sandstone dominates, forming a less fertile plateau. The Condroz comprises a succession of valleys hollowed out of the limestone between sandstone crests. Its northern part is structurally very similar to the Ardennes, because of a band of sandstone and schist. It is therefore known as 'Ardenne condruzienne'.

These regional variations in altitude, relief, and geology had a significant impact on crop growing, land use, settlement, landscapes, and field systems. A regional overview of field systems is provided in the collection *Architecture rurale de Wallonie* (Genicot et al., 1983; 1986; 1987; 1988; 1989a; 1989b; 1992). The main topic of these books is rural architecture, but each regional survey is accompanied by a well-informed overview of 'traditional' farming, land use, and landscapes, mainly based on evidence from the eighteenth century. According to these surveys, in Hesbaye, Condroz, and the Fagne-Famenne, a three-field system predominated, often associated with a settlement pattern of nucleated villages, and cultivators were obliged to practise a three-course rotation (*assolement triennal*). Major crops would have been spelt or wheat as winter grain (Genicot, 1947; Billen, 1989), and oats as spring grain. Although the rotations were similar in these three regions, the landscapes varied on opposite sides of the Sambre and Meuse Valley. In Hesbaye, much of the countryside was arable land, with extensive open fields (Genicot et al., 1983: 59; Genicot et al., 1986: 56). Settlement mainly took the form of medium-sized and large nucleated settlements, with only occasional isolated farms. The fields consisted of strips of land grouped into at least three, but generally more, blocks. In this open landscape, extensive meadows were often limited to river valleys. Pasture was mainly provided through rough grazing on the stubble that was left as fallow. This type of field system was very similar to those identified in contemporary northern France, on similar types of loess soils and flat or gently undulating relief (Sivéry, 1990: 38–39; Leturcq, 2004: 54–55).

South of the Sambre and Meuse valley, in the hilly parts of the Condroz and Famenne regions, the arable land was less extensive, and heaths and woodland more frequent. Nevertheless, here too the fields comprised strips, grouped into several blocks forming open fields, across which a three-course rotation was enforced (Genicot et al., 1988: 77; Genicot et al., 1989a: 44). Here the settlement pattern was dominated by a dense network of isolated farms, combined with large and medium-sized villages. The woods and heaths were common land and provided a range of resources: they were used as pasture, but also occasionally as temporary arable land (*sartages*), cropped for two or three years at intervals of six, ten, twenty, or even thirty years. In the Ardennes, with their poor quality soils and hilly relief, heath and woodland were also extensive. The settlement pattern was dominated by isolated farms and hamlets with only occasional nucleated villages. Farming communities cultivated small, continuously-cultivated arable fields that were regularly manured (*terre à champs*), and outer fields (*sartages*) that were fertilized only by grazing animals, or by slash and burn (Genicot et al., 1987: 38 and 58; Genicot et al., 1992: 40–42). On the continuously cultivated core, the rotation combined one year of rye with two or three years of oats followed by between four to nine years of grass fallow, depending on the quality of the land and the quantities of manure available (Schroeder, 2014). The outer fields and commons (*sartages*) were cropped for one or two years and then left fallow for longer periods (from six to thirty years).

Two important observations emerge from this overview of the current literature about field systems in the southern provinces of modern-day Belgium. Firstly, environmental conditions had a significant impact on settlement, farming and field systems. In the fertile and gently undulating central Belgian Plateau, communities developed extensive open fields, whereas in the hilly Ardennes, infield-outfield systems were dominant. A strong case has been made recently for the importance of 'natural' factors such as topography, physical geography, and climate in shaping the organization of the medieval landscape in Britain (Williamson, 2003; Williamson, Liddiard and Partida, 2013). To a certain extent, the southern Belgian case confirms the validity of this 'environmental approach' in explaining regional variations in field system types (also see below). Secondly, regardless of the specific forms they took, open fields *per se* existed everywhere: arable land was divided in strips and subjected to common pasturing in all regions; common rules about cropping were observed by all farmers; and many communities had areas of common land that were used as pasture and as temporary arable land. However, it is necessary to ask if this overview of open fields in southern Belgium is not excessively generalized.

The regional surveys presented in the collection *Architecture rurale de Wallonie* associate only one type of field system with each region. As one would expect, analysis of regional and local studies suggests a measure of diversity *within* regions, and local complexity in field system types. In an edited volume about rural communities published in 1987, several regions of southern Belgium were studied. In the Counties of Namur (Hesbaye, Condroz) and Laroche (Ardennes), the organization of common grazing and the management of open fields are described as general features of rural communities (Douxchamps-Lefevre, 1987: 329–30; Petit, 1987: 388–89 and 396–99). In the contributions concerning the Haute Semois (Gaume) and the principality of Stavelot-Malmedy (Ardennes), the picture is more nuanced. In these regions, where heathland and woodland were abundant, only some communities practised common grazing on their fields (compare Dorban, 1987: 416–18; Hansotte, 1987: 361–66 and Petit, 1987: 388–89). Unfortunately, a careful and thorough analysis of variations in field system types within southern Belgium has not yet been attempted. One of the main obstacles to such an essential work are the biases of many local studies. They tend to reproduce the generalization criticized by Joan Thirsk in 1966: 'much that was once accepted as evidence of common fields merely illustrates the presence of intermingled strips; it does not necessarily follow that such strips were part of a common-field system' (Thirsk, 1966: 143–44). In spite of such warnings, the model of the typical, extensive, 'three-field system' has often been projected on landscapes which, in reality, were more complex and varied in character, as recent research in both northern France and Flanders has demonstrated (Derville, 1984; Derville, 1994; Thoen, 1993; Thoen, 2010).

A good example of this bias is provided by the use that has been made of a fifteenth-century list of parcels of arable land situated in Éprave, belonging to the lords of Fexhe and Rochefort in Famenne, to demonstrate the existence of a three-field system with enforced crop rotations (Nemery, 1960). The list, which dates from 1493, is divided into three parts: parcels planted with winter-sown corn, parcels planted with spring crops, and parcels lying fallow. In several cases, the list shows that parcels were adjacent to land belonging to other landowners. Moreover, parcels sown with similar crops are roughly localised in the same areas of the township. However, nothing proves that these parcels were part of an extensive open-field system with common rules about cropping and the common grazing of the fallow. Nevertheless, in the volume of the collection *Architecture rurale de Wallonie* devoted to Famenne, Nemery's paper is cited as a clear example of a fifteenth-century *assolement triennal* (Genicot et al., 1988: 77). Another example of this tendency is provided by a frequently cited paper that claims that a village-based three-field system with enforced crop rotations was organized in Hesbaye in the late twelfth century (Rorive, 1978). In 1178, a charter was written in order to enact an arrangement about the payment of the tithe of a parcel of demesne land belonging to the abbey of Neufmoustier in Huy, situated between Chapon-Seraing and Seraing-le-Château. The tithe payment is clearly based on a three-year crop rotation: the monastery had to pay three *modii* of spelt in the year of winter-sown corn; one *modius* of spelt in the year of spring-sown corn; and nothing at all in the year when the land lay fallow. Rorive suggested that this charter reveals the existence of *assolement triennal* although there is absolutely no evidence that the field in question was part of a larger field system. His paper is, nevertheless, often cited as evidence for the existence at an early date of extensive open fields, with regular crop rotations, in Hesbaye (Kupper, 1981: 83; Suttor, 2006: 329; Wilkin, 2008: 541 is more critical). Such generalizations, as Thirsk emphasized, should be avoided, not least because individual landowners did not always have to practise rotations organized on a village base. Recent research on open fields in northern France shows that collective crop rotation could be a matter of consent between peasants (Leturcq, 2004: 55–56; Leturcq, 2007).

Figure 8.2 The settlement and field system of Wéris, 17th-18th century

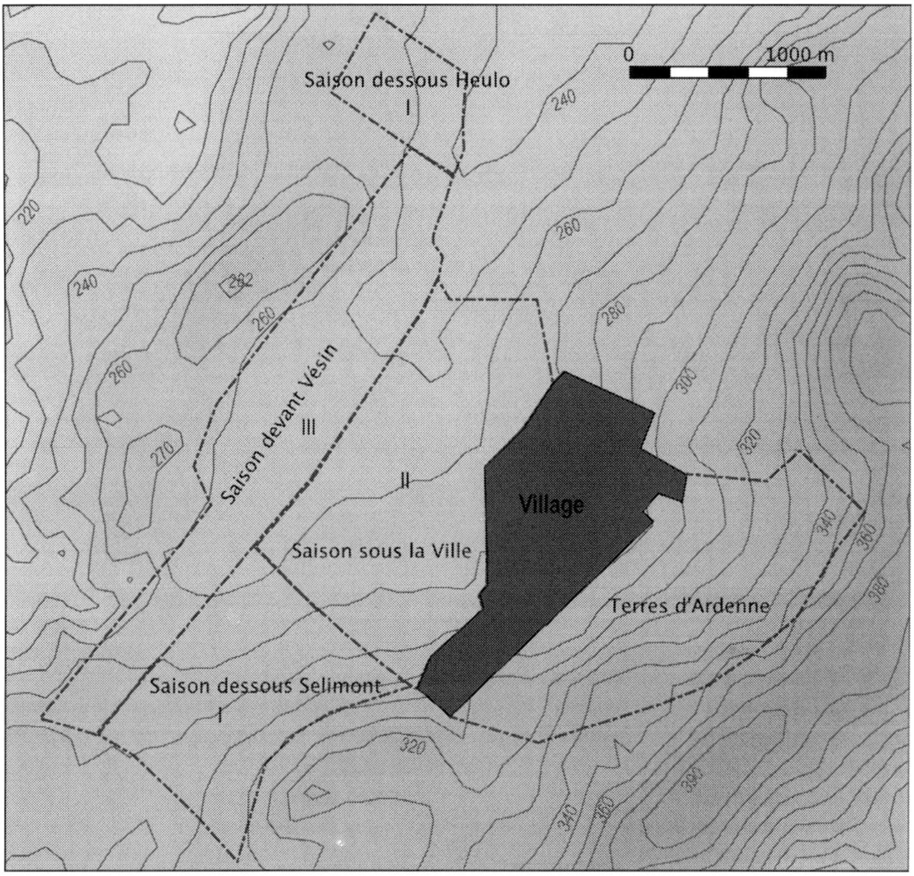

Source: N. Schroeder, based on Pirotte, 1974.

Further difficulties are raised by the complexity which can be displayed in the arrangement of local field systems. Fernand Pirotte showed that in the seventeenth and eighteenth centuries, the farmers of Wéris, a village in Famenne, organized a double open-field system (Pirotte, 1974). On the western side of Wéris lies the limestone band called the Calestienne, with its excellent lime-rich soils (Figure 8.2). On the east of the village rise the foothills of the Ardennes. Here, the subsoil consists of schists and sandstone. The land must have been inhabited since prehistory, as numerous dolmens have been found in the area. Pirotte used archives and maps from the eighteenth century to reconstruct the township and its field system. He showed that the arable land comprised intermingled strips, grouped in four fields (*saisons*), on the west of Wéris: here, from at least the seventeenth century, a compulsory three-year crop rotation based on spelt and oats was practised (Pirotte, 1974: 142). On the east of the village lay another field (*Terres d'Ardenne*) which was also divided in strips, but this was farmed under a different rotation, one typical of the Ardennes, featuring one year of rye, two or three years of oats, followed by two to five years of fallow. This example shows, once again, the impact of the environment on the organization of open fields. But it also demonstrates that at a local level, much more complexity and diversity have to be expected than is usually suggested in textbooks (another interesting example of a double open field system is Torgny: Lanotte, 1983: 21).

Rural historians and geographers have provided many monographs devoted to field-systems in southern Belgium (Raucq, 1951; Damas, 1960; Nemery, 1960; Bodson, 1965; Pirotte, 1974; Wieger, 1976; Rorive, 1978, to name but a few). Some of them are excellent, but others need to be treated with more caution. The mapping of strips and fields is not sufficient to demonstrate the character of the field systems of which they formed a part, because it does not reveal anything about their organization, or the presence or nature of rules concerning cropping and pasturing. Synthetic overviews, including the present contribution, will be at best half-formed until they are more firmly grounded in more local studies, avoid excessive generalizations, and fully reflect the true complexity of past agricultural arrangements.

III. The origins of open fields

Adriaan Verhulst's textbook on the rural economy of Belgium (Verhulst, 1990) and Léopold Genicot's fourth and last book about rural society and agriculture in the county of Namur (Genicot, 1995) are the most recent generalizing contributions to the question of the origins of open fields in southern Belgium. Verhulst's and Genicot's discussions of the chronology, but also of the causalities and the process at work in the development of open fields, are in broad agreement, although not identical. Some of the similarities and divergences of their approaches will be discussed critically in this section. Both historians were deeply influenced by wider European research, and this discussion will also take into account recent literature on French and German open fields. In the concluding paragraph, it will be suggested that new evidence (mainly archaeological and palaeoenvironmental) can and should be taken into consideration when studying the origins of open fields in southern Belgium.

According to Verhulst, the first examples of a three-year crop rotation can be found in the ninth century on the loess belt that covers parts of southern Belgium and northern France (Verhulst, 1990: 67–68). Three-course rotation was initially applied on *culturae*, big blocks of demesne land belonging to monastic estates. Extensive open fields with intermingled strips, common enforced rules, and three-year crop rotations appeared later, in the twelfth and thirteenth centuries, and only in the most fertile regions such as the Hesbaye (Verhulst, 1990: 67–68). The main reason for their development was the need to increase grain production in a time of rising prices. Big landowners were key actors in this process. The fragmentation of large estates and blocks of demesne land, combined with land clearances, allowed a reorganization of pre-existing arable land into extensive open fields. The widespread adoption of village-based three-field systems with enforced crop rotations (*assolement triennal*) required more labour, but also served to increase the proportion of the arable that was cropped each year, and (through fallow grazing) provided larger quantities of manure. The increase in oat production also had a positive impact on livestock numbers. Verhulst stressed the flexibility, adaptability, and variety of open fields. From the very beginning, some land was cultivated outside of the compulsory rotations. In Hesbaye, some fields were reserved to cultivate woad. Certain landowners wanted to produce more fodder (Verhulst, 1990: 119–20) and some small peasants were more interested in producing winter-sown corn than spring-sown corn (Verhulst, 1990: 124–25). These farmers managed to be exempted from the enforced crop rotations.

Léopold Genicot also noted that three-course rotations were practised on the demesne land of monastic estates as early as the ninth century (Genicot, 1995: 167). Using the polyptych of Lobbes (Devroey, 1986a), he even suggested that they might have been employed on arable land belonging to peasants. However, there is no specific evidence for co-ordinated rotations, or for the common grazing of fallows, until the thirteenth or fourteenth centuries. From this time forth, lease contracts specify that the leaseholders have to respect the same three-course rotation as the cultivators of neighbouring strips. Genicot dates this development to the thirteenth century, but the oldest examples of leasehold contracts he provides actually date from the fourteenth (see Genicot, 1943: 281). Rentals dating from this time show that strips and small fields were grouped into large fields called *champaigne* or *saison* on which a three-year rotation was collective and compulsory (Genicot, 1995: 169). To explain the development of these extensive open fields in the thirteenth century, Genicot highlights the importance of demographic growth, which both increased the size of the labour-force and the demand for cereals. The need for more fodder, the lack of land, urban expansion, and the development of commerce were also factors that contributed to the adoption of three-field systems. However, like Verhulst, Genicot shows that, from the beginning, part of the arable land remained outside the open-field system. This was mainly demesne land (*culturae*) and land belonging to the more powerful peasants.

In terms of chronology, Verhulst considered that extensive open fields originated in the second half of the high Middle Ages (that is, in the twelfth and thirteenth centuries), an opinion that was shared by many historians working on French and German open fields in the 1980s and 1990s (Rösener, 1985: 130; Sivéry, 1990: 39–40; Genicot, 1995: 167–69). However, more recent research suggests that although the process was initiated in the twelfth and thirteenth centuries, in certain regions, in others the new system was only adopted gradually, during the late Middle Ages and early modern period (for France: Leturcq, 2004: 54–55; for Germany: Schreg, 2009 and Schreg, 2011: 202–04). The main problem when elucidating the origins of south Belgian open fields is an absence of written evidence. The oldest reference that attests the existence of open fields in southern Belgium dates from the early fourteenth century (Genicot, 1995: 167–69; Van Derveeghde, 1955: 121). Similarly, in his recent study devoted to the management of the estates of the chapter of Saint-Lambert in Liège, Alexis Wilkin could not find any example of open fields in the late thirteenth century (Wilkin, 2008: 541–42). How are we to interpret this? Did open fields really only develop in the region after 1300, or are we dealing with an absence of evidence? In this context it should be emphasized that the type of evidence that reveals the presence of open fields, such as detailed rentals and local customary laws, only emerges in the fourteenth century. In particular, the development of leasehold (van Bavel and Schofield, 2008) gave rise to a new type of evidence – the lease contract – that is one of the most revealing documents we have for the study of open fields and crop rotations. A difficult question that must be asked is whether earlier evidence would, in fact, necessarily reveal the existence of open fields. Early and high medieval polyptychs and records of estate management focus on demesne land and rent, without giving much information about peasant farming. Units of land assessment (*mansus*, *curtis*) are frequently mentioned, but rarely described. As a result, for a long time the only way to obtain information about field systems was to look at demesne land. In the following paragraphs, the information given by this type of evidence will be analysed.

Carolingian polyptychs show that three-year rotations were organized on the demesne land of monastic estates in northern France and southern Belgium (Sivéry, 1990: 36–37; Morimoto, 1994; Verhulst, 2002: 64–65). The most revealing document is a fragment of the polyptych of Elnone (Table 1.1). Additions to the polyptych of Lobbes, dating from *c.* 889, suggest that in Biesmerée, Castillon, Lobbes, Berzée, Haine and Hyon, a three-course rotation was followed on the lands of the demesne (Table 1.2). The chapter of the polyptych of Prüm, written in 893, concerning the estate of Villance shows that in the Ardennes, oats were the predominant crop (Table 1.3). Rye was cultivated as well, but the polyptych does not reveal whether or not a regular crop rotation was applied on the demesne land. However, the chapter concerning the Villance estate is particularly interesting, because it reveals an important aspect of how labour was sometimes organized on demesnes. Some tenants had to plough and dung half a *jornalis* of land to prepare the field to be sown with rye. This type of work must have led several peasants to plough and dung parts of the same large field. Using the polyptychs of St Germain-des-Prés and St Remi de Reims, the Japanese historian Yoshiki Morimoto has insisted on the importance of the so called '*lot-corvée*' for the development of the three-year rotation and, perhaps, open fields (Morimoto, 1994). According to Morimoto, with the '*lot-corvée*' a plot of the demesne was regularly tilled by the same tenant. These plots were laid out in long parallel strips, very similar to those later found in open fields. The main difference is of course that with the '*lot-corvée*' all the land belonged to the lord. Morimoto suggested that progressively, the property of plots of land was transferred to the peasants who worked them. This process might have led to the creation of open fields. This model is partly hypothetical. It is not certain that peasants had to work the same plot of land every year (the best discussion of these questions remains Perrin, 1925). Moreover, the transfer of demesne land to peasants is not clearly recorded in the documentary sources (see below). Be that as it may, the evidence from large estates regarding '*lot-corvée*' and labour services shows that the technical knowledge (three-course rotation) and the concept (a large field made out of individual strips worked by several farmers) necessary to develop open fields already existed in the ninth and tenth centuries.

For the region studied here, we have examples of '*lot-corvée*' with a three-course rotation on two estates belonging to St Peter of Brogne, dating from the tenth century (Table 1.4). Whether farmers were applying these techniques individually, or collectively on their own land, be it allodial or tenurial, is unknown. Using mainly written evidence, Adriaan Verhulst asserted that 'the field complexes [...] did not yet form continuous open areas and were still separated by woods, heath, or uncultivated plots and possibly even enclosed by hedgerows or trees' (Verhulst, 2002: 64–65). Archaeological and palaeoenvironmental data confirms that the early medieval landscape was not as open as it was to become by the twelfth and thirteenth centuries (see below). It is, therefore, very unlikely that extensive open fields existed in southern Belgium in the early Middle Ages.

Table 8.1: Rotations and '*lot-corvée*' on demesne land according to early medieval estate records

1	*Mansus dominicatus [...] habens de terra arabili bunuaria* XXX. *Seminantur ad hibernaticum bunaria* X, *de modiis* XL, *et ad tremissem bunaria* X, *de modiis* LX. *Bunaria* X *interjacent.*	Polyptych of St-Amand-les-Eaux concerning Maire, 821–72 (Hägermann and Hedwig, 1990: 104)
2	*De terra arabili bunaria ubi possunt seminari ad hibernaticum modia CC, ad tremensium modia CXXX*	Polyptych of Lobbes concerning Biesmerée, *c.* 889 (Devroey, 1986a: 22)
	Sunt ibi de hibernatico seminata modia LXIII, de tremesio modia XLIII, de legumine modia II	Castillon (p. 23)
	[...] aspiciens ubi sunt seminata de hibernatico modia XLIIII, ad tremesium modia L	Lobbes (p. 24)
	De terra arabili bonuaria XXXIIII, sunt ibi seminata de hibernatico modia L, de tremesio modia XXXVIII, de legumine modia II et semis	Berzée (p. 26)
	Habet ibi seminatum inter hibernaticum et tremesium modia LXX	Haine (p. 27)
	Terra arabili ubi possunt seminari inter hibernaticum et tremensium de annona modia [...]	Hyon (p. 28)
3	*Est in Vilantia mansus indominicatus, aspiciunt ad ipsum culture.*VII. *Prima est in Rotunda Hasila, ubi potest seminari avena modios [...]; secunda est, que dicitur Merilonis Cruce, ad seminanda avena modios [...]; tercia ad Hulsiz, ubi potest seminare modios.*CCCC.*; quarta est in loco, qui vocatur Albieg, ad seminandos modios.*CCCC.*; quinta est in loco, qui vocatur Reimanbarba, ubi postest seminari modios.*CCL.*; sexta est secus fontem, qui vocatur Scaitla, ad modios.*LX.*; septima est in loco, qui vocatur Ruchengas, ad modios.*XXX. [...] *Arant et fimant de illorum fimum iornalem dimidium ad hibernaticam sationem, ad sigulum seminandum; ad tremensem in marcio et aprile arant iornales.*IIII.	Polyptych of Prüm concerning the estate Villance, 893 (Schwab, 1983: 201–02)

Table 8.1: (continued)

4	*In Romereis.IIII. mansi […] et unus mansus de terra sancti Lamberti.* […] *in martio.II. dies de bobus et bonuarium* […] *in maio de quindecim diebus bonuarium* […] *inter augustum et septembrem.X. dies falcis et.II. bonuarios, unum de spelta et.I. de avena*	Tenure description of St Gérard in Brogne concerning Romerée, tenth century (Devroey, 1982: 217)
	In maio de quindecim diebus bonuarium, de quindecim censum solvet […] *inter augustum et septembrem.X. dies falcis et.II. bonuarios, unum de spelta et.I. de avena*	Manise (Devroey, 1986b: 61)

Table 8.2: Size and status of the *culturae* of the monastery of St Truiden in 1253 (Pirenne, 1896: 360–61)

Location	Size (in *bonuaria*)	Status
St Truiden	47	Fragmented and transformed into tenures
St Truiden	20	Owned and exploited by the monastery
Stayen	40	Owned and exploited by the monastery
Borloo	76	Owned and exploited by the monastery
Mielen	27 and 16 *virgae*	Owned and exploited by the monastery
Meer	26,5	Owned and exploited by the monastery + a part that has been granted out in fief
Laer	53	Owned and exploited by the monastery
Villers-le-Peuplier	108	Owned and exploited by the monastery
Oreye	68	Owned and exploited by the monastery
Kerkom	32	Owned and exploited by the monastery + a few *bonuaria* that are disputed
Donck	20	Owned and exploited by the monastery
Houtsem	4	Owned and exploited by the monastery
Webbecom	24	Owned and exploited by the monastery
Sény	20	Owned and exploited by the monastery

Table 8.3: Location and status of demesne land belonging to the count of Namur in 1294 (Brouwers, 1910)

Fleurus	*prisiet a hiretage*
Golzinne	*Coustures qui gisent à Malerot et à Sorlées [...] sont tout achensi* *Coustures qui gisent à Boukaing et à Templos qui contienent entour 42 bonniers, dont on rent rentaulement par an 85 muis et 4 stiers d'espiaute au m. de Namur* *Coustures qui gisent à Goulesines qui Libers de Hines et Henris Chariaus tienent pour 52 m. d'espiautre par an au muids de Namur*
Vedrin	*Coustures [...] que Gerars de Frisey tient*
Champion	*Coustures, valent par an 100 m. d'espiautre*
Sclayn	*Li cuens y a ses coustures dont il y a entour 12 bonniers, valans par an 24 m. d'espiautre au mui de Namur*
Anthée	*Li cuens y a seur 42 bonniers de cousture gisans à Miauvoie le quarte partie, lequel il tient por li quant li autre parchoniers tiennent les leur parans: vaut se partie par an 12 m. et demi de grain au m. de Namur, c'est à savoir moitié espiaute, moitié avaine*
Flavion	*26 bonniers de coutures qui Colignons de Fluns tient, peu plus peu mains, en espiautre 7 m., en avaine 7 m.*
Floriffoux	*Coustures que Servais et Baudins et autres tiennent*
Floreffe	*Li cuens y a 38 bonniers de coustures qui li valent par an en espiautre 66 m. et demi au mui de Namur*
Libenne	*Li cuens y a se partie de 28 bonniers et 60 verges de coustures*
Moulins	*Li cuens y a coustures qui contiennent 35 bonniers et demi, et 22 verges, valent par an 35 m. d'espiautre*

However, tenth-century charters show that even in the Ardennes and Famenne, plots of cultivated land could lie adjacent to each other (Devroey and Schroeder, 2012: 54; Mignot and Schroeder, 2016). It is unclear whether they were fenced and cultivated in severalty or not, but the existence of small open-field systems with intermixed strips and some degree of common regulation cannot be excluded. Shifting cultivation in an infield-outfield system also implies some form of collective management of resources, and common rules for the use of heathland and woodland are certainly attested in the ninth and tenth centuries (Devroey, 2003: 83–85). The relevant sources do not explicitly mention intermittent cropping but this absence of evidence is not sufficient to exclude the existence of infield–outfield systems. If they existed, early medieval field systems must have been less formally structured and less extensive than they were to be in the high Middle Ages. We should note in this context that recent research has shown that the development of cities in the Meuse valley in the high Middle Ages had a significant impact on the surrounding countryside (Wilkin, 2008). Palaeoenvironmental data suggests a significant increase in arable land in the Ardennes between the tenth and the thirteenth centuries (see below). Clearances and the reorganization of field systems must have taken place in this period in order to respond to the increasing demand for cereals (Verhulst, 1990: 67–68; Renes, 2010: 55–59).

In Adriaan Verhulst's perspective, the demesne land of large early medieval estates had an important role in the development of open fields in the high Middle Ages (Verhulst, 1990: 67–68). He described how in Flanders large fields on the demesne (*Hofkouter*) were fragmented and converted into an open field with intermingled strips, common rotation and grazing rules. According to Morimoto, the '*lot-corvée*' might have been an intermediate stage in this fragmentation process (Morimoto, 1994). In other regions, such as Hesbaye, Brabant, and southern Flanders, Verhulst suggested that the three-course rotation organized on the demesne land of big landowners was progressively imposed on entire townships. This opinion was influenced by the work of Fossier and Sivéry on northern France, as Verhulst made clear in an earlier paper: 'les recherches de ces deux historiens ont en effet démontré qu'en Picardie et dans le Hainaut en particulier l'extension à tout le terroir du village de l'assolement et de la rotation triennale, devenus obligatoires vers le milieu du XIIIe siècle, a débuté sur une base volontaire au commencement de ce siècle, précisément à partir de terres de la réserve qui, à ce moment, ont toutes été mises en soles et qui dès lors ont joué un rôle pilote' (Verhulst, 1980: 25). A similar model has been suggested by German historians (Schröder-Lembke, 1978: 22). David Hall's recent attempt to explain the origins of extensive open fields in some areas of Britain is very close to this, although slightly different: 'block demesnes remained [...] confined to their own area, but with common rights extending throughout the township' (Hall, 2014: 189).

For some areas of southern Belgium, however, this interpretation is at variance with Léopold Genicot's observation that large fields of demesne land (*culturae*, *coutures*) frequently remained separate from, and outside, extensive open fields in the late Middle Ages and the modern period (see above). An excellent example of this is provided by the case of a village called Thisnes, situated in Hesbaye (Bodson, 1965). In the tenth century, the monastery of Andenne was granted an estate here and late medieval and modern archives from Andenne allowed Bodson to study the field system and its use in the fourteenth and fifteenth centuries. He drew a map of the township, using records of tithes and lease contracts (Figure 8.3). It shows that the township mainly consisted of arable land surrounding the village. It was divided into twenty-eight separate fields in which peasants owned intermingled strips. A three-year crop rotation was applied across these fields, and lease contracts from the fourteenth century show that it was compulsory. Each year, the amount of land planted with winter-sown corn was slightly higher than that planted with summer grain or left as fallow. The demesne land belonging to the estate of Andenne was composed of three fields situated on the western side of the village. They were called *couture* (a word derived from the Latin *cultura*) and did not form part of the village's open-field system. This example suggests that even after the entire land of the township had been converted into open fields, the demesne land still comprised discrete, undivided blocks which were not submitted to common rules. In other cases, the demesne land was scattered over the entire township in several parcels (for example the demesne land of the abbey of Waulsort in Anthisnes in Condroz or the demesne land of the abbey of Stavelot in Vieuxville, in Famenne: see respectively Wieger, 1976 and Damas, 1960). However, in neither case is it certain whether these rather large parcels were part of an extensive open field or not. The farmers who held them by lease had to respect the crop rotation imposed by the landowner, but not necessarily common rules about cropping or common pasturing. This evidence implies that the evolution of demesne land in the high Middle Ages might have been less important for the origins of open fields in southern Belgium

Figure 8.3 The settlement and field system of Thisnes, 14th-15th c.

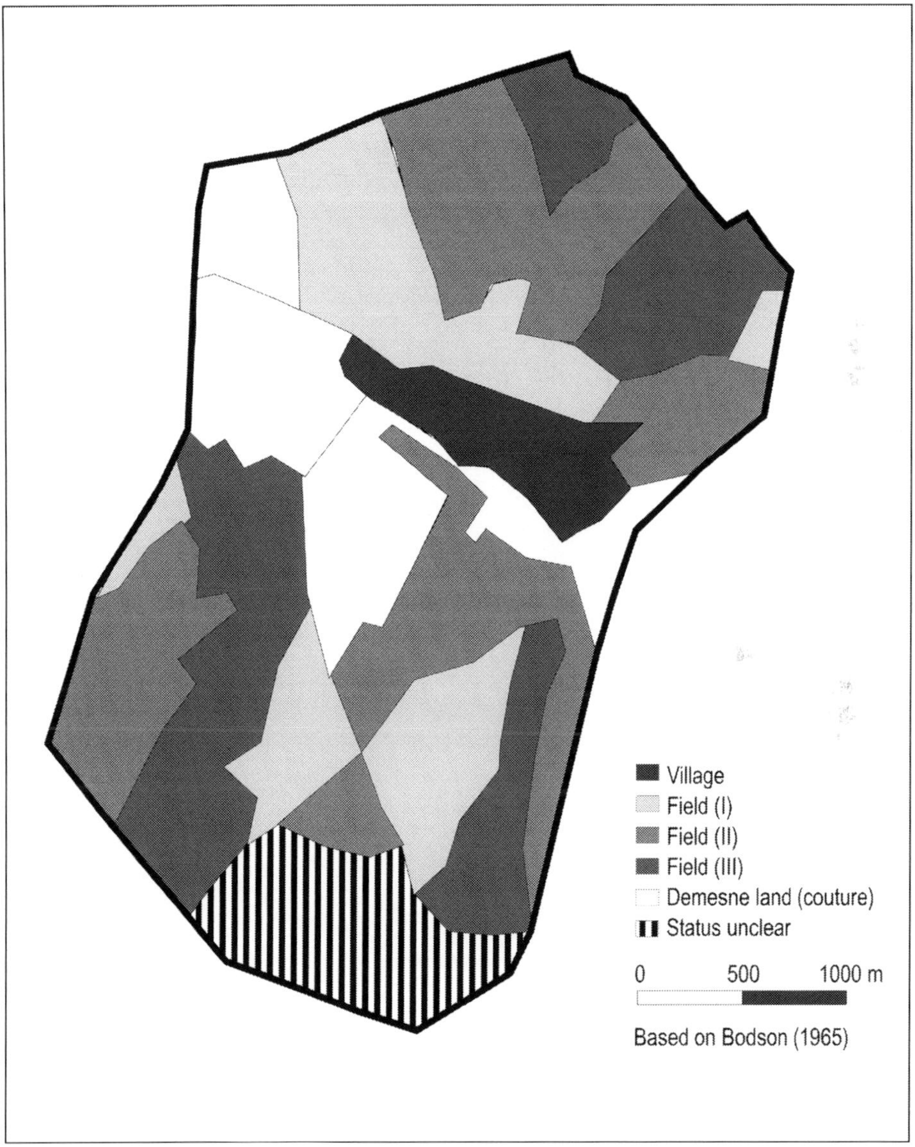

Source: N. Schroeder, based on Bodson, 1965.

than Verhulst suggested. However, this does not necessarily mean that the fragmentation of demesne land and its integration into open fields did not happen at all. Estate records from the thirteenth century suggest that it did occur in some cases, but not in others.

Many ecclesiastical and lay landowners prepared very detailed lists of their property in the thirteenth century and some of them have survived (St Truiden: Pirenne, 1896; St Lambert in Liège: van Derveeghde, 1958; the Count of Namur: Brouwers, 1910–1926). They rarely give information about peasant farming, beyond rent and abstract units of land assessment, but demesne land is often described in much detail. A comprehensive study of these documents cannot be undertaken here, but a brief analysis based on the existing literature is suggestive. Alexis Wilkin recently showed that most *culturae* of the chapter of Saint-Lambert in Liège were granted out in leasehold in the late thirteenth century (Wilkin, 2008: 522–27). These large block of land were thus preserved from fragmentation and still existed in the sixteenth and seventeenth centuries. Some division of the *culturae* into plots, granted out individually, took place, but it was rare (something which might also explain why Wilkin could not find any traces of open fields in his evidence – see above). A similar observation can be made for the *culturae* of the monastery of St Truiden in 1253 (see table 2). Out of fourteen *culturae*, only one had fragmented and been granted out in separate tenures; another was partly granted out as a fief; and the ownership of another was disputed. The others were all apparently still owned and directly exploited by the monastery. In 1262, in Oreye, the abbot tried to exchange portions of demesne land in order to reduce fragmentation and create more coherent blocks (Pirenne, 1896: 50). In the same year, the estate of Seny in Condroz was granted out to a farmer (Pirenne, 1896: 52). The estate still consisted of intact *cultura*e, on which a three-course rotation was practised. We do not know if the lease-holder had to respect any common rules about grazing and pasturing. According to Pirenne, from the second half of the thirteenth century onwards, most of these *culturae* were granted out on lease (Pirenne, 1896: xxiv–xxv). This helped to preserve them from fragmentation.

These examples suggest that much demesne land remained in large compact blocks, free from communal crop rotations and common grazing. However, the records concerning the estates of the count of Namur in 1294 provide an excellent counter example. Out of twenty-six estates, only eleven had demesne land (*culturae*). On these, the proportion of land devoted to *culturae* was quite low (Figure 8.4 and Genicot, 1943: 99–101). For the estates of Champion, Sclayn, Floreffe, Temploux, Libenne and Moulins, the survey only gives the amount of land and/or the quantity of grain produced every year (see table 3). For Fleurus, Malerot, Sorlées, Vedrin and Flavion it specifies that the land was granted out to a farmer. In Flavion, the farmer owed the same quantity of oats and spelt to the count each year, suggesting that a three-course rotation was followed on this land. In Goulesines, Anthée and Floriffoux, the demesne land was held by two or more farmers. These last examples suggest a situation in which demesne land was granted out to several farmers and subjected to a three-course rotation. This could have been a first stage towards the fragmentation of blocks of demesne land and their inclusion within an open-field system, as described by Verhulst. The fact that a significant amount of the count's estates did not have any demesne land in 1294 suggests that such a process of fragmentation, and granting out, had already happened in several places. This short comparative discussion of the history of demesne land suggests that, depending on landowners, geography, and socioeconomic context, the evolution towards open fields

Figure 8.4 Land allocated to *culturae* and tenures in the estates of the count of Namur in 1294 (*in bonnarii*)

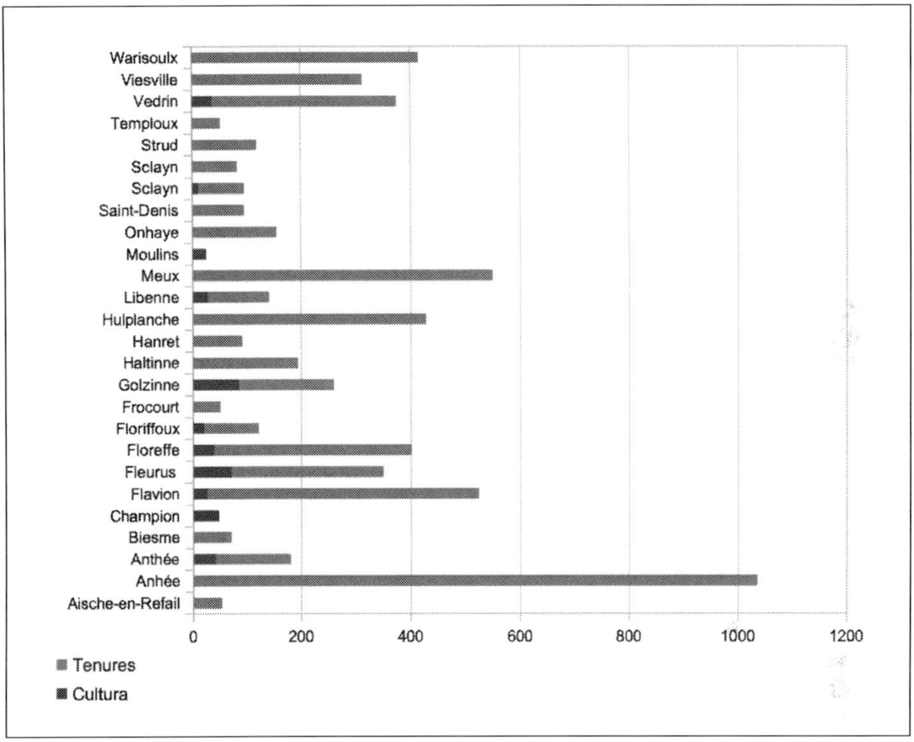

Source: N. Schroeder, based on Brouwers, 1910.

varied from place to place. In some cases demesne land played a part in the development of open fields, whereas in others it did not.

More local studies are needed before a general model concerning the origins of open fields in southern Belgium can be developed. One of the most problematic aspects is that peasant farming largely remains hidden from sight. Future research should integrate material evidence in order to compensate the biases of written documents. Research in British medieval field systems in recent years has made increasing use of archaeological and palaeoenvironmental data (Hall, 2014: 134–52 provides a recent and synthetic overview). This is hardly the case in southern Belgium. Extensive surveys and excavations of medieval settlements have only started recently (compare De Meulemeester, 1996; Vanmechelen, 2009 and Vanmechelen, 2013). There is a strong tradition among Belgian medievalists of using the evidence of pollen analysis (see Noël, 1972), but it has rarely been employed to study change in field systems. The study of alluvium deposits has also been developing over the last decade. Geoffrey Houbrechts and François Petit have shown recently that in the Ardennes, extensive sedimentation developed in the high Middle Ages (Houbrechts and Petit, 2006: 96). For the same region, pollen deposits show an increasingly open landscape, a process of clearance which, by the thirteenth century, had reached the upland moors (Damblon, 1994: 268). This evidence suggests a gradual, yet significant, increase in the extent of arable land use in the high Middle Ages (Schroeder, 2016: 182–84). However, these results are only valid at a broad, regional scale. Integrated multidisciplinary research on a local level remains an elusive goal (see Preiss et al., 2016 for a recent case-study that highlights the possibilities, and limitations, of such approaches in southern Belgium).

IV. The farmer's perspective? Economy, society and morality

Traditionally, research on southern Belgian open fields has been focussed on agrarian, economic, and legal aspects. Open fields have been seen as a solution to the problem of how to produce more grain when no more woodland or 'waste' remained to be cleared. It is often accepted that the identity and social organization of late medieval and modern village communities were related to the management of their open fields and commons (Genicot, 1995: 167–70). However, much more emphasis has been put on the legal and formal structures, such as lordship and parish, in order to explain the origins of village communities (Genicot, 1988). As we saw above, an important role has been assigned to demesne land in the development of open fields. Hence, the role and motivation of individual farmers and communities have been largely ignored. Recent research in northern France shows the importance of approaching open fields from a 'peasant's perspective' (Leturcq, 2007). It seems that a particular type of late medieval and modern evidence from southern Belgium is well-suited for such research. 'Statements of right' (*records de coutume*, *Weistümer*) have often been used to study the collective management of common resources such as woodlands or wastelands, but rarely the organization of arable farming. These documents are transcriptions of the laws of local villages, rules which were established by the villagers or their representatives in order to regulate the community and its interactions with its lord(s) and with other villages. Such rules were generally transmitted orally but sometimes a higher authority (a lord or a court of justice) committed them to writing, in whole or part, in order to conserve and reinforce them. Many laws were transcribed in the period from the fourteenth to the eighteenth century and have been preserved in various archives.

One example dates from 15 June 1612 and concerns the community of Membre, a village situated in the Ardennes (Hardt, 1868: 512–16 – Document I). Several chapters are relevant for our discussion. Firstly, members of the community had the right to cultivate in Membre if they gave a twelfth of their harvest to their lord and a tithe to the priest. Secondly, they had to pay fifteen pounds to the lord at Easter, in order to be allowed to own a plough, and to use it. However, 'ploughing alone' (*labourer seul*) was only permitted with the consent of the community (*par la permission des autres bourgeois*). Every year, on 24 February, the community gathered and its members had to 'find an agreement' (*s'accommoder les uns avec les autres*). Every farmer, 'the poor and the rich, who own a plough, an oxen or a horse' (*aussy bien les pauvres que les riches ayant bêtes et harnois, soit bœufs ou cheval*), had to join others in forming ploughing teams (*s'associer*). If a member of the community was excluded from the 'agreement', he was allowed to call the ploughs back from the field (*faire revenir toutes et chascunes les charues qui sont aux champs*) and the community had to debate until a new arrangement was found that would not exclude anyone. Harrowing was also submitted to common rules and the right/obligation to team up with other peasants. Such rules acknowledged the varying degrees of wealth within the peasant community and imposed a measure of levelling, based on the obligation to exchange both labour and the instruments of labour. The second document dates from 1660 (Genicot and Allart, 1981: 817–18 – Document II). It reveals that the community of Thuin in Hainaut was facing hard times because of war and the abuses of its richest members (*bourgeois et surcréans*) and outsiders (*étrangers*). The main problem was that the organization of common grazing was ineffective, advantaging some of the most powerful members of the community and people who were not even part of it, and causing its poorest members (*pauvres mannants et surcréans*) to struggle. To deal with this situation the representatives of the community made new rules about grazing on the commons, and also on the arable land when it lay fallow (*pâturage et champiage des bestes à laine sur les communes et héritages des particuliers*). These rules were supposed to support the weakest members of the community, who raised sheep 'for their subsistence and to nurture their family' (*pour la petitte subsistence et entretenance de leurs familles*). How are we to interpret such concerns? It might be naïve to take the rhetoric about social equality and morality that this document conveys at face value. Taking decisions inside the village community was certainly a question of power networks, influence and personal interests (see, for example, Popkin, 1979). However, it is not insignificant that the representatives of the community refer to moral concerns to justify their decisions. Morality and values such as solidarity might have been factors that had some importance in the management of open fields, even if equality was not actually achieved. Mark Bailey has recently claimed that with open fields, some 'communities [were] rating sustainability and equitable access to resources, above raising agrarian productivity' (Bailey, 2010: 170–71).

These two seventeenth-century documents reveal situations in which the lord is a distant figure, who receives rent and enforces the law by his authority. The ones who were both working the land, and taking decisions about how to do so, were groups of farmers. Their decisions were certainly led by an economic logic, but as the example of Thuin makes clear, they could also be informed by social and moral concerns. From the farmer's perspective, open fields allowed them to face the pressure of lords, markets, or particularly powerful farmers collectively. From the thirteenth century onwards in southern Belgium, village communities became stronger, and more formalized and institutionalized (Genicot, 1988; Genicot, 1995). This evolution has often been seen as a response to the need to protect communal woodlands (Genicot et al., 1988: 60–63). However, in a context of limited agrarian resources and increasing demand for cereals, farmers might well have tried to develop forms of common management of the arable land, leading to the increasing importance of communities in decision making. There has been a long-lasting 'top-down vs. bottom-up' debate about the origins of extensive open fields (see, for example, Nitz, 1988). It is certainly not my aim to reopen that argument here. However, the individual and collective decision-taking of farmers has not been significantly taken into account in the literature on southern Belgian open fields. The evidence provided by these 'statements of rights' offers an alternative view, one that stresses the importance of peasants in the management of open fields.

V. Conclusion

Open fields in the southern provinces of modern-day Belgium have been studied for many years. Their distribution and origins have been discussed in broad, generalized overviews. This chapter has provided a critical summary review of this literature, as well as proposing a number of new avenues of research. Previous studies have shown the regional diversity of open fields, with extensive open-field systems in Hesbaye and infield-outfield systems in the Ardennes. In Condroz and Famenne, mixed forms that combined relatively large areas of arable land, heathland, and woodland commons were predominant. These types largely reflect the physical geography of southern Belgium. However, such models appear to be overly general. They tend to smooth over intraregional diversity, local complexity, and temporal flexibility. Common rules about cropping and pasturing have often been assumed, rather than actually demonstrated.

Research into the origins of open fields has mainly focussed on demesne land. According to this model, in the twelfth and thirteenth centuries the fragmentation of blocks of demesne, land clearances, and the extension of crop rotation and grazing over entire townships led to the formation of extensive open fields. Although in some cases open fields may well have developed in this manner, this approach tends to overemphasize the importance of demesne land, which in many cases stayed outside the open field system.

The chronology of the origins of open-field systems is also likely to have been more complex than originally assumed (Renes, 2010: 55–56). Infield-outfield systems and small field systems with intermixed strips and some degree of common regulation may have existed in the early medieval period. It is nevertheless likely that the first extensive open fields only developed in the twelfth and thirteenth centuries as a result of the high medieval 'cerealization' connected with demographic and urban growth (Renes, 2010: 49–50). This evolution might be linked to the development of more formalized peasant communities. Further changes leading to the organization of new open fields or the reorganization of old ones in the late Middle Ages and the modern period cannot be excluded. The analysis of modern 'statements of rights' suggests that, besides economic efficiency, moral concerns may have been part of the incentives for a collective organization of production. This evidence also shows that peasant communities had an active role in the management of open fields, an aspect that earlier studies have not always taken into account.

Ideally, all these questions should be investigated in new studies that also take into account archaeological and palaeoenvironmental evidence. Above all, perhaps, the study of field systems in southern Belgium would benefit from the comparative analysis of more local case studies, taking into account such things as the character of the local environment, settlement patterns, manorial structures and property-structures, as well as the proximity of urban centres and the impact of commercial activity on agrarian production. Carried out systematically in selected areas, such an approach would allow us to establish the relative importance of such factors in the emergence and shaping of medieval field systems, as well as providing a clearer indication of relative degrees of regional homogeneity, and local diversity, in their character.

Document I: statement of rights of membre, the 15th of June 1662
(Hardt, 1868: 512–13)

Sur chaque labeur que les bourgeois peuvent labourer et cultiver sur le ban et seigneurie du dit Membre, payent terrage au seigneur à la 12e gerbeet au curé parochial l'on paye ensuite aussy la dixme à la douzième gerbe. [...]
Chacun bourgeois qui a charue pleine pour labourer et faire marsage, doit xv sols au seigneur, lesquels se payent aux festes de Pasque, et ne peut neantmoins le bourgeois labourer seul, encor qu'il auroit sa charrue à luy seul, si ce n'est par la permission des autres bourgeois, pour et autant qu'ils ont une coutume entre eux de loingtems observé de s'accommoder les uns avec les autres au jour st. Mathias, que l'on appelle fête de février, pour faire leur marsages, et ne peuvent labourer l'un sans l'autre, que prealablement ils ne soyent associés, aussy bien les pauvres que les riches ayant bêtes et harnois, soit bœufs ou cheval, et s'il advenoit qu'un desdits bourgeois ne fut associé avec les autres, comme est la coutume, il luy est permis sans autre forme faire revenir toutes et chacune les charues qui sont aux champs, et les faire cesser de labourer, tant et si longuement qu'il soit reunis et associé avec les autres. D'avantage si un bourgeois avoit une herse à rabattre la terre labourée et qu'il dit: "je veux être associé avec vous selon ma contingent et autant de bestial que porte la charue, j'irois autant de journées faire mon pouvoir", il auroit une journée de charue à son tour, et en cas de refus, il peut faire revenir d'aux champs ceux qui ont bêtes de harnais, car c'est la coutume, et partant n'y a que le mayeur qui peut labourer seul avec la charue.

Document II: statement of rights of Thuin, the 11th of April 1660
(Genicot and Allart, 1981: 817–18)

Nous les bourgemaitres, jurez et conseil de la bonne ville de Thuin, assemblez en notre lieu accoustumé ce onzième du mois d'avril 1660, délibérans sur les moyens de pouvoir remédier à divers abus et emprise tant des bourgeois et surcréans de la jurisdiction de cette ville que des étrangers glissez par le malheur du temps et la confusion causées des ruines et ravages de la guerre, laquelle a fait ressentir ses funestes effects par une si longue suitte d'années en ce quartier d'Entre-Meuse-et-Sambre, et faisans une particulière réflexion sur l'intérest et préjudice que le pâturage et champiage dereglez et indiscret des bestes à laine sur les communes et héritages des particuliers apportent de jour en jour, signament aux pauvres mannans et surcréans qui pourroint plus facilement nourrir bestes à corne pour la petitte subsistence et entretenance de leurs familles, désirans d'i pourvoir par les moyens les plus promptes et convenables, après avoir longtemps concertez et conférez, même pris avis de jurispérits là-dessus, avons d'un commun accord et unanimement conceu le règlement selon les points et articles suivants.

Bibliography

Bailey, M. (2010), 'Beyond the Midland system: the determinants of common rights over the arable in medieval England', *Agricultural History Review*, 58, 2, pp. 153–71.

Bakels, C. C. (2009), *The Western European loess belt. Agrarian history, 5300 BC-AD 1000*, Dordrecht – London – Heidelberg – New York.

Billen, C. (1989), 'L'épeautre en Wallonie du Bas Moyen-Âge jusqu'à l'époque contemporaine. Une mise au point' in: J.-P. Devroey and J.-J. van Mol ed., *L'épeautre (Triticum spelta). Histoire et ethnologie*, Treignes, pp. 177–90.

Bodson, M.-J. (1965), 'L'évolution d'un paysage rural au Moyen Âge. Thisnes en Hesbaye', *Bulletin de la société belge d'études géographiques*, 34, pp. 117–58.

Brouwers, D. D. (1910–1926), *L'administration et les finances du comté de Namur du XIII*[e] *au XV*[e] *siècle*, 1 (Cens et rentes du comté de Namur au XIIIe siècle), 2 t., 3. vols., Namur.

Damas, H. (1960), *Vieuxville, commune rurale de Wallonie: essai de géographie sociale*, s.l.

Damblon, F. (1994), 'Les dépôts tourbeux et l'histoire de la végétation sur le Plateau des Hautes-Fagnes (Belgique)', *Annales de la Société géologique de Belgique*, 117, 2, pp. 259–76.

De Meulemeester, J. (1996), 'L'archéologie rurale dans l'espace Belgo-Luxembourgeois. Un état de la question', in: J. Fridrich eds *Ruralia 1. Conference Ruralia 1, Prague, 8*[th]*-14*[th] *September 1995*, Prague, pp. 160–70.

Derville, A. (1989), 'L'assolement triennal dans la France du Nord au moyen âge', *Revue Historique*, 280, 2, pp. 337–76.

Derville, A. (1994), 'Verzelgung und Dreifelderwirtschaft in Nordfrankreich im Mittelalter' in: A. Verhulst and Y. Morimoto ed., *Économie rurale et économie urbaine au Moyen Âge*, Ghent – Fukuoka, pp. 127–33.

Devroey, J.-P. (1982), 'Documents inédits de l'abbaye Saint-Pierre de Brogne', *Bulletin de la Commission royale d'Histoire*, 148, pp. 205–27.

Devroey, J.-P. (1986a), *Le polyptyque et les listes de biens de l'abbaye Saint-Pierre de Lobbes (IX*[e]*-XI*[e] *siècles)*, Brussels.

Devroey, J.-P. (1986b), 'Les premiers domaines de l'abbaye Saint-Pierre de Brogne (X[e]-XIIe siècles)' in: *La seigneurie rurale en Lotharingie. Actes des 3*[es] *journées Lotharingiennes, 26–27 octobre 1984*, Publications de la section historique de l'institut Grand-ducal de Luxembourg, 102, pp. 51–67.

Devroey, J.-P. (2003), *Économie rurale et société dans l'Europe franque (VI*[e]*-IX*[e] *siècle)*, t. 1 (*Fondements matériels, échanges et liens social*), Paris.

Devroey, J.-P. and Schroeder, N. (2012), 'Beyond royal estates and monasteries: Landownership in the early medieval Ardennes', *Early Medieval Europe*, 20, 1, pp. 39–69.

Dorban, M. (1987), 'Introduction aux communautés rurales de la vallée de la Haute-Semois (XVIe-XVIIIe siècles) in: *Les communautés rurales. 5*[e] *partie. Europe occidentale et Amérique. Synthèse générale*, Recueil de la Société Jean Bodin pour l'histoire comparative des institutions, 44, Paris, pp. 409–41.

Douxchamps-Lefevre, C. (1987), 'Les communautés rurales du Namurois durant l'époque moderne' in: *Les communautés rurales. 5*[e] *partie. Europe occidentale et Amérique. Synthèse générale*, Recueil de la Société Jean Bodin pour l'histoire comparative des institutions, 44, Paris, pp. 317–34

Genicot, L. (1943), *L'économie rurale namuroise au bas Moyen Âge*, 1 (*Seigneurie foncière*), Louvain-la-Neuve.

Genicot, L. (1947), 'La limite des cultures du froment et de l'épeautre dans le Namurois au bas Moyen Âge', *Namurcum*, 22, pp. 17–23.

Genicot, L. and Allart, R. M. (1981), *Sources du droit rural du quartier d'Entre-Sambre-et-Meuse*, 2, Brussels.

Genicot, L. (1988), 'La communauté rurale en Belgique jusqu'au XIIIe siècle', in: *Les Structures du pouvoir dans les communautés rurales en Belgique et dans les pays limitrophes: (12e–19e siècle). 13e Colloque international, Spa, 3–5 sept. 1986*, Bruxelles, pp. 17–38.

Genicot, L. (1995), *L'économie rurale namuroise au bas Moyen Âge*, 4 (*La communauté et la vie rurales*), Louvain-la-Neuve – Brussels.

Genicot, L.-F. et al. (1983–1992, 7 vols.), *Architecture rurale de Wallonie – Hesbaye namuroise – Hesbaye Liégeoise – Ardenne centrale* Liège–*Fagne et Famenne – Condroz – Hesbaye brabançonne et pays de Hannut – Ardenne Herbagère*, Liège.

Hägermann, D. and Hedwig, A. (1990), *Das Polyptychon und die Notitia de Areis von Saint-Maur-des-Fossés. Analyse und Edition*, Beihefte der Francia, 23, Sigmaringen.

Hall, D. (2014), *The open fields of England*, Oxford.

Hansotte, G. (1987), 'Les communautés rurales au pays de Stavelot' in: *Les communautés rurales. 5e partie. Europe occidentale et Amérique. Synthèse générale*, Recueil de la Société Jean Bodin pour l'histoire comparative des institutions, 44, Paris, pp. 355–75.

Hardt, M. (1868), *Luxemburger Weisthümer, als Nachlese zu Jacob Grimm's Weisthümern*, Luxemburg.

Houbrechts, G. and Petit, F. (2006), 'Utilisation des microscories métallurgiques comme traceur de la sédimentation dans les plaines alluviales des rivières ardennaises' in: *L'érosion entre société, climat et paléoenvironnement*, Clermont-Ferrand, pp. 91–96.

Jacob, R. (1987), 'Les communautés rurales du Pays de Liège' in: *Les communautés rurales. 5e partie. Europe occidentale et Amérique. Synthèse générale*, Recueil de la Société Jean Bodin pour l'histoire comparative des institutions, 44, Paris, pp. 335–53.

Kupper, J.-L. (1981), *Liège et l'Église impériale XIe-XIIe siècles*, Brussels.

Lanotte, A. (1983), *La Lorraine: village/paysage. Ensembles ruraux de Wallonie*, Liège.

Leturcq, S. (2004), *La vie rurale en France au Moyen Âge: X^e-XVe siècle*, Paris.

Leturcq, S. (2007), *Un village, la terre et ses hommes: Toury en Beauce, XIIe-XVIIe siècle*, Paris.

Mignot, P. and Schroeder, N. (2016), 'Agrarian practices and landscape in the estate of Wellin (Belgium) from the early Middle Ages to the modern period: archaeology and history' in: J. Klápšte ed., *Agrarian Technology in the Medieval Landscape, Ruralia 10*, Turnhout, 2016, pp. 267–78.

Morimoto, Y. (1994), 'L'assolement triennal au haut Moyen Âge. Une analyse des données des polyptyques carolingiens' in: A. Verhulst and Y. Morimoto ed., *Économie rurale et économie urbaine au Moyen Âge*, Gent – Fukuoka, pp. 91–125.

Nitz, H.-J. (1988), 'Introduction from above: intentional spread of common-field systems by feudal authorities through colonization and reorganization', *Geografiska Annaler. Series B. Human Geography*, 70, 1, pp. 149–59.

Noël, R. (1972), *Les dépôts de pollens fossiles et l'histoire du Moyen Âge, Typologie des Sources du Moyen Âge Occidental,* 5, Turnhout.

Perrin, C.-E. (1925), 'De la condition des terres dites *ancingae*' in: *Mélanges d'histoire du Moyen Âge offerts à Ferdinand Lot*, Paris, pp. 619–40.

Petit, R. (1987), 'Les communautés rurales en Ardenne: l'exemple du comté de La Roche' in: *Les communautés rurales. 5e partie. Europe occidentale et Amérique. Synthèse générale*, Recueil de la Société Jean Bodin pour l'histoire comparative des institutions, 44, Paris, pp. 377–407.

Pirenne, H. (1896), *Le livre de l'abbé Guillaume de Ryckel (1249–1272). Polyptyque et comptes de l'abbaye de Saint-Trond au milieu du XIIIe siècle*, Brussels.

Pirotte, F. (1974), *La Terre de Durbuy aux XVIIe et XIIIe siècles*, Louvain.

Popkin, S. L. (1979), *The rational peasant: the political economy of rural society in Vietnam*, Berkeley – Los Angeles – London.

Preiss, S., Henrard, D. and Schroeder, N. (2016), 'Approche des pratiques agricoles durant le haut Moyen Âge en Hesbaye: étude de l'habitat rural de Lohincou/Villers-le-Bouillet (Province de Liège, Belgique)' in: M.-F. Dietsch-Sellami, Ch. Hallavant, L. Bouby and B. Pradat ed., *Plantes, produits végétaux et ravageurs. Actes des Xe Rencontres d'Archéobotanique, Les Eyzies-de-Tayac, 24–27 septembre 2014*, Aquitania, Supplément 36, Bordeaux, pp. 155–82.

Raucq, P. (1951), 'À propos de pratiques agraires anciennes et actuelles dans le Condroz oriental', *Bulletin de la société royale belge de géographie*, 74, 1–4, pp. 35–43.

Renes, H. (2010), 'Grainlands. The landscape of open fields in a European perspective', *Landscape History*, 31, 2, 2010, pp. 37–70.

Rorive, J.-P. (1978), 'Un cas précoce d'assolement triennal en Hesbaye hutoise (1178)', *Bulletin de l'Institut Archéologique Liégeois*, 110, pp. 1–6.

Rösener, W. (1985), *Bauern im Mittelalter*, München.

Schreg, R. (2009), 'Nach der Wüstungsphase: Umstrukturierungen des ländlichen Raumes in der frühen Neuzeit – eine umwelthistorische Perspektive' in: B. Scholkmann, S. Frommer, Chr. Vossler and M. Wolf ed., *Zwischen Tradition und Wandel. Ergebnisse und Fragen einer Archäologie des 15. und 16. Jahrhunderts*, Tübinger Forschungen zur historischen Archäologie, 3, Büchenbach, pp. 451–64.

Schreg, R. (2011), 'Die Krisen des späten Mittelalters: Perspektiven, Potentiale und Probleme archäologischer Krisenforschung' in: F. Daim, D. Gronenborn, R. Schreg ed., *Strategien zum Überleben. Umweltkrisen und ihre Bewältigung. Tagung des Römisch-Germanischen Zentralmuseums, 19./20. September 2008*, RGZM – Tagungen, 11, Mainz, pp. 197–213.

Schröder-Lembke, G. (1978), *Studien zur Agrargeschichte*, Stuttgart – New York.

Schroeder, N. (2014), 'Les cultures temporaires en Ardenne belge du Moyen Âge au XIXe siècle' in: R. Viader and C. Rendu ed., *Cultures temporaires et féodalité. Les rotations culturales et l'appropriation du sol dans l'Europe médiévale et moderne. Actes des 34es Journées internationales d'Histoire de Flaran, 12–13 octobre 2012*, Toulouse, pp. 101–15.

Schroeder, N. (2016), 'L'Ardenne: appropriation, exploitation et paysages du haut Moyen Âge à 1300' in: M. Pauly and H. Pettiau ed., *La forêt en Lotharingie médiévale. Der Wald im mittelalterlichen Lotharingien. Actes des 18es journées lotharingiennes. Luxembourg, 30–31 octobre 2014*, Publications de la section historique de l'Institut Grand-Ducal/ Publications du CLUDEM, 127/43, Luxembourg, pp. 163–91.

Schwab, I. (1983), *Das Prümer Urbar*, Rheinische Urbare, 5, Düsseldorf.

Sivéry, G. (1990), *Terroirs et communautés rurales dans l'Europe occidentale au Moyen Âge*, Lille.

Suttor, M. (2006), *Vie et dynamique d'un fleuve. La Meuse de Sedan à Maastricht (des origines à 1600)*, Brussels.

Thoen, E. (1993), 'Dries versus kouter. De wisselbouw in de Vlaamse landbouw van de Middeleeuwen tot de zestiende eeuw', *Heemkring Scheldeveld, Jaarboek*, 22, pp. 71–102.

Thoen, E. [with the collaboration of Lambrechts, Th.] (2010), 'Le paysage flamand et les structures de la société médiévale: nouvelles hypothèses sur l'origine et la fonction des kouters', in: J.-M. Yante and A.-M. Bublot-Verleysen ed., *Autour du "village": établissements humains, finages et communautés rurales entre Seine et Rhin (IVe-XIIIe siècles)*, Louvain-la-Neuve, pp. 379–99.

van Bavel, B. J. P. and Schofield, P. R. ed., (2008), *The development of leasehold in north-western Europe, c. 1200–1600*, Turnhout.

Van Derveeghde, D. (1955), *Le domaine du Val Saint-Lambert de 1202 à 1387*, Liège.

Van Derveeghde, D. (1958), *Le polyptyque de mil deux cent quatre-vingt (1280) du Chapitre de la Cathédrale Saint-Lambert à Liège*, Brussels.

Vanmechelen, R. ed., (2009), *Archéologie entre Meuse et Hoyoux. Le monde rural en Condroz namurois, du Ier au XIXe siècle. I. Les sites*, De la Meuse à l'Ardenne, 41.

Vanmechelen, R. ed., (2013), *Archéologie entre Meuse et Hoyoux. Le monde rural en Condroz namurois, du Ier au XIXe siècle. II. Contexte, analyses*, De la Meuse à l'Ardenne, 45.

Verhulst, A. (1980), 'Le paysage rural en Flandre intérieure: son évolution entre le IXème et le XIIIème siècle', *Revue du Nord*, 62, pp. 11–30.

Verhulst, A. (1990), *Précis d'histoire rurale de la Belgique*, Brussels.

Verhulst, A. (2002), *The Carolingian economy*, Cambridge.

Wieger, A. (1976), *Das Siedlungs–und Agrargefüge des Condroz und der Famenne in seiner historischen Entwicklung und in der Gegenwart*, Aachen.

Williamson, T. (2003), *Shaping medieval landscapes: settlement, society, environment*, Macclesfield.

Williamson, T., Liddiard, R. and Partida, T. (2013), *Champion: the making and unmaking of the English Midland landscape*, Exeter.

9 The open-field system and the persistence of communal land systems: lessons from the Andes

Hanne Cottyn, Ghent University, Belgium

I. The open field concept outside Europe

This chapter presents an historical assessment of communal land regimes in the Andes in relation to European open field systems. Entering the twenty-first century, the Andean region is characterized by relatively strong communal land systems, with both Bolivia and Peru recognizing community-based rights to over 30 per cent of their land area (Rights and Resource Initiative, 2015: 7). These communal systems combine collective with individual land rights, yet embrace complex regional variations ranging from classic open-pool resources to more restricted and differentiated systems of organization. As this chapter will explain, the persistence and complexity of Andean communal land systems in their internal organization and external embeddedness is the outcome of a long history of reform and resistance. An examination of communal forms of land use and tenure in different world regions, and their historical formation, can provide useful comparative insights into the organization and development of particular customary land right systems in rural Europe.

Dominated by research on Germany and England, the literature on open-field systems is strongly focused on European rural realities (McCloskey, 1991). Comparative literature on other parts of the world, where similar land systems once reigned or are still in place, is scarce (Godoy, 1991: 395–96; Campbell and Godoy, 1986). However, systems of common-field agriculture have developed in other world regions, including the Central Andes (Orlove and Godoy, 1986). Godoy points to important parallels with Spanish forms of common fields and the significance of fallow land, low population pressures and state institutions (Godoy, 1991). In European as in Andean practice, scattered holdings within an unenclosed landscape, grazing during fallow intervals, collective gathering rights and communal forms of decision-making over resource management are all present (Thirsk, 1964; Cole, 2002: 116–17).

In spite of fundamental differences arising from pre-colonial roots and particular historical transformations, a good case can be made, not only for the development of an open-field system in the Andes, but also for its preservation over the centuries (Godoy, 1991). This does not mean that a particular concept developed in research on European rural society can be copied and pasted on to other regions. The existing body of literature needs to be nurtured by critical comparative explorations into the historic development of communal land systems elsewhere. This chapter does not tell a completely new story, but it does open up a transregional dialogue by bringing in the particular trajectory of

Peasants and their fields. The rationale of open-field agriculture, ed. by Christopher Dyer, Erik Thoen and Tom Williamson, Turnhout, 2017 (*CORN* Publication Series, 16), pp. 207-232

BREPOLS PUBLISHERS DOI: 10.1484/M.CORN-EB.5.114274

rural communities in the Andean central highlands. In that way, it aims to foster a fresh look at the evolution of a key feature in rural European history.

Rather than providing a region-wide overview, the chapter focuses on Carangas, a rural highland area in the Oruro department of Bolivia. Through this case study a broader question is addressed, of how rural communities on the *altiplano*, the vast plateau at the heart of the Andean highlands, experienced and responded to privatization pressures in the context of post-colonial state formation. The changing relation between these communities and the state – first the Spanish colonial administration, subsequently the Bolivian republic – forms a key factor in the persistence of collective use and property rights. In the late nineteenth century, measures were taken in all Andean countries to formally abolish the community as such. The effect was the incorporation of a large share of highland communities into extensive private estates, the *haciendas*, and the conversion of their members into semi-servile tenant farmers. However, the land often continued to be managed and cultivated according to existing communal arrangements, and several regions even managed to escape the absorption of their lands in this manner, and maintained their communal organization. The case study presented in this chapter corresponds to the latter situation. The province of Carangas does not represent the trajectory of communal land systems in the entire Andean region, but strikingly reveals how the persistence of such arrangements cannot be explained simply by their geographical isolation. It was the outcome of consciously developed political negotiations, not a manifestation of indigenous isolation and passive victimhood.

This chapter first assesses what rural livelihoods in this particular region look like and how land use and tenure is organized. It then asks how and why these patterns of organization changed in the context of the restructuring of supra-local frameworks for resource management and sovereignty over the last two centuries. This trajectory will be assessed in relation to the abolition, defence, recuperation and fortification of communal land systems in the specific case of Carangas, which constituted a single provincial unit from the early colonial period until 1951. The analysis focuses on the latter phase in Carangas' 'land rights trajectory', from its exemption from late nineteenth-century disentailment measures to the post-revolutionary consolidation of communal land titles and their current accommodation within a pluralist legal framework. The reconstructed trajectory exposes the tensions and deviations through which communal land systems are incorporated into modern standardized land regimes.

II. The social agro-pastoral system of the Andean highlands: maximizing ecological complementarity

Covering the entire western part of the South-American continent, the Andes are geographically constituted by parallel north to south oriented volcanic mountain chains or cordilleras. These, relatively young in geological terms, serve to separate the Pacific coast from the Amazon and are hard to traverse. Yet it is exactly the interconnectedness between diverse ecological settings, from west to east, that constitutes the formative factor in Andean livelihoods. Altitudinal zonation splits the region into different belts or ecoregions comprising a very dense patchwork of ecological niches marked by a specific set of climatological, geological, vegetative and wildlife features. The ecoregion upon which this research focuses is known as the *altiplano*, the world's second highest plateau. Situated in the dry Central Andes, this vast plain covers contemporary Southern Peru, the western part of Bolivia, and the northern parts of Argentina and Chile.

On the *altiplano*, European immigration has remained very limited: to European eyes, the inhospitable terrain always appeared uninhabitable. As a result, the Andean highlands display a segregated society with an almost exclusively indigenous countryside, organized in communities, while whites and mestizos constitute a powerful minority confined to the cities and mining centres (Grieshaber, 1980: 260–61). Rural life in this part of the world is organized within the frameworks of what can be described as a specific social agrarian system.[1] A major factor in the constitution of that system in the Andes was the fact that there was an abundance of land, but few people, resulting in the strongly communal organization of labour. The agro-pastoral production modes that defined that system on the *altiplano* were not significantly altered by the introduction of new crops and livestock from the Old World. The potato and camelid species still count as the productive 'pillars' of the highland agrarian system. Because of the co-existence and complementary character of the extensive pastoralism and raised-field agriculture that has marked the Central Andes since pre-Inca times (Graffam, 1992), I define it as an agro-pastoral system.

In common with other mountain regions, such as the Himalaya or the Alps, characterized by a perennial '*struggle against geography*' (Mann, 2011: 203), south-central Andean post-Neolithic production modes have developed into an archetypal mountain system that has survived to this day. This system is marked by the ingenious adaptation of human settlement and production patterns to the challenging Andean environment. Examples of how human intervention has converted factors of apparent inhospitality into strengths are the development of terrace farming, the chewing of coca leaves to relieve altitude sickness and fatigue, and dry-freezing food storage techniques (Mann, 2011: 202). Ecological adaptation also includes the development of specific land tenure and resource management structures that testify to a durable human-nature co-evolution, reflected in corresponding systems of ethical conduct and religious-spiritual views (Cunill Grau, 1999: 18–21 and 89–150).

Within the Andean agro-pastoral system, the household constitutes the basic institution of social and economic organization (Custred and Orlove, 1980: 23; Bolton and Mayer, 1977; Mayer, 2002). However, despite the apparent autonomy of Andean households, their subsistence and reproduction can only be ensured by being embedded within larger corporate groups, the community (Custred and Orlove, 1980: 34; Mayer, 2002: 35–42). The 'community' refers to a multi-layered framework for group formation, task distribution and territorial control. The organizational axis behind this framework is land, and communal leaders at each level have the vital task of distributing usufruct rights to, and collecting usufruct contributions from, all members.

---

[1] Agrarian systems have been designed as concrete research tools by rural (history) scholars, associated with the Annales school and in particular with Slicher van Bath. A concrete example is the 'French' model of Agrarian Systems Analysis (Mazoyer, Roudart and Membrez, 2006). Although coming from quite a different regional setting, the social agrosystem framework as developed by the consortium for Comparative Rural History of the North Sea Area (Thoen, 2004) is instructive to the analysis of the Andean rural world for its strong emphasis on the role of social power.

The key concept in understanding the way in which land is used and managed is ecological complementarity, or how the proximity of starkly contrasting zones is exploited to the greatest benefit. Still counting as one of the most elemental ethnohistorical breakthroughs in Andean studies, John Murra has described this as the 'vertical control of a maximum of ecological niches' (Murra, 1975: 59–115). In his ground-breaking work, Murra defines how this resource management and governance strategy developed under early Andean processes of state-formation. Further elaboration of this vertical archipelago model has provided 'a dramatic demonstration of the continuing distinctiveness and viability of the Andean adaptive repertory,' not only under Inca rule but also at a local level.

> Many contemporary peasant communities solved the subsistence problem through a variety of kin-based strategies, as local peasantries tried to diversify agricultural and pastoral production in different ecological niches [...] and to coordinate communal work efforts (Larson, Harris and Tandeter, 1995: 13).

The survival of Andean rural households and communities still depends on the exploitation of this complementarity, both vertically and horizontally (Fonseca, 1973; Browmann, 1991; Medinaceli, 2010). Vertical control refers to direct community-based exploitation of multiple niches which allows households to combine access to dispersed plots located in various altitudinal zones. In the highest zones, above 4,000 m, peat lands known as *bofedales* generate excellent grazing conditions for llamas and alpacas. Descending the mountain slope, between 3,500 and 4,000 m, a wealth of potato varieties, root crops and grains such as quinoa are cultivated, despite the restrictions of elevation, aridity and frost, thanks to the opportunities offered by microclimates. Further down, the *altiplano* makes way for the maize production zone in the valleys lying between the mountains (Guillet, 1981: 141). The land is worked according to a communal tenure system which organizes the rotational use of open-field grazing lands, fallowing and crop rotation (Guillet, 1981: 143–44). This principle of complementarity allows for a heterogeneity of production options much more efficient than would arise from an all-encompassing control over the land. Horizontally, mobility and exchange patterns also provide an alternative strategy to the (in pre-colonial times absent) market system (West, 1981: 5–6), giving indirect access to complementary resources which circulated through long-distance inter-ecological barter networks on an altiplano-valley-coast axis. Over the centuries, both models have continued to complement each other, while increasingly meshing with the expanding sphere of capitalist market exchange.

III. Community-based land management in the Andes: Carangas

This section provides a brief examination of the ways in which land use and land management is organized at the community level in the Bolivian highlands. As the overview above shows, the ecological and hence social and political conditions in this environment are extremely different to those in which open fields flourished in medieval and modern Europe. Nevertheless, the still-surviving land systems in Andean communities share several characteristics with the open-field system (Thirsk, 1964), such as the dispersed character of cultivated parcels, the practice of common grazing on cultivated lands while these lie fallow, and the management of land through assemblies in which all community members participate.

An introduction to Carangas

The Carangas region, as already noted, is located on the central *altiplano*, characterized by a semi-arid cold climate, and by wet and dry grassland and scrubland vegetation. Entirely situated at over 3,600 m above sea level and roughly the size of Belgium, this immense open plain is currently inhabited by a little over 50,000 people (see map in figure 9.1). Here, pastoralism constitutes the prevalent economic activity, something which has particular implications for the local organization of community life, production and land administration. The region's landscape, culture and history are dominated by a desert-like soil with limited production options, a llama and alpaca population that largely outnumbers the inhabitants of the scattered rural communities, and relative geographical and administrative isolation (Figure 9.2).

Located within the Oruro department, the Carangas region borders to the west the regions of Arica and Tarapacá, which were originally Peruvian until they were occupied by Chile in 1879 during the War of the Pacific. In a departure from pre-colonial ethnic federation structures, the region was integrated as a single province under Spanish administration and maintained its unity until 1951, when it gradually splintered in the face of an administrative fragmentation process (Arteaga, 2006: 522). Today, over 90 per cent of the population identify themselves as members of the *Aymara*, Bolivia's second largest ethnic group (Instituto Nacional de Estadísticas, 2001 cited in SEDES, 2007: 16). In fact, Carangas has always had a multi-ethnic profile, but a strong historical process of discrimination has obscured the presence of the region's original inhabitants, the *Urus*. White and mestizo population have always constituted a minority group in numerical terms, yet have long formed a monopolistic class tied to the mining sector and the official juridical-administrative institutions of the region.

Originally, Carangas included a larger discontinuous territory that linked the *altiplano* heartland to distant valley lands to both east and the west. However, as communities were contained by delimited coherent political-territorial units under the Spanish administration, the population's inter-ecological mobility and its direct control over complementary territories was seriously restricted, forcing people into the market sphere to obtain the food they needed. Through these historical cycles of land tenure and territorial reorganization the *altiplano* communities were cut off from their ecologically complementary land property (Rivière, 1982: 19–24; Pauwels, 1983: 152; Hidalgo and Durston, 2004: 471–534; Ruz, Díaz and Fuentes, 2011: 24–25). The task of regulating a more enclosed communal space and securing self-sufficiency within that space hence became more demanding, with important implications for the shaping of field systems (Godoy, 1991: 404–05). It was, however, a slow divorce. As late as the early twentieth century we still find highland communities claiming effective land rights over private estates in the eastern valleys. An example is the community of Totora which possessed, in common with the Quillacas community of a neighbouring province, the 'La Chulla' estate in the valleys of Cochabamba. In the 1920s, the community started a campaign to recover its property which had been abandoned for several years, but the initiative had no success, while the 'Pisaqueri' estate, also owned by Totora, was sold in 1906.[2]

---

[2] ADRO, *Propiedades Provincias*, *Carangas*, 1922, no. 47, fs. 376–377^{v}; and AJC, *Civiles*, 1906, no. 2, f. 8–9^{v}.

Figure 9.1 Location of the Carangas region

Despite this severance of the communities of the Andean plateau from their valley lands, a remarkable continuity can be observed in altiplano-valley connections in the form of long-distance exchange contacts. Frequent inter-ecological travel and exchange continued into the second half of the twentieth century, mitigating the effects of territorial amputation. While control over valley lands was lost, territorial control over the *altiplano* heartland largely remained in the hands of the predominantly indigenous population as indivisible property, except for the mining concessions and the private plots in the rural towns. The sterile soil and the centrality of pastoralism have compounded the apathy and absence of a (white and mestizo) landlord class, and added to the strength of a communal tenure system.

Figure 9.2 Llamas grazing near the Sajama mountain, the highest of Bolivia (author's picture, Sajama, October 2008)

Figure 9.3 Meeting of indigenous authorities during the annual consultation visits of the Native Nation of Jach'a Karangas (author's picture, San Miguel, March 2012)

Figure 9.4 Overlapping official and indigenous territorial ordering in Carangas

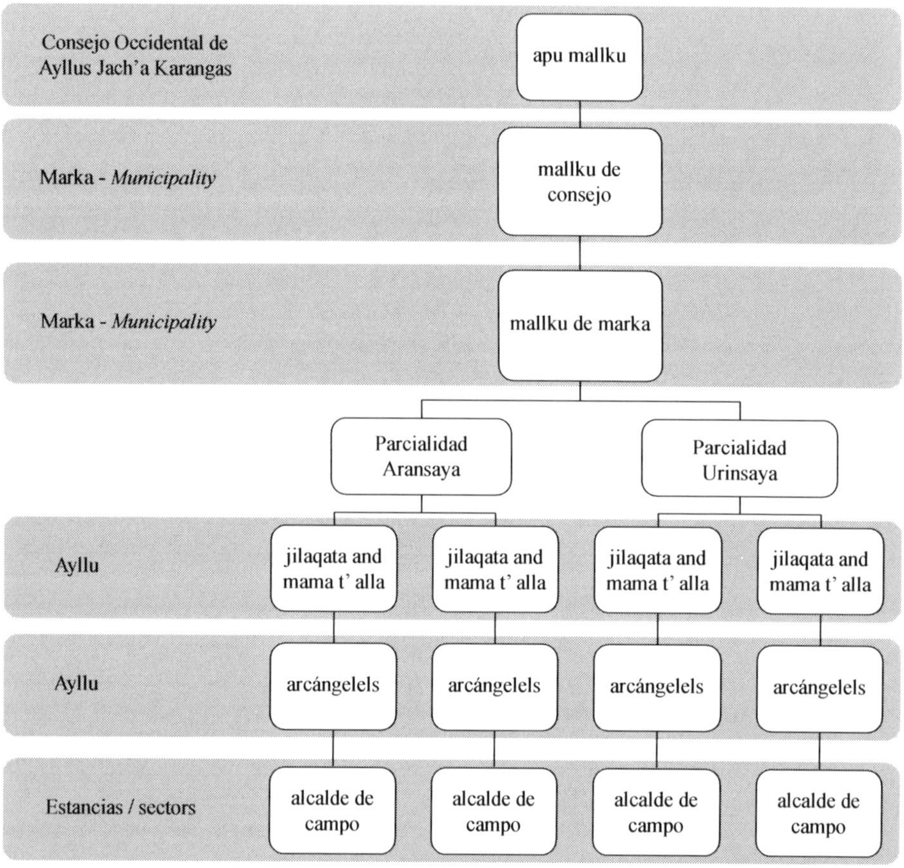

The foundation of this communal system is the nuclear households which cluster around one or more patronymics in a shared site of residence, referred to by the Spanish term of *estancias*. Varying according to ecological conditions, one *estancia* may comprise anything from tens to hundreds of individuals (Rivière, 1982: 238). *Estancias*, in their turn, are grouped together in *ayllus*, which embody a more institutionalized degree of group formation and correspond to what most people understand as the Andean *comunidad*, or community, although this is a somewhat superficial translation of a complex quechua-aymara term. Ceremonial-political representation, land administration, customary jurisdiction, tax collection and productive organization in the *ayllu* are community responsibilities, which are assigned through a rotating leadership system (Mayer, 2002: 40). At a higher level, *ayllus* group together in a 'larger community' or *marka*, which roughly coincides with the municipal units of the modern state administration. Headed by internal and external indigenous authorities, and sub-divided into a lower and an upper 'faction', the *marka* functions as the mediating sphere for indigenous ceremonial and administrative life. On a regional level, the *markas* of the former province have been gradually reintegrating into native federations or *suyus*. Since the 1980s, Aymara territorial and political control is being reclaimed by the indigenous council of the 'Native Nation of Jach'a Karangas', while the Uru population now formally constitutes the 'Native Peasant Nation Uru Chipaya' (Figure 9.3).

Eroded, transformed and complexly overlapping with state-administrative subdivisions, the twelve Aymara *markas* and the Uru community of Chipaya today display a remarkable level of continuity with respect to colonially installed rural administrative structures: not as an empty residual skeleton but as a workable (though delicate) framework for authority, resource-pooling, and cultural life (Figure 9.4).

Land use in Carangas

Challenged by the ecological character of the *altiplano*, households employ a dispersed settlement and land use strategy. Agriculture is based on fallowing and crop rotation, while pastoralism relies on the rotational use of vast grazing spaces. Both require well-organized systems of communal control (Guillet, 1981: 143–44). In Carangas the importance of cultivation is restricted, due to altitude and the dry climate. Most of the soil is so sterile that it hardly produces enough to sustain the food consumed by each household, and large areas of land are devoted exclusively to livestock herding (Lopez, 2003: 12). Mountain torrents, *bofedales*, lakes, rivers, and the climate shape excellent grazing conditions for llamas and alpacas (and sheep), which continue to constitute the most important productive asset of the local people. Thus, the most constant and reliable option in terms of risks and flexibility is offered by pastoralism, in spite of the fact that compared to agriculture, it might appear as an illogical strategic decision, if viewed simply in terms of energy input and profits (Mayer, 2002: 17). Until the arrival of motor transport in the area in the late 1970s, livestock also played a key role in obtaining cereals, fruits and vegetables, which had to be transported from the valleys.

In terms of land use, agriculture is mainly an additional though vital household activity. The main products farmed in Carangas are potatoes and barley, together with small quantities of quinoa,[3] beans, cereals, alfalfa, and some vegetables. Together, these products constitute the main items of consumption, dominated by the potato.[4] Crops are grown in small individually-owned plots called *canchones*, which used to be fenced by stone walls in order to prevent livestock from entering, and are organized into different terraces (Figure 9.5). Ploughing is done by oxen, although over recent decades there has been an increase in the use of tractors. In contrast to the maize zone where irrigation is crucial, the *altiplano*'s climate constraints enforce a system in which more land is left fallow than cultivated; and in which crop rotations mainly involve potatoes, native tubers and cereals. These are features both of individual plots and of communally managed

[3] The current boom in quinoa on the international market has increased the importance of agriculture in the region, although at the cost of collective grazing lands, soil depletion, and land conflicts.

[4] According to West's reconstruction of rural consumption in the Oruro countryside in 1977 (West, 1981: 117), an average nuclear family of six consumers needed each year approximately 455 kilogrammes of fresh potatoes, which were the year round stored in underground silos, and 137 to 182 kilogrammes of *chuno* (freeze dried potatoes) between two harvests. An average family therefore consumed about 12 kilos in varying proportions of fresh and dried potatoes per week.

Figure 9.5 Terraced cultivation plots in the outskirts of the village of San Miguel. The most fertile soils are located in the Northeast of Carangas (author's picture, March 2012)

(unfenced) terrains (Guillet, 1981: 143–44). Orlove and Godoy include the example of the Escara community in the Carangas region in their assessment of sectoral fallowing systems in the Andes (Orlove and Godoy, 1986).

As production is based on the management of multiple ecological niches, mountain slopes are converted into a dense collage of cultivation plots. *Thola* grows and quinoa is cultivated to a height of 3,800 m (Rivière, 1982: 234–35), and potatoes to around 4,100 m. The lands higher up are used for grazing and the gathering of fuel (*queñua* wood and *yareta* herbs). Due to the character of the soil, a significant proportion of the terrain remains uncultivated for periods which can extend for up to two decades. Since the introduction of practices and institutions based on Spanish models, highland agriculture and herding spaces have been drawn together within an increasingly enclosed 'amputated' system, facilitating the formation of something like an open-field system (Godoy, 1991: 404).

To understand properly communal land administration in Carangas we must look first at the modalities of livestock management. Andean camelids require much space to thrive, hence pasture land depends upon an abundant 'unfenced' resource. Long-standing (and still continuing) practices of 'free pasturing', without the use of animal housing, allow a pastoralist household to manage an ample workspace. Within this workspace familial flocks graze continuously on different pastures, but always within the same fixed area and according to a defined logic. In a sequence which runs from most to least selective forage consumers, cows are the first to enter a pasture, followed by llamas, and then sheep, to 'finish' it (Lopez, 2003: 37). Cows are herded separately by men, often left unwatched

Figure 9.6 An estancia with corrals keeping a llama herd together, along the road between Chuquichambi and Choquecota (author's picture, March 2012)

in the highest areas for several days (*ibid*, 36). Male llamas may also be left alone for several days on outlying pastures in the highest zones (over 4,200 m). Female llamas, however, are kept at lower altitudes (approximately 3,700 to 4,200 m), often accompanied by sheep. The latter are herded closer to the pastoralist dwelling by women, and also by children. The animals are spurred with whistles and a sling, which is also used to protect them against predators such as foxes, pumas and condors, and also other herds or stray llamas (López, 2003: 16–18). At night they are locked in corrals constructed around the *estancia* (Figure 9.6).

When lorries began to be used in the area in the 1970s and 80s, transporters and traders shifted to new means of capital formation, reducing the importance of livestock and hence of the grazing lands. However, the commercialization of llama meat from the 1980s countered this tendency and today the communities of the former Carangas province are widely renowned as centres of camelid breeding, with some communities possessing herds of over 100,000 llamas (Ministerio de Desarrollo Rural y Tierras, 2014: 132–33).

Land tenure in Carangas

The land-use system just described, well-adapted to the productive conditions of the *altiplano*, is managed through strong communal control over land tenure. In pasture zones such as Carangas, communal land administration is exerted in its most 'complete' form, based on indivisible use rights. Highland agriculture zones, dominated by tuber cultivation, are also communally administered but with divisible use rights, while land tenure in the lower zones, where maize and tubers are grown using irrigation systems, is characterized by private property (Guillet, 1981: 142–43; Urioste, Barragán and Colque, 2007: 79). The rationale behind this way of managing fields at the community level is largely to be ascribed to motives of risk prevention.

Quite different from theoretical and exclusive models in which collectivism overrules individualist principle, such as communism, the communal land administration in the Andes comprises several types of tenure that flexibly co-exist, and which may alter according to context and changing state and market imperatives. In the Carangas region, most of the land was, and still is, held as indivisible property, yet tenure entails varying degrees of collective ownership. Rather than an all-encompassing control over the land, each household combines access to dispersed plots of pasture, cultivation, and fallow land, preferably of as varied a quality as possible.

The *estancia*, the core site of residence and operation of a family, usually comprises the plots containing the best of the cultivated land. In addition, each household has access to small, scattered plots lying a short distance away, and to a number of outlying fields, mainly used for grazing, which are provided with temporary dwellings (Llanque, 1995 cited in Rodríguez and Quispe, 2009: 35). This heterogeneity of options guarantees a diversified system of production in a fragile environment, and epitomises the principle of complementarity, which until the severance of Carangas' valley lands was upheld on a much larger scale.

The *ayllu* is the basic level of organization for the distribution of land among community families. All land pertaining to an *ayllu* is divided into four interlinked categories, allowing households to control different ecological levels (see Figure 9.7). Closest to but not identical with private property are the individual inalienable usufruct rights assigned to nuclear families. These rights endow each household with a *sayaña*, which contains several cultivation fields, differently located within the territory of the *estancia*. Rivière adds that the Aymara verb *sayaña* means 'get up, stand up', hence referring to the place where a person 'raises' or 'builds' his space. According to Bertonio's Aymara dictionary, the verb entails the meaning of 'delimiting, establishing the limits of a property' (Bertonio I. N. E. L., 1979: 95, cited in Rivière, 1982: 258).

The counterpart of these privately used lands is the *saraqa*, which are communal terrains over which one *estancia* (the extended family) holds collective usufruct rights. These terrains comprise a geographically discontinuous territory, including fallow, water, pasture and woodlands dispersed over the *ayllu* to which the *estancia* belongs, sometimes lying at several days walking distance (Rivière, 1982: 237). Lying still further from the permanent settlements, and much larger in extent, are the *qallpa*, for cultivation, and the *aynuqa*, used as pasture or lying fallow. These are communal lands, administered at either the *estancia* or *ayllu* level by distributing temporal, collective non-adjudicated usufruct rights to member households (Galindo, 2007: 85–86) and functioning as a reserve for future internal redistributions.

Figure 9.7 Altitudinal varied land use in the Sabaya community

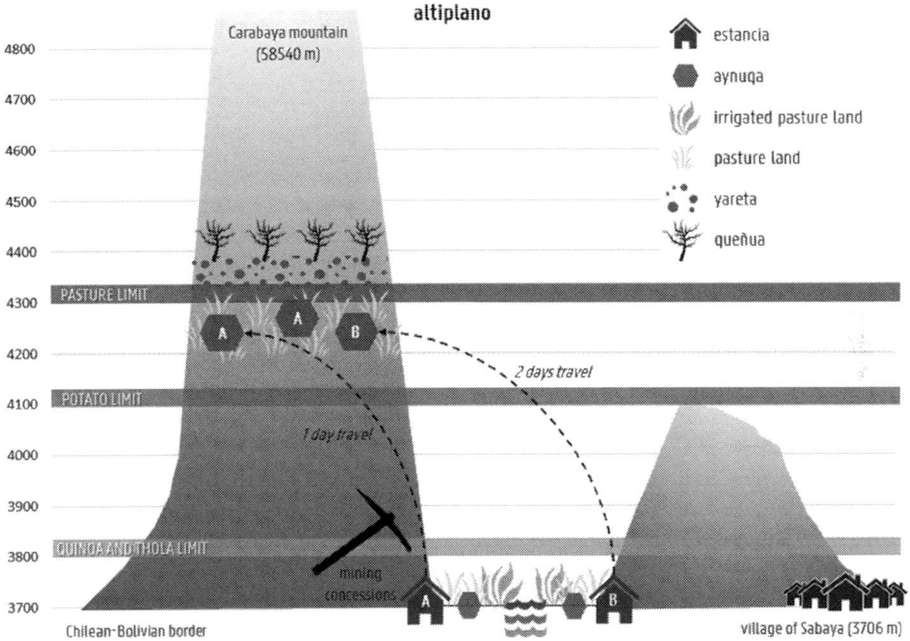

Source: Adaptation of Rivière, 1982: 234

> As a provincial inspection report of 1910 demonstrates, this system has not changed significantly over time:
> The property of the land is maintained communally by aillos [*ayllus*], distributed in ainocas [*aynuqa*], and no one of the co-proprietors can extend its cultivations at will. The aillos are divided into 10 or more ainocas, destined one for each year, in a way that the community must sow in one single ainoca, leaving the others for the pasturing of their livestock. This [is a] system of rotation, that is applied in the cultivation lands (Bacarreza, 1910 in Pauwels, 2006: 394).

Sayañas are found everywhere, on the slopes, in the high *bofedales* and spread over the plain. Communal cultivation lands are usually located on the highland plains, at an altitude of around 3,700 m, and are either non-irrigated or are associated with *bofedales* for young and female llamas, and provided with permanent dwellings and corrals. A little higher, between 4200 and 4300 m, *aynuka* grazing lands are used for the (male) herd.

Access to land is conditioned by a family's affiliation to a particular *ayllu*, but is not equally distributed among all community members. In the community of Turco, for example, an equal division of land would provide each family with 314.1 ha., but in reality holdings range in size from barely 40 ha. to as much as 1,800 ha. (HAM Turco, 2007: 47). While population (as well as the livestock) increases, the only way to multiply the land base is by subdividing the *sayañas*. This explains why the search for ways to reinforce the security of individual or familial land rights occurs essentially within the framework of the community, and why this search is intrinsically entangled with internal community struggles.

The social control over the size, use and distribution of the communally held lands, together with the allotment of individually cultivated lands to all members of the *ayllu*, is entrusted to the *jilaqatas* (Urioste, Barragán and Colque, 2007: 79). In order to be granted *sayaña* land rights and obtain the associated status of *sayañero*, but also to obtain usufruct tights in collective lands, households are required to make a yearly contribution to the community and take up their responsibilities in the fiesta-cargo cycle (although indigenous authorities were exempt from this contribution while executing their function).

Community authorities are entrusted with the distribution of land among all members, with the collection of their contributions, with negotiating community control over the *ayllu* lands with external actors, and with the mediation of conflicts over land. The existence of individual and collective rights within the *ayllu*, combined with fluid land boundaries, leads to frequent disputes between households, *estancias* and *ayllus*. Cases concerning disputed property rights, trespass, and violent occupation of land abound. Land boundaries are usually indicated by stones (*mojones*), representing imaginary, fluid lines, and can hence be easily moved (see also Goodale and Sky, 2000). The collective use of grazing lands adjacent to individually cultivated plots also leads to frequent accusations of land invasion, or attempts to establish *de facto* ownership over communal land (West, 1981: 88–89). Land disputes and usurpations are a constant feature of life in Carangas. The Uru population – the first inhabitants of the region – has fallen most strongly victim to its disturbing effects. The arrival of the Aymara in the region is often highlighted as the moment when strict land regulations were imposed upon the Uru population. The latter was gradually pushed into a marginal position and remained at the shores of the lakes. While the water, the lake, is indeed essential to Uru livelihood and culture, this should not detract from the fact that land is also vital to their existence (see Wachtel, 2001; Pauwels, 1999; de la Barra et al, 2012). To insist on their status as 'people of the water' is, in fact a false excuse, a device employed to negate their basic rights, and still seems to be employed in arguments supporting their current precarious access to land.

Over time, Spanish influence introduced patrilineal inheritance principles in the allocation of *sayaña* lands, which are now invariably passed from father to son (Rivera and Platt, 1978). Moreover, inter-community and inter-household competition over land has become increasingly accompanied by competition with 'outside', non-community actors. This has resulted in the creation of several privatized enclaves within the Carangas territory. These comprise the small plots in the 'urban areas' of the rural towns, and the trade settlements and the mining districts in the province. For centuries, the mining districts (located on subsoil that – being state property – had been granted by formal concession), and the urban plots, remained the only marketable privatized terrain in the Carangas region. But for community lands, internal rules for land tenure are still in place, and no individual or institutional proprietor, only the community as a whole, exerts right of ownership over them.

IV. Land rights, community control and the state

To understand the survival of communal land systems in the Andes, the relations of local populations with the state must be taken into account. Particular communal arrangements over land use and tenure were developed in response to challenging ecological factors, yet over time they have also been shaped by, and became embedded in, a state-managed framework of land control. The community is essentially

> an institution of land tenure which regulates access of individuals to land. It is also a creation of the state, recreated over centuries by the Inca, the Spanish, and the modernizing governments (Lehmann, 1983: 22).

As states expand and consolidate their control over territories, peoples and resources, the alternative – of free access to certain resource pools held by certain population groups – is usually reduced in an attempt 'less to make them [these groups] productive than to ensure that their economic activity was legible, taxable, assessable, and confiscatable or, failing that, to replace it with forms of production that were' (Scott, 2009: 5). Indigenous communities were thus integrated as 'legible' units within a centralized land system.

In the Andes, the indigenous population was classified by the Spanish administration as *indios*, something which was accompanied by the gradual imposition of a juridical-fiscal order that established the terms of participation of differently classified sectors in society and political life, and of their access to resources. The internal bond between *sayañeros* and their communal authorities, regulating access to and use of land, was incorporated and monetized by the colonial state apparatus, paving the way for the formation of a centralized land system. This process started after the first chaotic episode of conquest, under the early colonial administration of viceroy Toledo in the 1570s. Toledo's reform constitutes a fundamental but unsettling step in the creation of the indigenous communities that we know today (Mumford, 2012). Indigenous hamlets were concentrated into Spanish-style villages within a demarcated territory for which formal property titles, known as *títulos de composición,* were to be purchased from the Spanish Crown at a cost of 1000 *pesos fuertes*. In the Carangas region colonial property titles were obtained around 1645 when each community paid that sum to the royal inspector José de la Vega Alvarado during his visitation (Pauwels, 1983: 150). This forced the indigenous population into a contractual relationship with the state, in which the purchased land titles served as the 'contract'. This contract was to be regularly renewed by paying a half-yearly tax, known as the *contribución* or *tributo indígena*, the rate of which was updated through regular record-keeping operations, the *Revisitas*. Over time, the yearly contribution to be paid by all community members shifted from non-remunerated services and agricultural yields, to money.

Spanish interventions seem to have contributed to the development of common-field agriculture in terms of labour allocation, while communities appear to have come to rely more heavily on this system to comply with tribute obligations and the effects of depopulation (Godoy, 1991: 408). In the same manner that the transformation of land-use systems seems to have supported both Spanish extraction and communal 'survival', it also drew Spanish Crown and indigenous communities into a mutually enforcing relationship. In return for the communities' purchase money and fiscal contributions, the state granted and protected communal land rights, including the right to communal organization without state interference. This established an asymmetrical but vital relationship between the *ayllus* and Spanish Crown. The mutual reliance on this kind of contractual relationship has been described by Tristan Platt as a reciprocity pact (Platt, 1982: 100),[5] in which the Spanish Crown guaranteed the protection of communal tenure by granting collective land titles in return for a two-yearly payment of a head tax and non-remunerated services by each commoner.

Community land became both an important source of revenue and loyalty for the state, and a lever of negotiation for the communities. While for the former securing a major source of fiscal revenues, for the latter it granted autonomous control over the customary organization of communal tenure. Entailing both individual and hereditary usufruct rights, communal tenure was preserved as un-marketable and non-alienable. The *composición* titles, taxation and land inspections were the central elements of the pact, and crucially articulated the relations between indigenous communities and the government. After independence, this arrangement remained in place and would continue to regulate the relations between the indigenous population and the 'pact-making' state (Barragán, 2002).

At the same time, this tributary pact was 'incorporated' at the heart of community organization. Access to community land and to the community hierarchy was determined by an individual's fiscal status as a *contribuyente* (taxpayer) and by his or her ability to pay a head tax at the appropriate time. In Carangas there were three different communitarian categories; the *originarios* or 'original' inhabitants with strong access rights; the *forasteros*, also known as the *agregados*, or newly incorporated members with weaker access rights; and the Urus with the weakest rights. The haciendas categories of *forasteros* with a tenant status, and the *yanaconas* who were semi-slaves, remained absent (Grieshaber, 1980: 267; Rivière, 1982: 73; Platt, 1982: 52–53; Irurozqui, 1994: 14).

The virtual pact resettled the conditions for community membership, which now became a fiscal category defined by the state. However, in several places in the Andean highlands, these tributary ties were internalized, extending the principles of reciprocity and redistribution, defined by Murra as the Andean ideals that supported both Inca imperial coherence and community-state relations (Platt, 1982). Moreover, this reciprocal bond also regulated usufruct and authority relations among hacienda Indians (Grieshaber, 1980). The implication for the community leaders who became responsible for land distribution and tax collection was their empowerment as intermediaries between local communities and the state. As the mutual pact with the state was internalized by indigenous communities, the maintenance of tributary ties became, in the eyes of the Indians, a guarantee for the viability of native land tenure systems (Langer, 1989: 4).

V. How and why did communal arrangements alter?

While common lands disappeared in Europe, community land thus continued to be a decisive component in the constitution of Andean land systems. It is possible to view

[5] See also Platt, 1984 and 1987; and Albornoz, cited by Langer, 2009: 537. Rivera and THOA rather refer to a "truce" instead of a pact (Rivera, 1991: 625).

the assertion of indigenous territorial demands and the preservation of a communal land system as manifestations of stagnation and isolation in the midst of a globalizing world. Although superficially Carangas may appear to correspond to this picture, it was in fact embedded in one of the first world regions to be incorporated into the capitalist world-economy as this emerged about 500 years ago (Moore, 2010). Through the mobilizing force of llama caravans, transporting minerals and supplies between the mines of Potosí and the ports of the Pacific, this pastoralist region became tightly linked to the colonial circuits of extraction and commerce (Medinaceli, 2006: 514).

From late colonial times onwards, a protracted process of marginalization weakened Carangas' position, yet without completely neutralizing its interconnection with state and market structures. Meanwhile, and despite this longstanding link, most land in Carangas has remained outside the privatization pressures which are driving the expansion of the world-economy and related state-building processes. These pressures have intensified over the last two centuries. The following analysis focuses on the implications of nineteenth-century anti-community measures for the Carangas communities and their 'land rights trajectory' during the land reforms which followed in the twentieth and twenty-first centuries.

Apart from the role of (limited) market factors, Guillet points to the intrinsic connection between sectoral fallowing and communal regulation, the importance attributed to permanent access rights over concepts of private property and population and ecological factors in averting Hardin's 'tragedy of the commons' (Guillet, 1981). To these external and endemic factors must be added the importance of mutual community-state relations, grafted onto the issues of labour and tax extraction. As Godoy puts it, 'it is not a mere coincidence that common-field agriculture attained its most complex form and survived the longest in the same area in the Andes that faced the heaviest labor-tribute liabilities during the colonial era' (Godoy, 1991: 409).

The foundations of the colonially inherited reciprocity pact started to crumble in the late eighteenth century, as Enlightenment and liberal ideas emphasized the importance of private property. The community came to be viewed as an obstacle to the indigenous population's integration as citizens, independent economic actors and private landholders within a 'modern' society (Yambert, 1980: 65). In Bolivia, however, plans to dismantle corporative landownership were only applied to indigenous communities about a century later. Widespread resistance to Simon Bolívar's measures to abolish the *tributo* and the treasury's dependence upon *tributo* revenues impeded the new regime's attempts to break radically with the colonial agrarian policy in the first decades of independence (Antezana, 2006: 90–93). The declaration of community lands as state property in 1843; and the radical, although short-lived, decree of 1866 that exposed communal land titles to sale transactions; heralded the breakthrough of this major enclosure operation (Ovando Sanz, 1985: 60 and 123). It was only when a revival in mining provided the National Treasury with an income source that dethroned the *tributo* as the prime contributor to the country's finances, and when the Pacific War (1879–1883) subsequently led to a need for extra income sources, that the law would finally be implemented.

The new land policy proclaimed by the 'Disentailment Act' (*Ley de Exvinculación*) of 1874 and the corresponding deployment of fiscal instruments to 'commodify' the communal ownership of land in Bolivia would seriously undermine the security of communal land tenure. In the process, as in other Andean countries, the government developed a particular interest in what came to be categorized as *tierras sobrantes*. This referred to the collectively used lands; lands that lay fallow, which were kept as reserve by the community, or were located in distant archipelagos; as well as the lands usurped by non-indigenous proprietors. The 1874 law classified these communal 'reserve' lands as 'residuary estate' or 'vacant' and declared them to be the property of the state. The official argument was that these lands exceeded the minimal amount required for the subsistence of local inhabitants (Langer and Jackson, 1990: 24). In the eyes of the state, they were excess land, wastelands that were to be brought into the productive cycle and additionally incorporated as a revenue source for the national treasury, as they were easily marketable through auctions (Larson, 2004: 216).

Aimed at converting collectively-owned lands into individually-held marketable plots, the law led to the large-scale expropriation of highland communities between 1880 and 1930, particularly in the La Paz area, where numerous communities were incorporated into private estate land and became tenant farmers within the expanding hacienda system (Klein, 1993: 157). Despite this massive exercise in land grabbing, indigenous resistance and networking – both within and beyond the limits of legal action – as well as the government's own financial constraints, led to the incomplete implementation of the new system, and especially to a measure of regional variation in the degree to which it was implemented (Barragán, 2012; Irurozqui, 1999). For the few communities of the Carangas province which still possessed lands – now rented as 'private' estates – in the valleys, the land reform brought about their final severance from these distant properties. Internally, the sort of open-field arrangements that regulated land use within communities also came under severe attack. The way in which property regulation in the Carangas province was described by a provincial inspection report of 1910 well reflects the attitude of contemporaries towards local open-field systems as undesirable obstacles to the expansion of the agricultural frontier:

> Ainocas [are] destined one for each year, in a way that the community must sow in one single ainoca, leaving the others for the pasturing of their livestock. This system of rotation, that is applied in the cultivation lands, restricts the area to which agriculture could extend (Bacarreza, 1910 in Pauwels, 2006: 394).

However, such views also demonstrate the short-sightedness of representatives of government both to the way in which the system was well adapted to ecological circumstances, and its internal complexity. Moreover, they ignored the strength of the communal organization backing this system. The ecological conditions of the region, combined with communal resistance, prevented the law from making any progress within the province. No lands were sold or divided.[6] In effect, the tributary pact, inherited from colonial times, continued to regulate community-state relations at the departmental level, as communities continued to pay the indigenous contribution and hence confirmed their status as community members whose collective land rights were to be protected by the state.

[6] Although some remarkable cases of sales and transfers of *sayañas* in Carangas were registered during this period, these did not necessarily go accompanied by effective monetary transactions to purchase plots. In several cases in which a *sayaña* is attributed a concrete cost, this only concerned an expression of the monetary value of the plot but the transfer would be free. ADRO, *Propiedades Provincias*, *Carangas*, 1930, no. 6.

The fact that enclosure had been kept at bay in Carangas can largely be attributed to the lack of incentives for the expansion of haciendas in the region, due to its unattractive soil and the strength of indigenous self-government. Governmental representatives repeatedly pushed for a more intensive use of land in the province, ignoring the fact that the system of rotational use of communal grazing lands was carefully designed to prevent over-grazing (West, 1981: 59; Lima, 1921: 41–50). While ecological and demographical factors are indeed key (Grieshaber, 1980), it must be stressed that, even in Carangas, the preservation of communal land control was never a 'given'. Its defence implied the deployment of a series of juridical, legislative and on occasions more violent resistance strategies to force the state to abandon its plans and retreat from the region. A more detailed assessment of these strategies goes beyond the scope of this text (see Cottyn, 2014), but it is important to note, in relation to open-field systems, that regardless of whether land continued to be controlled by communities or was absorbed into haciendas, it continued to be divided into individual and collective plots. It was only in those areas where a class of independent smallholders emerged, particularly in fertile valley lands, that peasant households cultivated and transferred lands free from communal supervision (Larson, 1998: 172).

Through reforms, constitutional changes and new land policies introduced during the twentieth century, communal arrangements which had been openly denied by nineteenth-century liberal legislation were restored. First, the constitution of 1938 recognized the legitimacy of the corporative character of the indigenous communities.[7] The most important breakthrough, however, was the land reform of 1953, carried out as part of the broader National Revolution of 1952, which abolished the hacienda and confirmed the communal ownership of lands by granting 'undivided titles' (*título proindiviso*). Yet, while the state demonstrated an effort to respect communitarian autonomy, this recognition was based on a dichotomist understanding of communal land systems, giving indigenous ex-hacienda peasants the option of either individual *or* undivided land titles. This imposed a dual private vs. communal system that effectively denied the traditional combination of individual and collective arrangements *within* communities, and eventually stimulated a process of individualization (Urioste, 2005: 17). In fact, for both community lands that had 'escaped' the pre-revolutionary hacienda expansion, and for those that were effectively parcelled out, the 'real' enclosure happened in the decades after 1953: not so much through the implementation of a reform law but as the outcome of increasing demographic pressure (Klein, 1993: 131). Together with ecological degradation and new inheritance practices that gave a more individual meaning to *sayaña* rights, this inflamed conflicts over land use, boundaries and access, incited a rural exodus and contributed to subdivision, individualization and differentiation of access to land. The reformed land system failed to halt land fragmentation, extreme poverty, and the marginalization of the highlands, and continued to favour capitalist exploitation of the land (Kay and Urioste, 2007; Urioste, Barragán and Colque, 2007).

---

[7] *'The State recognizes and guarantees the legal existence of the indigenous communities.'* Own translation, article 165 of the political constitution of Bolivia of 30 October, 1938, Busch, 1938 in Lexivox. See also Choque and Ticona, 2010: 29.

Figure 9.8 Map of Carangas (detail)

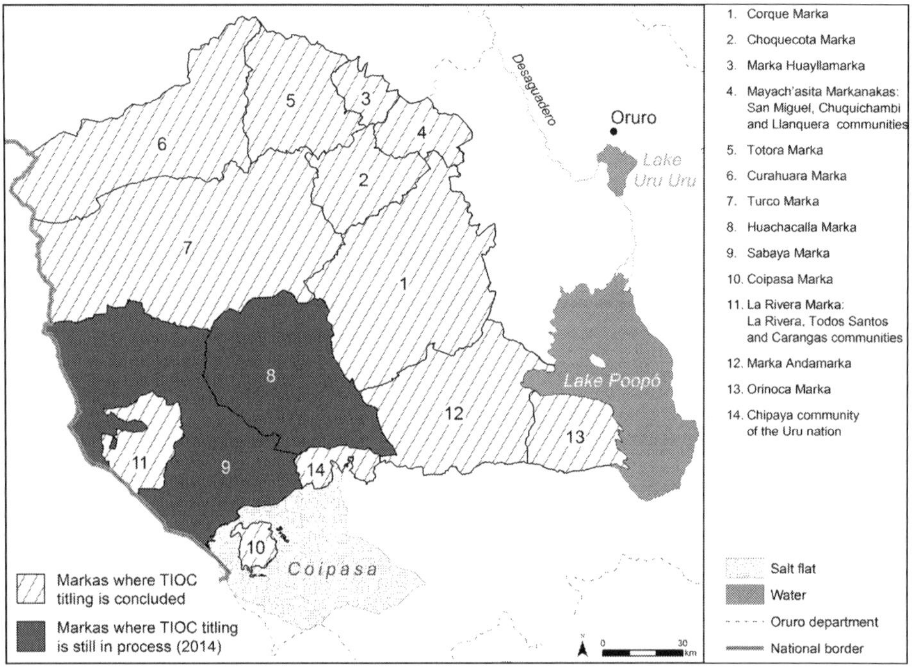

Quantitative data which might confirm the process of land fragmentation are lacking precisely because of the character of the new land policies. The allocation of 'undivided' property rights represents a deliberate decision on the part of the central state not to intervene at a local level. Communal land property was allocated at the *ayllu* level without specifying the internal property limits between plots accessed individually by families. This lack of intervention is the ambiguous outcome of both an institutional deficiency in effective governmental land management and a reinforced struggle by indigenous communities for communal self-determination. The result was the non-involvement of the state within communal structures of land management, something which was consolidated through the subsequent land reforms implemented at the turn of the twenty-first century.

As a result of pressure from both peasant and indigenous organizations, a new constitution was enacted in 1994, introducing the concept of 'Native Community Land' or *Tierra Comunitaria de Origen* (TCO) to recognize explicitly indigenous rights to communal resources, customs and decision-making (Campero, 1999: 479). The 1996 land reform (Law 1715, also known as Ley INRA) subjected all post-1953 land property documents and titling procedures to a technical-juridical regularization operation, and adopted the TCO as communal property title (INRA, 2008: 85). The collective TCO title, which left interior boundaries, ownership and inheritance patterns to internal customary regulations under indigenous authority supervision, was a crucial innovation. However, obtaining that title (at the *marka* or *ayllu* level) forced peasants to explicitly renounce their family property rights (Sanjínes, 2005: 42). Under the Morales government, the 2006 land reform 'updated' this regularization, and converted the TCO into 'Indigenous Native Peasant Territory' or *Territorio Indígena Originario Campesino* (TIOC), also recognized by the new constitution of 2009.

By February 2014, the property title regularization was almost completed for the entire Carangas region, with procedures still running in the Sabaya and Huachacalla provinces/ *markas* (Regalsky, Pablo et al, 2015: 180). Most of these lands are titled as TIOC (Servicio Estatal de Autonomías, 2012: 3), with exception of the state owned salt flats and minor pieces of individualized land of private persons or groups of families who chose to be excluded from the community as 'third party', and who obtain a private property title. The TIOC titling map (Figure 9.8) depicts a simplified picture of the advance of the titling procedure by February 2014 on the level of *markas*, although *marka* boundaries do not strictly coincide with TIOC delineations and might include several TIOCs titled on the *ayllu* level and 'third party' privatized enclaves (see the database on http://territorios.ftierra.org). While security of tenure over communal lands – which comprise virtually all land in Carangas – has been seriously eroded over the last few centuries, the community's role as the primary landowning institution, as well as the inalienability of community lands, have survived. The recently created legal modalities pretend to strengthen them, by incorporating them within the framework of the centralized state. This remains a continuous tightrope-walking challenge, involving the balancing out of the collective-individual complementarity of communal land systems and heightened expectations of autonomy and security (Urioste, 2008: 8; Rojas, 2012; Chumacero, 2012).

VI. Concluding remarks: change and continuity

In Carangas, a significant measure of community-based land and resource management has endured over centuries of shifting central state land policies. The region can be envisaged as a 'fissure' in modern, standardized land regimes. After more than 500 years of capitalist expansion, to which rural zones such as Carangas have been actively contributing since early colonial silver routes first developed, titling procedures have effectively embedded communal land tenure systems within national legal frameworks. Although subjected to successive land reform initiatives, communal land tenure systems have remained in place, albeit challenged, and altered. In this respect, communities such as those in Carangas present a fascinating laboratory to explore the dialectics of communitarian self-determination and capitalist expansion.

The case of Carangas provides ample support for those who believe, in the words of Fenelon and Hall, that communal ownership poses 'the most fundamental challenge to capitalism, [...] because it denies the overarching dominance of private property rights' (Fenelon and Hall, 2009: 6). The survival of communal control systems in the Central Andes, despite centuries of close connections with the capitalist world-economy, is an example of how Hardin's theorem is countered by communal action. There are, however, several factors to take into account when we assess the full significance of this particular example, and especially the way in which the community-based land systems in the Andes were backed and reinforced by particular ecological, demographic and social factors.

In the case of Bolivian *altiplano* communities, the limited possibilities for arable agriculture and the optimal conditions for pastoralism necessitated the adoption of rotational systems and the maintenance of open, communal grazing lands. This way of organizing land use was driven by the need to reduce risk in a fragile environment. The preponderance of Andean pastoralism is a key factor in explaining the persistence of collective land rights, and privatization had a much greater impact in regions where the ecological conditions were more suited to arable farming. This said, the persistence of communal land management systems cannot be reduced to the dictates of ecology. Too great an emphasis on soil poverty and isolation as the determinants of historical processes in Carangas risks ignoring underlying local dynamics, of negotiation and confrontation: between local communities and agricultural entrepreneurs, between communities and state powers, and between communities.

This brief overview of how land is managed in certain parts of the Andes demonstrates how, even today, rural households can access both individually and collectively held land, whether cultivated, pasture or fallow. The question is not so much whether the concept of an open-field system can or cannot be applied to the Andes. It is rather the extent to which, through comparing the experience of different regions, important light can be thrown on factors encouraging the emergence, and survival, of communal land systems more generally.

As the Bolivian economy becomes ever more closely tied to the imperatives of the world market, community land and collective forms of resource regulation come under increasing pressure. Nevertheless, Bolivia is entering the twenty-first century with a legal framework that allows indigenous communities to uphold collective mechanisms to manage and use land, and other natural resources. The communities of the former Carangas province do not represent an isolated case. The development of their land rights, as explained briefly in this chapter, shows clearly how communal land systems, managed by indigenous and peasant communities, are not inevitably replaced by private smallholder tenure, but can instead take new forms within modern state regimes.

Bibliography

Archival sources

Archivo de los Derechos Reales de Oruro (ADRO). Corte Superior de Distrito, Oruro. *Properties in the provinces*, 1894–1930.

Archivo Judicial de Corque (AJC). Corte de Corque, Corque. *Historical archive*, up to 1900. Volume 18 (1860–1862) to volume 44 (1898–1900). *Semi-historical archive*, 1900–1950. *Civil* (1900–1930); *Criminal* (1900–1930).

Publications

Antezana Ergueta, L. (2006), *La política agraria en la primera etapa nacional*, La Paz.

Arteaga, A., Ramiro, M. R. et al (2006), 'Estudio de caso Oruro. Reseña histórica y consulta sobre regionalización del Departemento de Oruro' in: F. X. Barrios Suvelza ed., *Procesos de desconcentración prefectural en Bolivia*, La Paz, pp. 507–57.

Barragán, R. (2012), 'Los títulos de la Corona de España de los indígenas: para una historia de las representaciones políticas, presiones y negociaciones entre Cádiz y la república liberal', *Boletín Americanista*, 65, pp. 15–37.

Bolton, R. and Mayer, E. (1977), *Andean kinship and marriage*, Washington D. C.

Brownman, D. (1991), 'Llama caravan fleteros: their importance in production and distribution' in:

Campbell, B. and Godoy, R. (1985), 'Common-field agriculture: the Andes and medieval England compared' in: *Proceedings of the Conference on Common Property Resource Management, April 21–26, 1985*. National Research Council (U. S.). Panel on Common Property Resource Management. Washington: National Academy Press.

Choque Canquí, R. and Ticona, E. (1996), *Sublevación y masacre de 1921 (Jesús de Machaqa: la Marka Rebelde 2)*, La Paz.

Chumacero, J. P. (2012), *Desafíos de la problemática agraria en Bolivia*, La Paz.

Cole, D. H. (2002), *Pollution and property: comparing ownership institutions for environmental protection*, Cambridge.

Cottyn, H. (2012), 'Ruptura sin reforma? El proceso de reforma agraria en Turco' in: X. Medinaceli ed., *Turco Marka. Hombres, dioses y paisaje en la historia de un pueblo orureño. Historia-Arqueología-Arquitectura*, La Paz, pp. 157–86.

Cottyn, H. (2014), *Renegotiating communal autonomy. Communal land rights and liberal land reform on the Bolivian altiplano. Carangas, 1860–1930*, unpublished PhD thesis, Ghent University.

Cunill Grau, P. (1999), 'La geohistoria' in: M. Carmagnani, A. Hernández Chávez and R. Romano ed., *Para una historia de América. Vol. 1. Las Estructuras*, Mexico D. F., pp. 13–159.

Custred, G. and Orlove, B. S. ed. (1980). *Land and power in Latin America: agrarian economies and social processes in the Andes*, New York.

de la Barra, S. Z., Marcelo Lara Barrientos, G. and Oscar Coca Cruz, R. (2011), *Exclusión y subalternidad de los urus del lago Poopó. Discriminación en la relación mayorías y minorías étnicas*, La Paz.

Fenelon, J. and Hall, T. D. (2009), *Indigenous peoples and globalization*, Boulder.

Fonseca Martell, C. (1973), *Sistemas económicos Andinos*, Lima.

Galindo, M., Cruz, B., Pardo, E. and Bueno, R. (2007), *Visiones Aymaras sobre las autonomías: aportes para la construcción del estado nacional*, La Paz.

Godoy, R. (1991), 'The evolution of common-field agriculture in the Andes: a hypothesis', *Comparative Studies in Society and History*, 33, 2, pp. 395–414

Goodale, M. R. G. and Sky, P. K. (2000), *A comparative study of land tenure, property boundaries, and dispute resolution: examples from Bolivia and Norway*, Wisconsin, 2000.

Graffam, G. (1992), 'Beyond state collapse: rural history, raised fields, and pastoralism in the south Andes', *American Anthropologist,* 94, 4, pp. 882–904.

Grieshaber, E. P. (1980), 'Survival of Indian communities in nineteenth-century Bolivia: a regional comparison', *Journal of Latin American Studies*, 12, 2, pp. 223–69.

Guillet, D. (1981), 'Land tenure, ecological zone, and agricultural regime in the central Andes', *American Ethnologist*, 8, 1, pp. 139–56.

HAM Turco (2007), *Plan de desarrollo municipal 2008–2012, Municipio de Turco, Provincia Sajama*, Turco.

Hidalgo Lehuedé, J. and Durston, A. (2004), 'Re-constitución étnica colonial en la sierra de Arica: elcacicazgo de Codpa, 1650–1780' in: J. Hidalgo Lehuedé ed., *Historia Andina en Chile*, Santiago, pp. 507–34.

INRA (2008), *Breve historia del reparto de tierras en Bolivia. De la titulación colonial a la reconducción comunitaria de la reforma agraria: certezas y proyecciones*, La Paz.

Irurozqui Victoriana, M. (1994), *La armonía de las desigualdades*, Lima.

Irurozqui Victoriana, M. (1999), 'Las paradojas de la tributación. Ciudadanía y política estatal indígena en Bolivia, 1825–1900', *Revista de Indias* 59, no. 217, pp. 705–40.

Kay, C. and Urioste, M. (2007), 'Bolivia's unfinished agrarian reform: rural poverty and development policies' in: A. H. Akram-Lodhi, J. M. Borras Jr and C. Kay ed., *Land, poverty and livelihoods in an era of globalization: perspectives from developing and transition countries*, London and New York, pp. 41–78.

Klein, H. (1993), *Haciendas and Ayllus. Rural society in the Bolivian Andes in the eighteenth and nineteenth centuries*, Stanford.

Langer, E. D. (1989), *Economic change and rural resistance in southern Bolivia, 1880–1930,* Stanford.

Langer, E. D. (2009), 'Bringing the economic back in: Andean Indians and the construction of the nation-state in nineteenth-century Bolivia', *Journal of Latin American Studies,* 41, pp. 527–51.

Langer, E. D. and Jackson, R. H. (1990), 'El liberalismo y el problema de la tierra en Bolivia (1825–1920)', *Revista de Historia*, 5, pp. 9–32.

Larson, B., Harris, O. and Tandeter, E. (1995), *Ethnicity, markets, and migration in the Andes: at the crossroads of history and anthropology*, Durham NC.

Larson, B. (1998), *Cochabamba, 1550–1900: colonialism and agrarian transformation in Bolivia*, expanded edn, Durham NC and London.

Larson, B. (2004), *Trials of nation making: liberalism, race, and ethnicity in the Andes, 1810–1910*, Cambridge.

Lehmann, D. (1983), *Ecology and exchange in the Andes*, Cambridge.

Lexivox. Portal jurídico libre. http://www.lexivox.org/

Lima, E. A. (1921), Etnografía de la Provincia De Carangas. Su iportancia para el Bolivia. reformas sociales e institucionales, La Paz.

López Garcia, M. K. (2003), *Pastoreo Andino*, Oruro.

Mann, C. C. (2011), *1493: Uncovering the new world Columbus created,* New York.

McCloskey, D. (1991), 'The open field system' in: J. Eatwell, M. Milgate and P. Newman ed., *World of economics*, London.

Mayer, E. (2002), *The articulated peasant: household economies in the Andes*, Boulder.

Mazoyer, M., Roudart, L. and Membrez, J. H. (2006), *A history of world agriculture: from the neolithic age to the current crisis*, London.

Medinaceli, X. (2005), 'Los pastores andinos: una propuesta de lectura de su historia', *Bulletin de l'Institut Français d'Études Andines*, 34, 3, pp. 463–74.

Medinaceli, X. (2010), *Sariri*, La Paz.

Ministerio de Desarrollo Rural y Tierras (2014), *Atlas de riesgo agropecuario y cambio climático para la soberanía alimentaria*, La Paz.

Moore, J. (2010), '"Amsterdam is standing on Norway" part I: the alchemy of capital, empire and nature in the diaspora of silver, 1545–1648', *Journal of Agrarian Change*, 10, 1, pp. 33–68.

Mumford, J. (2012), *Vertical empire: the general resettlement of Indians in the colonial Andes*, Durham NC.

Murra, J. V. (1975), 'El control vertical de un máximo de pisos ecológicos en la economia de las sociedades andinas' in: J. V. Murra ed., *Formaciones económicas y políticas del mundo Andino*, Lima, pp. 59–115.

Orlove, B. and Godoy, R. (1986), 'Sectoral fallowing systems in the central Andes', *Journal of Ethnobiology*, 6, 1, pp. 169–204.

Ovando Sanz, J. A. (1985), *El tributo indígena en las finanzas bolivianas del siglo XIX*, La Paz.

Pauwels, G. (1983), *Dorpen en Gemeenschappen in de Andes: socio-culturele veranderingen bij Boliviaanse Aymara*, unpublished PhD thesis, Catholic University of Leuven.

Pauwels, G. (1999), 'Santa Bárbara de Chillagua. Aporte a la recuperación de la etnohistoria Uru en Carangas' in: *XIII Reunión Anual de Etnología. Identidades, Globalización o Etnocidio*, La Paz, pp. 72–78.

Pauwels, G. (2006), 'Carangas en el año 1910. El informe de Zenón Bacarreza' in: M. Cajías de la Vega et al ed., *Ensayos históricos sobre Oruro*, La Paz, pp. 337–403.

Platt, T. (1981), *Estado Boliviano y Ayllu Andino: tierra Y tributo en el norte de Potosi*, Lima.

Regalsky, P. et al. (2015), *La problemática de la tierra. A 18 años de la ley INRA. Territorios, minifundio, e individualización*. La Paz.

Rights and Resource Initiative (2015), *Who owns the world's land? A global baseline of formally recognized indigenous and community land rights*, Washington D. C.

Rivera Cusicanqui, S. and Platt, T. (1978), 'El impacto colonial sobre un pueblo Pakaxa: la crisis del Cacicazgo en Caquingora (Urinsaya) durante el siglo xvi', *Avances, Revista Boliviana de Estudios Históricos y Sociales*, 1, pp. 101–20.

Rivera Cusicanqui, S. (1991), 'Pedimos la revisión de límites: un episodio de incomunicación de castas en el movimiento de caciques-apoderados, 1919–1921' in: S. Moreno and F. Salomon eds, *Reproduccion y transformacion de las sociedades andinas, XVI–XX*, Quito.

Rivière, G. (1982), *Sabayas: structures socio-economiques et representation symbolique dans les Carangas, Bolivia*, unpublished PhD thesis, Ecole des Hautes Etudes en Sciences Sociales, Paris.

Rodríguez, C. T., and Quispe, J. L. (2009), 'Domesticated camelids, the main animal genetic resource of pastoral systems in the region of Turco, Bolivia', in: K. A. Tempelman and E. A. Cardellino ed., *People and animals,* Rome, pp. 33–45.

Rojas Calizaya, J. C. (2012), 'Agrarian transformation in Bolivia at risk', *BIF Bulletin*, 22.

Ruz, R., Díaz, A., and Fuentes, R. (2011), *Timalchaca. Fiesta, tradición y costumbre en el santuario de la Virgen De Los Remedios*, Arica

Salzman, P. C. and Galatey, J. G. ed., *Nomads in a changing world*, Naples, pp. 408–55.

Sanjinés Delgadillo, E. (2005), *Ayllu Jila Taypi uta Collana: una visión de la Tierra Desde La Norma*, La Paz.

Scott, J. C. (2009), *The art of not being governed: an anarchist history of upland southeast Asia*, New Haven and London.

SEDES (2007), *Atlas de Salud 2005. Departamento de Oruro*, Oruro.

Servicio Estatal de Autonomías and Observatorio Bolivia Autonómica (2012), 'Territorios Indígena Originario Campesinas potenciales para el acceso a las AIOC en tierras altas', *Boletín de Estadísticas y Análisis,* No. 3.

Thirsk, J. (1964), 'The common fields', *Past and Present*, 29, pp. 3–25.

Thoen, E. (2004), "Social agro-systems" as an economic concept to explain regional differences. An essay taking the former county of Flanders as an example (Middle Ages-19th century)' in: B. van Bavel and P. Hoppenbrouwers eds, *Landholding and land transfer in the North Sea area (late Middle Ages-19th century)*, Turnhout, pp. 47–66.

Urioste, M. (2005), *Bolivia: la reforma agraria abandonada. Valles y altiplano*, La Paz.

Urioste, M. (2008), 'Nuevo ciclo de reformas agrarias en América Latina?', *Pulso Semanario* 23.11.2008–29.11.2008: 4.

Urioste, M., Barragán, R. and Colque, G. (2007), *Los nietos de la Reforma Agraria. Tierra y comunidad en el altiplano de Bolivia*, La Paz.

Wachtel, N. (2001), '*El regreso de los antepasados: los Indios Urus de Bolivia, del siglo XX al XVI'*, Mexico City.

West, T. L. (1981), *Sufriendo nos vamos: from a subsistence to a market economy in an Aymara community of Bolivia*, unpublished PhD thesis, New School of Social Research, New York.

Yambert, K. A. (1980), 'Thought and reality: dialectics of the Andean community' in: G. Custred and B. S. Orolove, ed., *Land and power in Latin America: agrarian economies and social processes in the Andes*, New York, pp. 55–78.

10 The scattered and intermingled field system of Japan compared to the open-field systems of Europe

Junichi KANZAKA, Soka University, Japan

I. Introduction

In many agricultural regions, land fragmentation exists in which 'a single farm consists of numerous discrete parcels, often scattered over a wide area'. We find this phenomenon 'in most parts of the world where cultivation is long established' (Binns, 1950: 5), including India (Heston and Kumar, 1983; Bonner, 1987; Niroula and Thapa, 2005), Africa (Blarel et al., 1992), China (Tan, Qu, and Heerink, 2005), and Europe. 'Open fields' in Europe are regarded in these studies as an example of fragmentation, but they offer no examination of the unique properties of such field systems. When an international institution, such as the United Nations Food and Agriculture Organization (FAO), carried out programmes to abolish fragmentation in developing countries in the period after the Second World War (Meliczek, 1973), the process of consolidation in eighteenth–and nineteenth-century Europe acted as a precedent, although as Farmer cautioned, 'arguments from the experience in Western agriculture' are 'heard in South and South-east Asia, usually without sufficient thought about their relevance to local conditions' (Farmer, 1960: 226).

In Japan, fragmented agricultural lands have long been a feature of the landscape. In the Tokugawa period, 1603-1868, an individual family holding 'was generally split up into many parcels scattered over a wide area' (Fukutake, 1967: 83). In some regions, such as Toyohara-mura (Yamagata), land-owning peasants cultivated consolidated blocks of paddies (Toyohara Kenkyu-kai, 1978); while in newly-developed villages property was not dispersed. However, historians, geographers, and agricultural economists principally characterize the Japanese fields as '*reisai bunsan sakuho*', that is, as comprising 'tiny, scattered, and intermingled parcels'. Today, fragmentation is regarded as an obstacle to Japanese agricultural development. Kawasaki (2010) for example states that 'fragmentation increases production costs and offsets economies of size', and that 'although fragmentation does reduce production risk, its monetary value is far below the cost of land fragmentation'.

Since the nineteenth century, attempts have been made in Japan to consolidate fragmented parcels, similar to those which are currently underway in many developing countries. However, the Japanese experience is unique in three ways. First, while in many developing countries international institutions, such as FAO, or outside advisors such as European economists, have initiated programs of land consolidation, the Japanese by themselves learnt about European field systems and carried out their owns schemes of land consolidation. In particular, in the 1880s and 1890s Japanese officials and scholars eagerly studied European agriculture in order to formulate the Arable Land Readjustment Law. Second, the Japanese did not strictly follow European models. They realized that Japanese agriculture was markedly different from that of Europe in the early twentieth

Peasants and their fields. The rationale of open-field agriculture, ed. by Christopher DYER, Erik THOEN and Tom WILLIAMSON, Turnhout, 2017 (*CORN* Publication Series, 16), pp. 233–255

BREPOLS PUBLISHERS DOI: 10.1484/M.CORN-EB.5.114275

century, and therefore revised the Readjustment Law and constructed an infrastructure for irrigation and drainage, rather than simply consolidating scattered parcels. Since Bentley criticizes the government officials of developing countries for being 'eager to attempt consolidation schemes but unwilling to study farm fragmentation' (Bentley, 1987: 61), this reorientation of policy on the part of the Japanese government is instructive in today's world. Finally, and principally in the period after World War II, Japanese scholars examined the origin of their scattered and intermingled field system and compared them to European open fields. In this process, they showed the rationale for land fragmentation; furthermore, they compared the structures and functions of European village communities, based on open-fields, with those of Japan, founded on communal water regulation. I discuss each of these three points in turn in the following sections, which describe the development of Japanese agriculture from 1880 to 1899, from 1900 to World War II, and after World War II, respectively.

II. The study of European agriculture by the Japanese from 1880 to 1899

Early attempts at land improvement

Even before learning of the improvements in European agriculture, the Japanese made some attempts to consolidate scattered parcels; for example, in 1849, Kumenosuke Mori, a great landlord, created large blocks of one *tan* twelve *bu*, or 0.103 ha., by removing the ridges separating different holdings (Ogawa, 1953: 165–66). Such indigenous improvements developed into the nationwide movement for 'straightening up paddy field plots' (*denku-kaisei*) in 1887–97 (Ogura, 1963: 239). Agricultural innovation, so called 'Meiji Farming', which mainly involved the cultivation of new varieties of plants, ploughing with horses, and inputs of fertilizer, encouraged the 'improvement' of paddies and fields. In 1887, for instance, the villagers of Kamonishi-mura (Shizuoka), under the vigorous leadership of their headman (*Kucho*) Urahachi Suzuki, worked together to build straight roads, irrigation and drainage canals and levees. The project allowed the inhabitants 'to carry fertilizer and other goods by carts, horses or oxen, to plough fields by horses or oxen, to improve lands through good irrigation and drainage, and to let sunshine into the fields' (Ogawa, 1953: 184).

New knowledge of enclosure and land consolidation in Europe, however, also promoted agricultural improvements. Some Japanese were keen to import Western agricultural techniques. For example, in 1876, Sen Tsuda and seven friends founded the Gakunosha Agricultural School 'to build a foundation for the strength of our nation by studying diligently European books on agriculture, combining European and Japanese agricultural techniques, and enhancing prosperity of the Japanese' (Susuda, 2010: 5). Important figures in this movement include Josaburo Watanabe, who joined the Agricultural School and later, in 1886, became the head of Ishikawa-district Model Farm. The Model Farm supported the land consolidation project in Kami-yasuhara-mura (Ishikawa) during 1887, in which Kyubei Takata, a diligent agriculturalist, managed to persuade the villagers to cooperate in improving their paddies and fields (Susuda, 2010: 4–6, 13). Thus, the spread and development of innovative agricultural techniques encouraged the rearrangement of existing land parcels.

The Japanese Government was also eager to import 'advanced' techniques for land improvement. It invited many European scholars in several disciplines, including Udo Eggert, a German professor who taught public finances at Tokyo Imperial University from 1887 to 1893. In 1890, Eggert encouraged the consolidation of scattered fields: 'One of the most important questions, now agitating the minds of the Japanese, is the consolidation of dismembered area'. In his book, he further explains that such an amalgamation of small productive units would allow time, which 'the cultivator now wastes in going from one of his dismembered portion to another', to be saved. In addition, 'by the fields being laid out in a less irregular way, much land would be gained for cultivation, which is now taken up by necessary embankments and watercourses' (Eggert, 1890:21–22).

Japanese officials' research expedition to Europe and Roichi Hida

The Meiji government not only invited European scholars to come to Japan; it also sent some of its own officials to Europe to investigate agriculture, as well as manufacturing, and commerce. In 1886, the minister Tateki Tani, accompanied by officials of the Ministry of Agriculture and Commerce (MAC), travelled to Europe. They produced a detailed report of their research, which commented on important aspects of contemporary European agriculture. For example, when they asked Eugène Tisserand, the inspector general of the French Agriculture Ministry, about the consolidation of scattered plots, he replied: 'agriculturalists and government officials have noticed the inefficiency of scattered plots for many years. The first attempt to solve the problem originated in France'. He then described the late eighteenth-century land consolidation projects directed by François de Neufchâteau. Nevertheless, Tisserand showed the Japanese officials, not examples of French legislation, but a Saxon law of 1861, whose effectiveness was reflected in the land consolidation projects undertaken in Hohenheida, Saxony, in 1865. The Japanese officials translated all the articles of the Saxon law (MAC, 1888: 371–94). It is not clear why Tisserand presented Saxon law, although Tayama supposed that Tisserand regarded 'the achievement in Saxony as the fruit of the French idea' (Tayama, 1988: 316). However, it should be noted that the 'French idea' did not bear fruit in France itself, where a law passed in 1865 requested the 'unanimous consent of landowners associated in a *syndicat libre*' to determine consolidation and enclosure; it was not until after World War I that enclosure became widespread in France (Grantham, 1980). Tisserand may himself have examined the German laws, in order to promote land consolidation in France.

During this visit, Roichi Hida, one of the officials, observed land improvement projects in the Camargue of France and in Luxembourg. The latter was the more interesting to him. Although the details of the projects in Luxembourg are unclear (Sato and Hirota, 1999), I presume that they were implemented by the land improvement syndicates established in 1874. Luxembourg's government installed a German engineer, J. Enzweiler, to coordinate the work of these syndicates, and Hida states in his book on agricultural improvement that 'J. Enzweiler led me to several construction places and showed me design drawings and machines' (Hida, 1894: 3). The tasks of the syndicate included the regulation of watercourses, the installation of new drainage and irrigation channels, and land consolidation (Schaich, Karier and Knold, 2011: 249). All these works must have been of considerable interest to Hida, given the situation in his homeland. First, Japan had many rivers that required improvement; indeed the Japanese government adopted the first modern 'river law' in 1896. Next, irrigation and drainage were indispensable for rice paddies. Lastly, Japanese peasants had just started to consolidate their scattered plots. Hida speaks of the projects in Luxembourg as a model for Japan, since they included 'the improvement of irrigation, drainage, and field paths, and the exchange of scattered plots' (Hida, 1894: 3). Although Hida was criticized on the grounds that he was simply attempting to build 'modern European farms' in Japan (Ogawa, 1954: 206), in reality he tried to derive a land improvement plan suitable for Japanese agriculture from his research on European projects (Susuda, 2010: 43–97). As he himself put it, 'we should build our own plan by referring to the projects of Europe, although we cannot adopt these plans directly, since there are many differences in European and Japanese agriculture' (Hida, 1894: 4). After he returned to Japan in 1887, he was dismissed from the MAC, along with his superior Masana Maeda, in 1890. Thereafter, Hida facilitated the land improvement movement as one of the leaders of two civil organizations, the Agricultural Society of Great Japan (ASJ) (*Dai-nihon Nogyo-kai*) and the National Agricultural Association (NAA) (*Zenkoku Noji-kai*).

The arable land readjustment law and Tsuneaki Sakou

After Hida left the MAC, Tsuneaki Sakou was appointed as secretary of that ministry in 1898. He was in Europe from 1889 to 1891, where he mainly studied agricultural institutions in Germany. In his work on the readjustment of plots, he argued that '*Flurzwang* (the compulsory regulation of tilling) and easement prevent the development of agriculture by oppressing voluntarily improvement' (Sakou, 1893: 4). To overcome these obstacles, some peasants in those parts of Germany where enclosure was not yet prevalent consolidated their own scattered strips. In the latter half of the nineteenth century, many German states made this form of land consolidation, or *Verkoppelung*, legally secure (Mayhew, 1973: 187). In this regard, Sakou was aware of the following legislation: the 1872 act in Prussia, the 1861 act in Saxony, the 1856 act in Hanover, the 1861 and 1886 acts in Bavaria, the 1856 act in Baden, the 1886 act in Württemberg, and the 1871 act in Hessen-Darmstadt (Sakou, 1893: 30–36). He thus offered his readers the newest information on German land consolidation legislation (Tayama, 1988: 333–35). Furthermore, Sakou's book includes maps showing Neckarbischofsheim in Baden before and after land consideration (Figure 10.1), along with other maps that indicated the changes produced by the above-mentioned land project in Kami-yasuhara-mura (Figure 10.2).

Sakou presented 'ten benefits' of land readjustment. He emphasized, in particular, the increased acreage of plots which would result, and the government's decision to leave the land tax on consolidated land unchanged for thirty years (Ogawa, 1954: 209). He noted the technical gains made possible by larger, geometrically regular fields, such as the ability to plough with horses or oxen. Moreover, the removal of tight and tortuous paths allowed for the movement of implements, fertilizer, and produce more easily, since animal-drawn carts could more easily access the paddy fields. Such consolidation also made possible the more efficient supervision of the workforce, which was now grouped together in larger units of production. Improved channels and culverts allowed for the proper irrigation and drainage of paddy fields, and both they, and the new access paths, were easier to maintain. The overall effect of these alterations was not only higher productivity but also a rise in the value of land, which promoted agricultural trade and finance and produced higher profits for tenants, who were also less plagued by land and water disputes because of the advent of reliable mapping (Sakou, 1893: 5–6). In addition, Sakou insisted on the necessity for

Figure 10.1 Maps of Neckarbischofsheim in Baden

Before consolidation

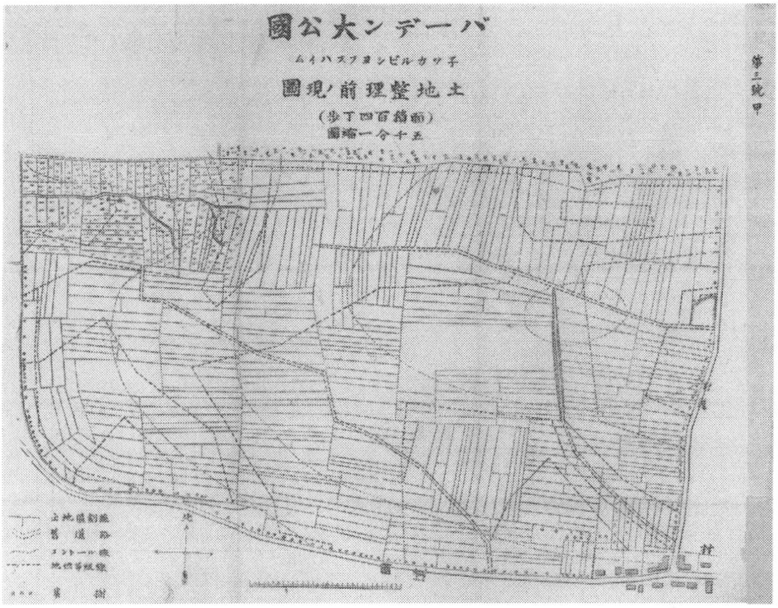

After consolidation

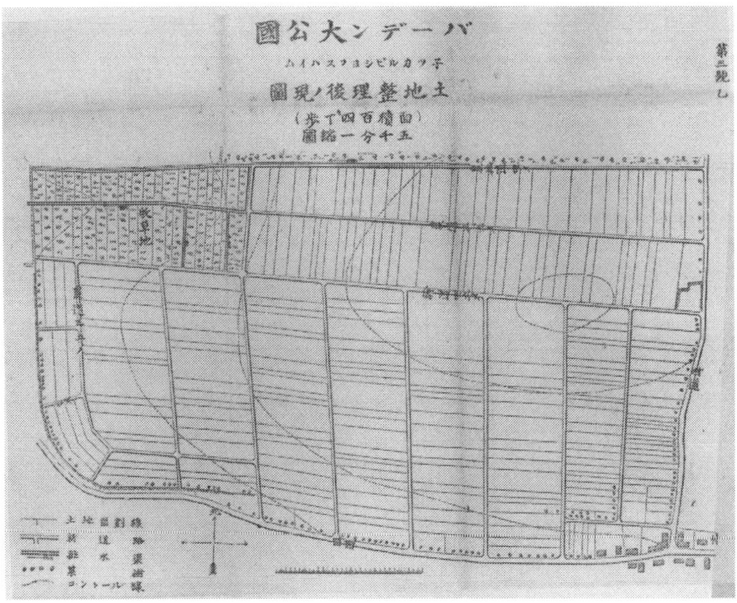

Source: Sakou, 1893

Figure 10.2 Maps of the land readjustment project in Kami-yasuhara-mura in Ishikawa

Before consolidation

After readjustment

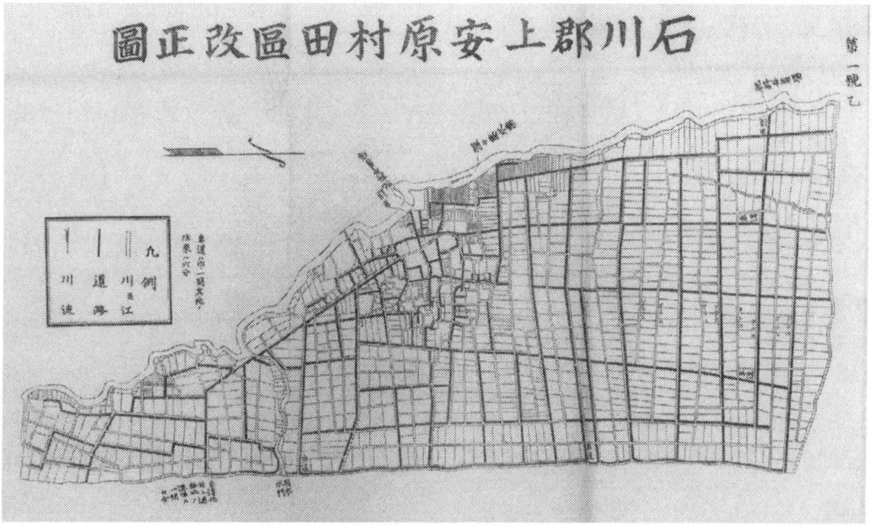

Source: Sakou, 1893

legislation to enforce the readjustment of plots. He described how 'In projects that refigure plots [...] it is necessary to force the objecting minority to follow the majority's decision. Therefore, we require a legal basis to carry out quickly these projects' (Sakou, 1893: 10).

Indeed, it was extremely difficult to obtain a consensus of all landowners. The leaders of the enclosure movement often faced severe protests. For example, in the case of the scheme at Kamonishi-mura, noted earlier, someone planned to assassinate Suzuki, the head of the project, and construction of the new works could only proceed under police guard. In the 1894 meeting of the NAA Genji Kitamura, the representative from Ishikawa, stated that such projects were regularly prevented by landowners' objections, troublesome procedures, and a lack of money (Ogawa, 1953: 231). In 1896, the NAA passed a resolution that requested legislation to deal with the problems caused by the 'so-called objecting minority, which is the serious obstacle against land improvement projects' (Susuda, 2010: 63).

In 1899, the Arable Land Readjustment Law was passed, 'for improving the utility of land, [permitting] landowners to exchange cooperatively or consolidate their plots, alter land shapes, and modify paths, ridge, canals, and ditches'. Sakou played a significant role in this legislation, which was explicitly based on the German land consolidation laws in Württemberg, Bayern, and Baden. In response to requests, the law allowed entrepreneurial landlords to fend off the 'objecting minority' if the following three requirements were satisfied: '1) The consent of more than two-thirds of landowners in the area concerned should be obtained; 2) the acreage of land belonging to the assentients should be more than two-thirds of the total acreage of land to be replotted; and 3) the aggregate amount of the price of lands belonging to the assentients should be more than two thirds of the total amount of the prices of the lands in the area;' (Ogura, 1963: 240–41). The first and third conditions are almost the same as those contained in the 1856 law of Baden. Furthermore, the second condition on acreage probably follows the tradition of the Bayern legislation (Tayama, 1988: 337–38).

Thus, in the late nineteenth century, the Japanese eagerly studied European field systems and land consolidation schemes, and in 1899 they enacted the Arable Land Readjustment Law which was firmly based on German precedent, in order to encourage Japanese landowners to consolidate their scattered plots.

III. The reorientation of policy from consolidation to water management in the early twentieth century

The stagnation of land consolidation

The Arable Land Readjustment Law, however, did not lead to much in the way of land consolidation, as it resulted in only isolated projects. For example, in 1902 landlords and owner-cultivators in Konosu-machi (Saitama) improved the productivity of 387 ha. of rice paddies and fields by making larger, regularly-shaped, and well-drained fields, which facilitated the use of horses for ploughing (Nagata, 1984). Eisaburo Ueno, a lecturer at the Agricultural High School attached to the ASJ, praised this project in a published textbook (Ueno, 1905). However, in the period before 1905 such projects were few in number (Figure 10.3).

Figure 10.3 Acreage of land affected by 'readjustment' projects (in ha.)

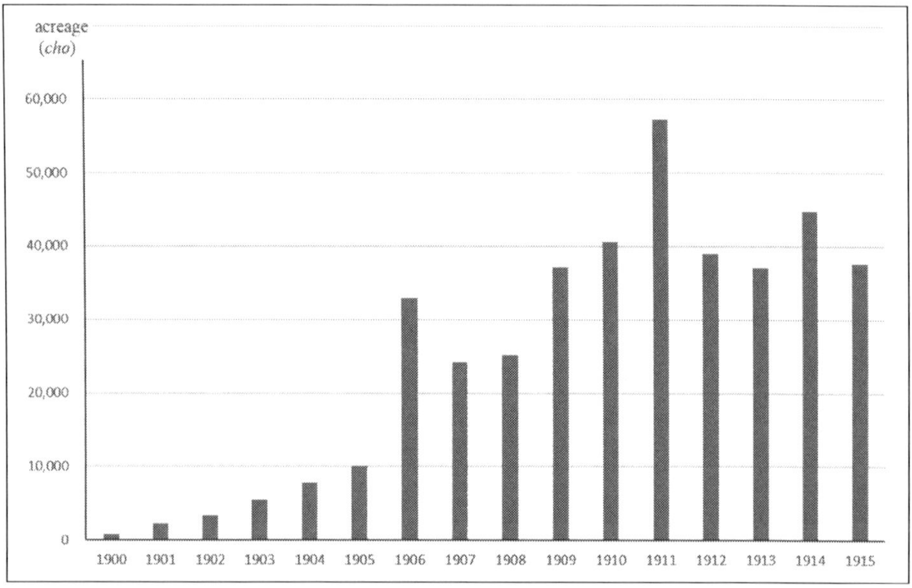

Source: Ogawa, 1954

Landlordism was a major obstacle to land consolidation. The development of landlordism was an important feature of Japanese rural society. At the end of the 1880s, at least 40 per cent of rice paddies and fields were cultivated by tenants (Gordon, 2003: 96). It was the proprietors of land, including landlords and owner-cultivators, who carried out land readjustment; but consolidation did not always seem profitable to landlords, many of whom had ceased farming and were more interesting in increasing the rents paid by their tenants than in making farm management more efficient. When strong demographic pressure raised the demand for tenancies, non-cultivating landlords preferred to divide and sublet land. In such cases, tenants usually cultivated rice paddies and fields consisting of as small plots as 0.1 ha. This was at a time when Ueno, in his analysis of 'cultivation profit', insisted that 'the standard acreage of Japanese agricultural plots' which constituted tenant's holdling should be between 0.2 and 0.3 ha. or, more optimally, 0.5 ha (Ueno 1905: 255). Shinzawa (1969) observes that Ueno's calculation do not reflect the normal situation where Japanese agriculture was based on landlordism.

The idea of the perpetual family (*iye*) was also a major impediment to land-consolidating transactions among the peasantry. The peasants of the Tokugawa period assumed that family property was the possession not of one person, the present head of the family, but rather of the perpetual family. Therefore, 'although the head of a family has dominion over his family, he cannot divide, sell, or alienate the family estates at his own will. It was believed that the head of a family is responsible for inheriting the family estates of his ancestors and conveying them to his offspring without any change' (Watanabe, 2008: 49). Meiji peasants had the same attitude and were therefore unwilling to exchange their land to achieve consolidation.

Revision of the arable land readjustment law

In the face of this stagnation, the Meiji government revised the law and added 'irrigation and drainage' to the aim of 'land readjustment' laid down in 1905. Consequently, water projects dramatically increased. The improvement of irrigation and drainage infrastructures raised the productivity of land without consolidation. Thus, 'land improvement' became the central part of 'arable land readjustment' (Yokoi, 1921: 1), and many water projects left scattered and intermingled plots untouched (Ogawa, 1954: 211). The Japanese were able to enhance the management of irrigation and drainage without land consolidation, since they had developed an efficient system for controlling water, based on close-knit village communities. As described below, land fragmentation was perfectly compatible with communal water management, in contrast to the situation in places like Haryana in India, where each peasant supplied his own water from wells, and where land consolidation therefore improved the efficiency of water management (Bonner, 1987: 112).

The Arable Land Readjustment Law, modelled on German laws, was thus soon revised along lines more in line with the actual cultivation practices of Japan. This process reveals the unique features of Japanese agriculture, namely, the importance of water management and the high intensity of labour. Indeed, Tokiyoshi Yokoi, a professor of agronomy at Tokyo University's agricultural college, insisted on the significance of these features, and criticized government policy that had encouraged land consolidation. In his published analysis of land readjustment, he stated that good irrigation and drainage facilities are indispensable for raising the yields of rice paddies. He further indicated that a failure to appreciate the importance of water management had led to many failures in land readjustment projects: '[M]any believed that the readjustment of small, distorted, and scattered plots would result in some profit'; but this could not be realized 'without increasing productivity through irrigation and drainage' (Yokoi, 1921: 34–35). Yokoi contended that labour-intensive cultivation, which was closely tied to improvements in water management, required reform. He classified land improvement schemes into two types. In 'productivity improvement' irrigation, drainage, soil dressing, and reclamation raised productivity; in 'economic improvement', workers saved time and labour by consolidating arable plots. The type of improvement adopted should, he argued, be dictated by the productive factor that is scarce: if labour is scare, 'economic improvement' is profitable, but if land is lacking, 'productivity improvement' is required (Yokoi, 1921: 53, 57). In Japan, where the labour/land ratio was high, 'productivity improvement', and especially the enhancement of irrigation and drainage systems, was thus the first priority, rather than the consolidation of land parcels. Forty years later, Farmer was to observe that in countries under strong demographic pressure 'there is no inefficiency in the sort of subdivision of paddy-lands' (Farmer, 1960: 230). They needed instead to carry out projects aimed at 'productivity improvement'.

It is also noteworthy that Yokoi described the rationale for scattered plots. He stated that the plots of cultivators should be scattered in order to divide, in an equitable manner, risks of disaster among the different rice paddies and fields. This is the reason why many land 'readjustment' projects failed. He wrote: 'You cannot expect to obtain common consent on land exchange for consolidation since it is extremely difficult to guarantee the equality in distance from messuages to plots, quality of soil, facility of irrigation and drainage, and risk of flooding and inundation' (Yokoi, 1921: 78). Further, he argued that the readjustment of arable plots, in simple imitation of Western models, was not 'appropriate to Japanese agriculture': 'Many policies in our country have their origin in stupid ideas. For example, arable land readjustment was begun by the blind worship of foreign countries and promoted by the misunderstanding of their policies' (Yokoi, 1921: 5, 26; see also Franks, 2006: 160–66).

The Japanese did not, therefore, continue to apply European models to their agriculture. By understanding the difference between European and Japanese agriculture, they changed their land improvement policy in the early twentieth century. After the revision of the Arable Land Readjustment Law, many projects – as indicated above – were simply aimed at the improvement of irrigation and drainage facilities. Government officials and scholars thus became more interested in water management. The MAC conducted a comprehensive survey of the 'customary system' of irrigation and drainage, beginning in 1913 (MAC, 1916). In addition, geographers and historians carefully researched water management (e.g. Kitamura, 1950) and did not focus on scattered and intermingled field systems.

IV. The study of the origins of fragmentation after World War II

Land reform and the persistence of land fragmentation

After World War II, while landlordism declined, land fragmentation persisted. In 1947, the Supreme Commander for the Allied Powers (SCAP) initiated an important phase of land reform in Japan; landlords were required to sell their holdings to tenants as family-sized plots at 1945 prices (Gordon, 2003: 235). Landlordism thus came to an end and cultivators obtained property rights over their land. However, this did not change the field systems of small, scattered and intermingled plots. Koubou and Yamazaki (1954) state that one household usually held a property made up of sixteen plots and that the sum of the distance between each of the plots was approximately eight kilometres. Moreover, the new owner-cultivators refrained from trading plots. Koubou and Yamazaki attribute this reluctance to an obstinate adherence to their own lands: 'Since farmers do not want to relinquish their ancestral land, they hesitate to exchange lands despite the understanding that it makes good sense' (Koubou and Yamazaki, 1954: 169, 184). As mentioned above, landlordism and the idea of the 'perpetual family' have widely been regarded as obstacles to land consolidation. Koubou and Yamazaki insisted that the latter institution continued to preserve land fragmentation after the abolition of landlordism.

With the persistence of fragmentation, some historians began to enquire about its origins and rationale. In particular, Shun Yamada showed that arable land had been divided into scattered, intermingled plots from at least the sixteenth century and argued that the main reason for this was a desire to minimize risk. 'When a one-family household managed a farm, the household had to disperse risk by scattering the plot to survive self-sufficiently' (Yamada, 1956: 65). Yamada demonstrated the importance of risk aversion to Tokugawa peasants by highlighting their strategy of planting several different types of rice and vegetables. However, he does not adduce direct evidence on the connection between the initial development of scattered plots and risk aversion. Instead, he refers only to contemporary instructions to hold lands in several parts of the arable in order to avoid risk (Yamada, 1956: 107; Koubou and Yamasaki, 1954: 192). In addition, Yamada quoted the explanation of Marc Bloch for French open fields: 'If the plots were dispersed, then everyone shared the same risk and enjoyed the same opportunity' (Bloch, 1931/1966: 55).

Hisao Otuka's criticism of the village community

At that time, however, only a minority of Japanese scholars regarded land fragmentation and open fields as a rational means to diffuse risk. Most understood it as being closely connected to wider systems of communal organization that restricted the free and independent activities of each individual. In fact, in the Japan of the 1950s and 1960s, 'village community' did seem to be obstacle to both 'modernization' and 'democratization'.

In 1952, a surprising event occurred in the countryside of Sizuoka. During an election campaign, Satsuki Ishikawa, a high school girl, sent the following letter to a newspaper office:

> When I came back home from school, my mother said, '*kumicho* (one of the leaders of the village) had collected the admission tickets to the polling station'. When I asked him, he stated that he had obtained many tickets of absentees and the elderly people 'to prevent abstention'. I cannot ignore this obvious election fraud. Please investigate it.

As a result, some village officials were arrested. However, the villagers did not praise Satsuki Ishikawa for promoting democracy; instead, they punished her. A villager said to her: 'You are a high school student. You have to be mature enough to understand whether it is good or not to enjoy incriminating others. You have brought disgrace on the village where you live'. Moreover, the villagers ostracized (*mura hachibu*) her family, and prevented them from joining in the social, political and economic life of the village (Ishiwawa, 1953)

Such events made many Japanese think that they had to eradicate severe communal restraints hampering the freedoms of speech and action in order to 'modernize' or 'democratize' their society. The study of the village community thus had a practical significance. In Dewey's words, 'where nations were still composed of myriads of village communities, the fate of civilizations seemed to depend on their rulers' estimates of the village communities' worth' (Dewey, 1972: 292). He gives India and Russia as examples of such societies: Japan can be added to them, since almost half of the Japanese still lived in villages in 1950. Many contemporary Japanese scholars, familiar with Western society and history, were hostile to the whole tradition of village community. In particular, Hisao Otsuka (1955) argued that the 'dissolution' of village communities in Europe provided a model for exterminating communal restrictions in Japan. He stated that European villages used to have communal constraints called *Flurzwang*, which Otsuka translates as *kouku-kyousei*, or literally 'compulsion in a furlong'. The furlong constituted the unit of cultivation on which village households worked, under obligation, at specific times of the agricultural year. Otsuka refers to a passage from Weber: 'Under such a system of husbandry, it was impossible for any individual to use methods different in any way from those of the rest of the community; he was bound to the group in all his acts' (Weber, 1927: 7). In contrast, Otsuka idealized the English yeomanry, independent from such communal restrictions, as 'the modern type of human being' (Otsuka, 1948).

Otsuka's theory had a strong influence on Japanese agricultural history. For example, when he explained the scattered field system, Toshio Furushima suggested that the Tokugawa peasants dispersed their holdings in 'several blocks like furlongs in England'. They had to irrigate and drain simultaneously all the plots of one block; as a consequence, those who shared land in one block usually lived in proximity to each other, and formed a mutual working organization, or *yui*; the timing of tasks was strictly regulated by the requirements of irrigation and by the rules of the organization (Furushima, 1963: 581–82). Yoden (1961) similarly compared the *Mizo-kakari Den*, or a group of paddies sharing a canal in Japan to a furlong in Europe.

> Each peasant held several plots in the groups of paddies sharing a canal. I call the scattered field system the *Mizo-kakari* system. The holders of each plot shared the communal facilities, such as canals, gutters, and pipes; they usually managed them communally [...] Communal organization and several restrictions based on the public character of the communal facilities constrained individual labour process; then peasants had to carry out several agricultural tasks communally (Yoden, 1961: 210).

In rice paddies, each plot was enclosed and levelled to keep the water evenly spread. In this sense, physically 'open' fields did not exist. Nevertheless, since plots shared canals and were closely connected through water management, peasants had to cooperate and follow communal restrictions, just like those in European open-field villages.

Different concepts of 'equality' in Europe and Japan

In these ways, Otsuka and other Japanese scholars highlighted the similarities between European and Japanese village communities; but they also showed the differences. Otsuka, referring to Weber's theory of 'formal equality' (*formale Gleichheit*), emphasized that the holdings of the peasants in open-field husbandry were almost equal in size (Otuska, 1955: 93). However, the principle of equality differed in Europe and Japan. In Europe, 'originally there was in theory strict equality among the members in this economic organization' (Weber, 1927: 8). The basic unit of holding was the *hide* in England, the *Hufe* in Germany, the *manse* in France, and the *mantal* in Sweden. Later, in England, the standard holding was reduced to a virgate or yardland (a quarter-*hide*); the virgate was, in turn, often further divided into halves or quarters. However, holders of these small parcels were still distinct from cottagers, who usually did not have lands in the open fields. In Germany, 'there arose a division between the peasants and another class of village dwellers, called in South Germany hirelings or cottagers (*Seldner, Häusler*), and in the north "*Brinksitzer*" or "*Kossäten*"' (Weber, 1927: 8–9). Although the original equality among peasants holding land in open fields was eroded over time, the distinction between large and small holders continued.

Japan did not have a standard landholding unit. Table 10.1 shows the acreage of each of ninety peasant holdings in Tada-mura of Shimosa (Chiba) in 1677. The average holding size was about 110 *une*, or 1.1 ha.; a 'relatively large holding among the rice growing villages in Southern Kanto' (Kimura, 1969: 371). But this average obscures a significant range of holding size, from more than 500 *une* to less than one, although the gradation was subtle and continuous. There was no sharp distinction between the holders of a standard unit and smallholders, and the latter held plots intermingled with those of other villagers. In this situation, the principle of equality functioned through requests for *equal contributions* to the maintenance of irrigation and drainage facilities. All the villagers, both rich and poor, had to cooperate. As Furushima puts it, 'even though a

Table 10.1 Paddies and fields in Tada-mura of Shimosa (Chiba) in 1677

	acreage (*une*)	degree of scatteredness			ratio of reclaimed land
		no. of plots A	no. of blocks B	Plots per block A/B	
1	535.07	89	32	2.78	0.19
2	257.73	43	21	2.05	0.26
3	255.63	45	24	1.88	0.19
4	230.03	37	23	1.61	0.24
5	208.97	33	22	1.50	0.25
6	204.10	31	23	1.35	0.22
7	200.17	33	22	1.50	0.22
8	198.37	30	18	1.67	0.33
9	188.93	26	18	1.44	0.29
10	188.87	33	16	2.06	0.20
11	183.70	37	22	1.68	0.19
12	181.37	27	16	1.69	0.18
13	170.97	24	19	1.26	0.28
14	170.70	27	19	1.42	0.23
15	165.20	30	15	2.00	0.15
16	161.30	22	13	1.69	0.16
17	153.80	25	15	1.67	0.16
18	153.23	32	18	1.78	0.27
19	152.97	28	15	1.87	0.21
20	151.93	22	14	1.57	0.09
21	151.87	29	17	1.71	0.22
22	142.87	16	12	1.33	0.20
23	139.53	15	11	1.36	0.29
24	138.63	19	15	1.27	0.18
25	138.17	18	11	1.64	0.17
26	127.70	19	13	1.46	0.28
27	126.97	20	14	1.43	0.17
28	124.53	26	15	1.73	0.25
29	124.07	18	13	1.38	0.15
30	121.57	26	16	1.63	0.28
31	119.33	16	11	1.45	0.16
32	119.23	16	11	1.45	0.27
33	119.23	16	12	1.33	0.00
34	118.97	20	14	1.43	0.18
35	116.83	17	11	1.55	0.15
36	116.33	17	12	1.42	0.20
37	113.87	16	10	1.60	0.00
38	112.30	14	12	1.17	0.20
39	110.50	15	13	1.15	0.19
40	109.90	18	13	1.38	0.31
41	109.30	18	12	1.50	0.33
42	109.30	13	13	1.00	0.26
43	108.10	16	14	1.14	0.17
44	107.37	14	13	1.08	0.27
45	106.53	17	14	1.21	0.15

Table 10.1 Paddies and fields in Tada-mura of Shimosa (Chiba) in 1677 (*continue*)

	acreage (*une*)	degree of scatteredness			ratio of reclaimed land
		no. of plots	no. of blocks	Plots per block	
46	106.27	18	12	1.50	0.19
47	105.53	18	14	1.29	0.18
48	103.47	17	11	1.55	0.15
49	101.60	21	16	1.31	0.18
50	100.47	14	13	1.08	0.19
51	100.20	16	12	1.33	0.19
52	97.60	16	12	1.33	0.17
53	96.20	16	14	1.14	0.14
54	96.00	19	12	1.58	0.31
55	95.50	14	11	1.27	0.18
56	92.53	14	12	1.17	0.19
57	91.87	17	11	1.55	0.29
58	89.20	17	13	1.31	0.14
59	88.90	17	12	1.42	0.25
60	88.30	17	13	1.31	0.11
61	87.47	12	9	1.33	0.19
62	87.13	13	10	1.30	0.14
63	85.73	21	10	2.10	0.22
64	85.43	18	9	2.00	0.18
65	85.23	15	11	1.36	0.09
66	85.20	15	13	1.15	0.27
67	85.07	16	12	1.33	0.16
68	78.43	14	11	1.27	0.20
69	73.80	13	11	1.18	0.12
70	73.47	11	9	1.22	0.16
71	67.67	16	12	1.33	0.12
72	63.63	16	13	1.23	0.33
73	63.03	13	9	1.44	0.29
74	61.83	17	13	1.31	0.02
75	57.47	8	8	1.00	0.06
76	51.83	7	7	1.00	0.21
77	46.87	10	7	1.43	0.22
78	46.30	10	9	1.11	0.18
79	45.10	6	6	1.00	0.16
80	44.77	12	9	1.33	0.83
81	42.77	6	5	1.20	0.16
82	41.10	6	6	1.00	0.12
83	31.23	3	3	1.00	0.00
84	30.00	3	2	1.50	0.00
85	20.43	4	4	1.00	1.00
86	20.37	3	3	1.00	0.39
87	16.27	5	1	5.00	0.00
88	8.87	2	2	1.00	1.00
89	7.33	1	1	1.00	0.00
90	6.60	1	1	1.00	1.00

1 une = $100m^2$ = 0.025 acre

Source: Kimura and Takashima, 1969: 367-370.

peasant lost the title of formal landholding, he was required to fulfil the obligation of managing water' (Furushima, 1963: 584). In addition, Kazuhiko Sumiya (1963) argued the rules for allotting water and the obligation of labour services for maintaining water facilities were as compulsory as *Flurzwang*. The *Flurzwang* was enforced on all the peasants in the village. In other words, only those who made this contribution were accepted as members of the village community. Fukutake emphasized that 'a household which moved in from outside could only be recognized as a constituent household of the hamlet' after meeting certain conditions, including performing 'their share of hamlet labour tasks'; 'until then the newcomers were thought of as resident strangers' (Fukutake 1967: 118–19). In Japan, in other words, the village community consisted of peasants who owed contributions to the village.

Such cooperation was necessitated both by the considerable labour required to clean and repair canals, and by the large externalities of production. Even though the paddy plots were small, mismanagement of one plot could adversely affect the entire village. Since the land of each household was usually divided into small plots, which were scattered and mixed with those of others; and since water was usually distributed through different paddies in turn; one household's negligent water management could obstruct the water supply of its neighbours. Furthermore, as the paddy plots were located close to one another, issues arising from water mismanagement, disease, and harmful insects in one paddy quickly spread to the other paddies (Watanabe, 2008: 81–85). Therefore, villagers severely punished those who caused such damages by ostracizing them. While, as mentioned before, Otsuka considered ostracization (*mura hachibu*) as an obstacle to 'modernization', some modern economists praise it as an efficient measure to facilitate the management of water and the maintenance of other aspects of the public good (Hayami, 2001; Aoki, 2001: 44–46).

In addition, cooperation in villages included assistance to poor and stricken households. While regarded as a moral obligation for the rich, the support was rational in that it assisted in the maintenance of water facilities. The mutual interdependence among villagers, both the rich and the poor, is an important feature of the Japanese scattered field system; the village guaranteed 'the subsistence of the peasant in severe conditions' (Furushima, 1963: 584). The village donated or lent money to them from its own funds. Indeed, many villages had mutual financing associations (*mujin* or *tanomoshiko*), from which the poor often borrowed money (Watanabe, 2008: 132–34). Leagues of villages developed 'a famine relief fund' (*bikou-chochiku*) to provide money or crops for disaster assistance. Wealthy landholders, who were usually village officials, contributed generously to these relief activities. The principle of mutual help worked well in the context of the landlord-tenant relationship: 'All landlords, regardless of the size of their holdings, were expected to protect their tenants from the impact of poor harvests by reducing rents' (Waswo, 1977: 30).

While in European open field-villages, peasants who had a standard unit of holdings were distinct from cottagers, in Japan all villagers, including the rich and the poor, were closely interconnected, mainly through water management. All were required to provide 'equal' contributions to maintain public goods, such as irrigation canals, and supported each other when they were in trouble.

Scattering plots for optimizing the distribution of labour

From the end of the 1960s, Japanese historians ceased criticizing the 'village community' as an obstacle to 'modernization'; instead, they began to analyse the functions of community, without ideological preconceptions. Unfortunately, however, little research has been undertaken into the scattered field systems with which such communities were intimately associated. This state of affairs is explained by, first, the interest of historians and geographers in irrigation systems, and, second, by the strong influence of Marxist theory on the former, which has led them to concentrate on analysing the 'social structure' of 'feudal society'. Nevertheless, from the end of 1960s, Teisaku Hayama (1969), Motoi Kimura (1969, 1978), and Azio Fukuta (1996) carefully analysed the historical documents of Tokugawa Japan.

Hayama's research is marked by its outstanding reproduction of the field systems of sixteenth and seventeenth century Japan. Based on a picture map of 1676, a note in a land register of 1677, a land register of 1688, and a land survey map of 1890, he has reconstructed the shape and location of each plot in Saraike and Shima-izumi mura, Kawachi (Osaka) during the Empo (1673–81) and Bunroku periods (1592–96) as Figure 10.4 shows. Interestingly, these maps show how each plot was irrigated. For example, Saraike village in the Empo period was divided into four irrigation blocks, according to its water sources and canals. The first, second, and third blocks obtained water from the *Shin-ike* pond; these blocks were called *Uwa-hinori, Naka-hinori, and Soko-hinori*, respectively. The fourth block, called *Fujii-hinori*, used water from a spring. The irrigation blocks correspond to Yoden's *Mizo-kakari Den*, or a group of paddies sharing a canal. In each block, water usually flowed directly from one plot to the next (*Tagoshi-kangai*, or over-levee irrigation). Furthermore, by marking each plot belonging to a each household on the map, Hayama reveals that a wealthy owner-cultivator, Yoshibee, who held about one hectare of paddy land, dispersed it into a number of different blocks.

Hayama states that the owner-cultivator purposely distributed his lands in this manner; it was a strategy to ensure that all could be cultivated using just the labour of his own household members. In rice cultivation, the labour demanded for rice planting was extremely high; once water flowed into a paddy several tasks, such as 'ploughing with oxen seven times coarsely and once carefully, puddling once by human beings, plastering ridges, and planting' had to be completed within a short period of time. It was impossible for these tasks to be performed at the same time on a single large holding. However, if the owner-cultivator dispersed his lands in different irrigation blocks, he could work on them sequentially, since water flowed into irrigation blocks at different times. In this way, peasants were able to distribute their household labour evenly over a number of days. Hayama therefore concludes that the scattered and intermingled field system was the result of the peasants' careful planning (Hayama, 1969: 246–47). This argument is similar to Fenoaltea's explanation of the rationale for scattered land in medieval England. He states that open fields were 'designed to achieve economic efficiency by optimizing the distribution of labour'; the constraints caused by the necessity of utilizing large amounts of labour in a short period was 'partially but significantly lessened by diversification into a balanced mixture of fallow, winter crops, and spring crops—and, within each cropping unit, by the heterogeneity of the land itself' (Fenoaltea, 1976: 142, 144).

Figure 10.4 Saraike and Shima-izumi mura of Kawachi (Osaka) in the late seventeenth century

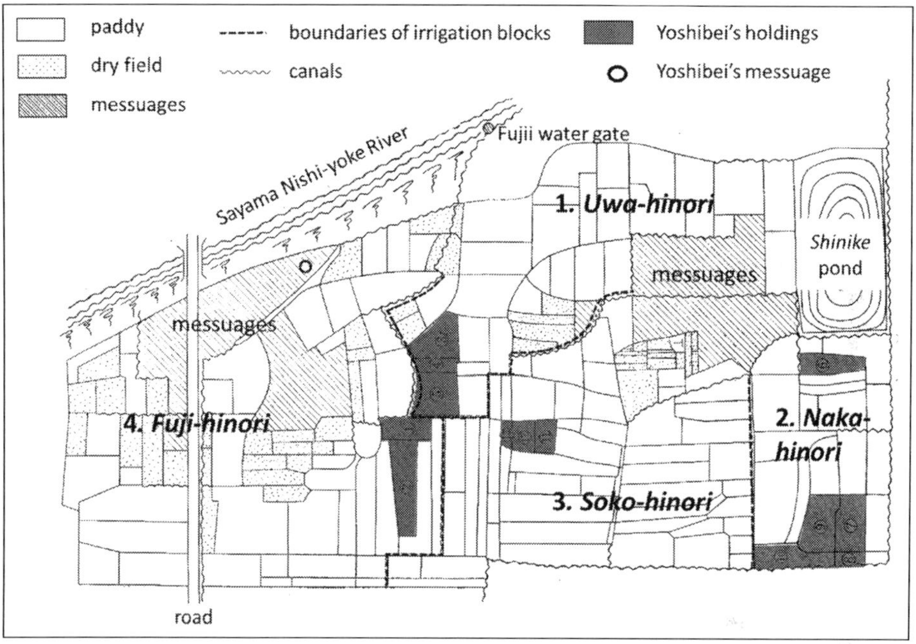

Source: Hayama, 1969

The evolutionary development of the scattered field system

Some historians, however, are sceptical about the idea that peasants designed scattered field systems, whether for diffusing risk (Yamada, 1956) or for allocating household labour equally (Hayama, 1969). As with the debate on the origin of open fields in Europe (Lewis, Mitchell-Fox and Dyer, 2001: 148–49), the planning of such systems is an important issue. Most historians accept the fact that in Tokugawa Japan the holding of a peasant household was usually divided into several scattered plots. For example, Kimura (1969) state that in Katorisya-ryo of Shimosa (Chiba) scattered field systems existed in all the villages. Land surveys in Tokugawa Japan usually registered the holders, the sizes, the quality, and the locations of plots, allowing their location to be reconstructed. We are therefore able to see how far a holding was scattered, or consolidated. Tada-mura thus consisted of four settlements, namely, *Motomura, Shinbashi, Yamada*, and *Shinden*; furthermore, its paddies and fileds were divided into more than one hundred blocks, or *ko-aza*. The holdings of ninety households consisted of several plots, or *fude*, which were usually dispersed in the blocks.

Unlike Hayama, however, Kimura does not regard this scattered field system to be the result of peasant planning, since in newly-developed villages of the Tokogawa period, household holdings almost all covered the same area, and were consolidated. For example, as Figure 10.5 shows, in Santomi-sinden of Musashi (Saitama), each peasant household held 5 ha. of rectangular strips, consisting of messuages, fields, grasslands and woodlands (Kikuchi, 1986: 73). These 'newly developed villages' were laid out at a stroke: yet rather than scattering their land in dispersed plots, the peasants chose to have consolidated holdings instead. This was the case in other countries. Binns states that fragmentation took place in those regions in which peasants had cultivated fields for a long time, while there was no fragmentation in recently settled areas (Binns, 1950: 10).

If this is so, why did old villages have a scattered field system? Kimura supposes that it was the result of the development of small conjugal family households in the late sixteenth and seventeenth centuries. At the beginning of the Tokugawa period, relatively large, extended households, which cultivated their land with the labour of subordinate family members, changed into small conjugal family households, composed of the direct relatives of a couple. While in large-scale families, a majority of subordinates remained unmarried, after this transformation, 'it became possible for all agricultural workers, so long as they reached adulthood, to marry and establish families' (Hayami and Kito, 2004: 218–19). In the seventeenth century, two types of household can be found. First, there were those that had their origins in the Middle Ages (the early sixteenth century or before) and, while they still had relatively large holdings, they no longer maintained these with the labour of subordinates. Second, we find newly-developed households, many established by former subordinates. Field systems reflected this distinction. The first type of household concentrated some plots and scattered others; Kimura supposes that the 'upper class peasants had concentrated the plots at the core of their holdings and obtained other scattered lands through reclamation and sales'. They usually had a number of plots in each block, or *ko-aza*. For example, as Table 10.1 shows, household No. 1 held 89 plots in 32 blocks; on average, 2.78 plots in each block. In contrast, newly developed households commonly had only one plot in one block (households No. 75, 76, 79, 82, 83, 85, 86, 88, 89, and 90 in Table 10.1), since they had to create plots in the places that the wealthy peasant had not already occupied (Kimura, 1969: 372–75).

Furthermore, it is noteworthy that the development of new, smaller households often accompanied land reclamation and settlement nucleation. In Tada-mura, eighty-two out of ninety households held newly reclaimed land; the ratio of reclaimed land was larger for smallholders. This reduced the difference in the total size of holdings among households. Moreover, since newly developed households had relatively similar sized plots, they usually made their residences around large houses in the Middle Ages; thus, nucleated settlements were created. As Kimura states, 'the major change in settlement landscape, i.e. from scattered to nucleated, or from loose-knit to dense-knit, was based on the expansion of newly cleared land. This must be the most frequent pattern of modification' from the medieval to the Tokugawa period (Kimura, 1978: 33).

In later centuries, land exchanges and inheritances served to disperse plots: 'It is not deniable that the large holdings of upper class peasants in the beginning of Tokugawa period were gradually divided' (Kimura, 1978: 150). In Renkoji-mura of Musashi (Tokyo) during the later seventeenth century, Fukuta (1996) found that 'heirs, in order to equalize the production conditions, divided land equally; then the small and scattered field system was established'. It is true that land transactions did not always result in scattered plots.

Figure 10.5 A new village in San-tomi-sinden, Musashi (Saitama)

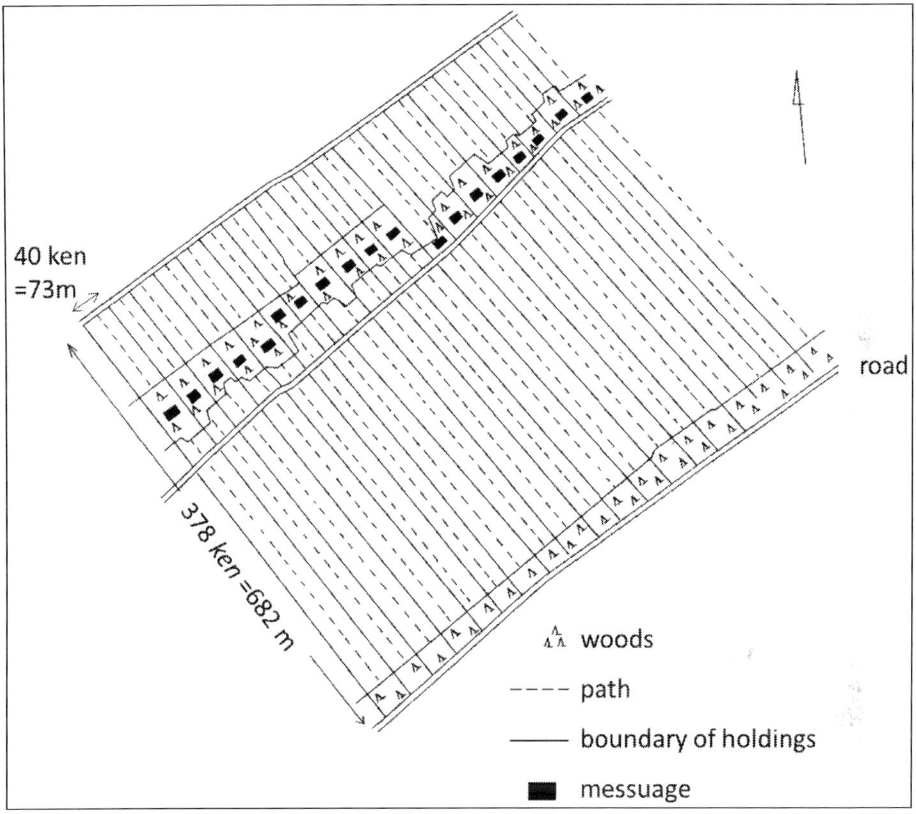

Source: Kikuchi 1986, 73

In particular, in the later centuries of the Tokugawa period, land exchanges sometimes led to the consolidation of wealthy peasant holdings. Fukuta (1996) indicates that in Renjoji-mura, from the middle of the eighteenth century, some peasants began to consolidate scattered plots. Furthermore, in Kakegawa Goake-mura in Toutoumi (Shizuoka), Gorobee, a great landlord who held almost half land of the village, had sixteen groups of consolidated plots in 1865 (Kimura, 1978: 35–36). However, this tendency was not common, which explains why late nineteenth-century village leaders and government officials often strongly encouraged peasants to consolidate their holding, as I noted in the beginning of this chapter.

Kimura argued that each Tokugawa peasant wished to have 'arable land of equal condition near his/her messuage' (Kimura, 1978: 181). However, they did not, in general, 'design' scattered field systems. Plots were dispersed as the result of the peasants' search for greater equality, in terms of the independence of former subordinates, partible inheritances, and land transactions. In this process, they dispersed their plots to equalize land quality and risk.[1] As to the question of whether the open field system was the product of 'a slow, evolutionary process' or 'a very systematic allocation of land' (Lewis, Mitchell-Fox and Dyer, 2001: 148–49), it is likely that the former model of development applies to Japan.

V. Conclusion

Japanese peasant holdings were small and divided into scattered, intermingled plots. In the later part of the nineteenth century, when the Japanese began to study European agriculture, they noticed that their scattered field system was similar to the open-field system of Europe. It is not, therefore, surprising that in following the European model, the Japanese tried to consolidate these dispersed plots. Referring to the German laws of Württemberg, Bayern, and Baden, the government passed the Arable Land Readjustment Law in 1899. Soon after, however, scholars and administrators realized that land consolidation alone was not sufficient to improve the agricultural system: unlike Europe, irrigation was crucial to rice farming in Japan. The law was revised to focus on building and improving canals and drainage. Since water management was crucial, the principle of equality functioned as the request for equal contributions in maintaining irrigation and drainage facilities. Historians give several reasons for the development of the Japanese field system, which has its origins in the sixteenth and seventeenth centuries. Some argue that peasants scattered their plots on purpose, in order to spread risk and distribute more equally labour-intensive tasks. Others argue that scattered and intermixed parcels were the result of an evolutionary, piecemeal attempt to equalize the quality and risk of landholdings.

Bibliography

Aoki, M. (2001), *Toward a comparative institutional analysis*, Massachusetts.

Bentley, J. W. (1987), 'Economic and ecological approaches to land fragmentation: in defence of a much-maligned phenomenon', *American Review of Anthropology*, 16, pp. 31–67.

Binns, B. O. (1950), *The consolidation of fragmented agricultural holdings*, Washington.

Blarel, B., Hazell, P. Place, F. and Quiggin, I. (1992), 'The economics of farm fragmentation: evidence from. Ghana and Rwanda', *The World Bank Economic Review*, 6, 2, pp. 233–54.

Bloch, M. (1931/1966), *French rural history: an essay on its basic characteristics*, trans. J. Sondheimer, London.

[1] Brown (2011) notes that some Japanese villages had *warichi*, or 'joint ownership' systems, where the 'allocation of cultivation rights was linked to family composition and a fixed order of rotation, or lottery'; joint ownership was an expression of 'parity, equity, and/or fair shares as important community values toward land'. However, the villages that carried out *warichi* were far fewer than those where small parcels were dispersed. Therefore, the 'community value' can be understood more clearly by analysing *warichi* in the context of land fragmentation.

Bonner, J. P. (1987), *Land consolidation and economic development in India: a study of two Haryana villages*, New Delhi

Brown, P. C. (2011), *Cultivating commons: joint ownership of arable land in early modern Japan*, Honolulu.

Dewey, C. (1972), 'Image of the village community: a study in Anglo-Indian ideology', *Modern Asian Studies*, 6, pp. 291–328.

Eggert, U. (1890), *Land reform in Japan: specially based on the development of credit associations*, Tokyo.

Farmer, B. H. (1960), 'On not controlling subdivision in paddy-lands', *Transactions and Papers of the Institute of British Geographers*, 28, pp. 225–35.

Fenoaltea, S. (1976), 'Risk, transaction costs and the organization of medieval agriculture', *Explorations in Economic History*, 13, pp. 129–51.

Franks, P. (2006), *Rural economic development in Japan*, London.

Fukuta, A. (1996), 'Sakuho-sei-kochi no Keisei to Kinsei Sonraku (The early-modern village and the evolution of "patchwork" farmland distribution)', Kokuritsu Rekishi-minzoku-kan Kenkyu-hokoku, 66.

Fukutake, T. (1967), *Japanese rural society*, trans R. P. Dore, Oxford.

Furushima, T. (1963), *Kinsei-nihon-nogyo no Tenkai* (The development of early modern Japanese agriculture), Tokyo.

Gordon, A. (2003), *A modern history of Japan: from Tokugawa times to the present*, Oxford.

Grantham, Q. W. (1980), 'The persistence of open-field farming in nineteenth-century France', *Journal of Economic History*, 40, pp. 515–31.

Hayama, T. (1969), *Kinsei-nogyo-hatten no Seisan-ryoku-bunseki* (An analysis of the productivity development of early modern agriculture), Tokyo.

Hayami, A. and Kito, H. (2004), 'Demography and living standards' in: A. Hayami, O. Saitô and R. P. Toby ed., *The emergence of economic society in Japan, 1600–1859*, Oxford.

Hayami, Y. (2001), *Development economics: from the poverty to the wealth of nations*, 2d ed., New York.

Heston, A. and Kumar, D. (1983), 'The persistence of land fragmentation in peasant agriculture: an analysis of south Asian cases', *Explorations in Economic History*, 20, pp. 199–220.

Hida, R. (1894), *Kochi-kukaku Kairyo-hoan* (A plan for improving arable field components), Tokyo.

Ishikawa, S. (1953), *Mura-hachibu no Ki* (A record of ostracization), Tokyo.

Kawasaki, K. (2010), 'The costs and benefits of land fragmentation of rice farms in Japan', *Australian Journal of Agricultural and Resource Economics*, 54, pp. 509–26.

Kikuchi, T. (1986), *Zoku shinden-kaihatsu* (The reclamation of new paddy-fields, part 2), Tokyo.

Kimura, M. (1978), *Nihon-sonraku-shi* (A history of Japanese villages), Tokyo.

Kimura, M. (1969), 'Kinsei-sonraku no Kakuritu (The establishment of early modern villages)', in M. Kimura and R. Takashima, *Kouchi to shuraku no rekishi* (A history of fields and settlements), Tokyo.

Kitamura, M. (1950), *Nihon kangai-kankou no shiteki-kenyu* (A historical study of irrigation customs in Japan), Tokyo.

Koubou, K. and Yamazaki, F. (1954), *Suiden to hatake* (Paddy and field), Tokyo.

Lewis, C., Mitchell-Fox, P. and Dyer, C. (2001), *Village, hamlet and field: changing medieval settlements in central England*, Macclesfield.

MAC (Ministry of Agriculture and Commerce) (1888), *Oubei-junkai torishirabe-sho* (A report of investigation in Europe and the United State), 7 vols., Tokyo.

MAC (1916), *Nogyo-suiri-kanko* (Customs of agricultural water management), Tokyo

Mayhew, A. (1973), *Rural settlement and farming in Germany*, London.

Meliczek, H. (1973), 'The work of FAO and experiences in land consolidation', *Land Reform, Land Settlement and Cooperatives*, 1, pp. 50–64.

Nagata, K. (1984), 'Tochi-kairyo-jigyo no tenkai to Nogyo-seisan-ryoku no keisei-mekanizum (The development of land improvement projects and the mechanism of agricultural productivity growth) in: S. Tamashiro ed., *Suiri no shakai–kozo* (The social structure of water management), Tokyo.

Niroula, G. S. and Thapa, G. B. (2005), 'Impacts and causes of land fragmentation and lessons learned from land consolidation in Asia', *Land Use Policy*, 22, pp. 358–72.

Ogawa, M. (1953), 'Kochi-menseki no zodai to kochi-seiri-jigyo no Taidou' (The increase of arable land and the beginning of land readjustment projects) in: Nihon-nogyo-hattatsu Shosakai ed., *Nihon-nogyo-hattatsushi* (A history of the development of Japanese agriculture), vol. 1, Tokyo.

Ogawa, M. (1954), 'Chisui suiri toch-kairyo no taikeiteki-seibi (Systematic preparation for water control, water management, and land improvement)' in: Nihon-nogyo-hattatsu Shosakai ed., *Nihon-nogyo-hattatsushi* (A history of the development of Japanese agriculture), vol. 4, Tokyo.

Ogura, T. ed. (1963), *Agriculture development in Modern Japan*, Tokyo.

Otsuka, H. (1948), *Kindaika no ningenteki kiso* (The human foundation of modernization), Tokyo.

Otsuka, H. (1955), *Kyodotai no kiso-riron* (A basic theory of community), Tokyo.

Sakou, T. (1893), *Tochi-seiri ron* (A theory of land readjustment), Tokyo.

Sato, Y. and Hirota, J. (1999), 'Waga-kuni kochi-seiri-ho no Seiritu to Doitsu Kochi-seiri-ho no eikyo' (Influence of German legislation on enactment of the Arable Land Readjustment Act in Japan)', *Doboku-gakkai-shi*, 67, 6, pp. 817–20.

Schaich, H., Karier, J. and Knold, W. (2011), 'Rivers, regulation, and restoration: land use history of floodplains in a peri-urban landscape in Luxembourg, 1777–2000', *European Countryside*, 4, pp. 241–64.

Sumiya, K. (1963), *Kyodotai no Shiteki-kozo-ron* (A study on the historical structures of communities), Tokyo.

Susuda, R. (2010), *Hida Roichi to kochi-seiri* (Hida Roichi and Arable Land Readjustment), Tokyo.

Shinzawa, K. (1969), 'Tochi-shoyu-sei no suiden-keitai heno eikyo' (The influence of landownerships on the forms of the paddy field), *Nogyo-doboku-gakkai Ronbun-shu*, 27, pp. 50–56.

Tan, S., Qu, F. and Heerink, N. (2005), 'What drives land fragmentation?: Theoretical approaches and empirical analysis', in: P. Ho ed., *Developmental dilemma: land reform and institutional change in China*, Oxford.

Tayama, T. (1988), *Nishi-Doitsu nochi-seibi-hosei no Kenkyu* (A study of Land Readjustment Acts in West Germany), Tokyo.

Toyohara Kenkyu-kai (1978), *Toyohara-mura* (Toyohara Village), Tokyo.

Ueno, E. (1905), *Kouchi-seiri Kougi* (A lecture on Arable Land Readjustment), Tokyo.

Watanabe, H. (2008), *Hyakusho no chikara* (The power of peasants), Tokyo.

Weber, M. (1927), *General economic history*, trans F. H. Knight, New York.

Yamada, S. (1956), *Nihon-hoken-sei no Kouzou-bunseki* (An analysis of the structure of Japanese feudalism), Tokyo.

Yoden, H. (1961), *Nogyo-sonraku-shakai no ronri-kozo* (A logical structure of agricultural village society), Tokyo.

Yokoi, T. (1921), *Keizaigawa no kouchi-seiri* (An economic aspect of Arable Land Readjustment), Tokyo.

11 Conclusion: the rationale of open fields

Christopher Dyer, Erik Thoen and Tom Williamson

When that outspoken advocate of agricultural improvement, and apostle of Enlightenment ideas, Arthur Young, took the road out of Cambridge to the west in 1791, he commented in the strongest language on the 'fallow system' that he saw: in other words, land cultivated in open fields. The land beside the road was covered with both thistles and dung, 'the worst husbandry in Great Britain', and Young launched into criticisms of the ignorance, backwardness and sloth which he supposed to prevail in the 'beggarly village' where cultivators of such fields lived (Young, ed. 1932: 203–04). The authors of the essays gathered in this book live in an age when agriculture is much more scientific and productive than any 'improved' system praised by Young, yet we can appreciate that in the centuries before 1791 purposeful and rational cultivators saw much merit in open fields. For so many farmers, fields of this kind provided the framework within which agriculture was practised for many centuries, from southern Finland to central France, and from eastern Scotland to eastern Europe. The many millions of people who engaged in open-field husbandry over that time and space cannot all be regarded as ignorant, lazy and backward.

The questions with which this book began were all directed to understanding the rationale behind the open fields. Having established a definition of open fields and discussed the sources that can be used to investigate the subject, and having considered the local and regional variants of such fields within their environmental context, our collective aim was to examine the chronology of the development of (open) fields, their social background (including the role of lords and peasant communities in decision making), their technical significance, the changes that were made to them during their often long histories, and the reasons for their eventual demise.

I. Sources

To answer the questions, our authors deployed a wide range of sources. The most explicit and valuable of documents is of course the detailed local map which depicts the cultivated fields and their internal divisions, the houses and closes of the settlements, and any associated woods and wastes. These maps were not plotted and drawn in any quantity until the sixteenth, seventeenth and eighteenth centuries but, through comparisons with early texts and place names, they can be used to throw light on earlier periods as well, when the boundaries they depict would often have first been fixed. Historians working on the early modern period can also use estate surveys, court records from the archives of the state, as well as the comments made by observers such as Arthur Young. Those investigating open fields before the sixteenth century must interpret administrative documents such as charters, rentals, surveys, accounts and local court records, often depending on incidental references to fields and their management. This evidence usually comes from the archives of lords, and reflects their concerns, giving only indirect help in understanding the peasants' perspective on the fields in which they worked.

Peasants and their fields. The rationale of open-field agriculture, ed. by Christopher Dyer, Erik Thoen and Tom Williamson, Turnhout, 2017 (*CORN* Publication Series, 16), pp. 257-275

BREPOLS PUBLISHERS DOI: 10.1484/M.CORN-EB.5.114276

Many sources provide few or no references to fields, not because they were unimportant, but because they belonged in the fabric of everyday practices, which were managed by word of mouth. Writing only gradually became part of the routine of administration, so when historians report that the first references to open fields in a particular place date only from the thirteenth or fourteenth centuries, they do not know for how long they had been functioning before that time. Many of those working on the history of field systems will also expect to use the evidence from observation of the landscape (at ground level and from the air), from archaeological work on settlements connected to the fields, from palaeoenvironmental investigations, especially pollen analysis, archaeological field surveys and the approaches of geo-archaeology (or 'archéogéographie' as it is often called today (see e.g. Turner, 2011)). The two chapters we have included on twentieth-century open fields, in Bolivia and Japan, can draw on abundant public archives relating to policies on land reform, as well as on the oral testimony of living farmers.

II. Defining and characterizing open fields

The term 'open fields' carries a range of meanings but the contributors to this book broadly agree on how they should be defined. Fields of this kind comprised areas of cultivated land which contained the intermingled plots of a number of different cultivators, without upstanding physical boundaries. Such areas of land were, in almost all cases, subject to common rotations agreed by the cultivators, and enforced by some form of consent or compulsion. In the most highly developed forms of open field, after the harvest or during a period of fallow the livestock of all the cultivators would graze together on the fields. In such systems, those who had a stake in the arable would also usually have had access to other types of land such as meadow and pasture. The intermingled lands usually took the form of narrow strips, although sometimes they were laid out as rectangular or square plots. Strips of land were convenient to cultivate, for ploughs could operate most effectively where they could be hauled for at least 100 metres before turning. It is sometimes suggested that the use of a heavy wheeled plough, with a mouldboard and pulled by six or eight oxen, was a particular incentive towards holding parcels in strips, and there may be some truth in this: but as some of the contributions to this volume make clear, strips were also the basic form of landholding in some areas of medieval Europe where lighter ards were in use, or even where spade cultivation was practised. Property was probably easier to lay out as a sequence of strips, while if – as many have suggested – intermixed properties originally arose through the subdivision of land by inheritance or sale, they may simply have developed through the repeated division of originally rectangular fields from end to end. In some cases, it has been argued that the basic framework of the open fields developed directly from patterns of Roman or even prehistoric land-division.

The small size of open-field parcels, and especially (in most cases) their narrowness, can explain why they lacked physical boundaries. The maintenance of fences, walls or hedges around a large number of small, narrow plots would have been expensive and time-consuming, and hedges in particular would have taken up a lot of the surface area and damaged the crop by shading it or robbing it of nutrients. Where cultivation was carried out in small, intermingled plots, it was inevitable that some degree of communal farming – especially involving common pasturing of livestock after the harvest or when land lay fallow – should be practised. This is turn would require the acceptance of some degree of common rotations, at least of the practice of laying groups of strips to fallow at the same time.

All contributors to this book agree in recognizing an 'ideal type' of open-field arrangement which had developed in parts of western Germany, northern France, and midland England by the high Middle Ages. In this 'ideal type', a nucleated settlement (a village or large hamlet) was associated with two or three extensive fields, managed on a two-or three-course rotation, in which the strips – often as small as a fifth of a hectare – might be arranged in an orderly rotation cycle, repeated throughout the fields. Such cycles are best exemplified in the arrangements termed *solskifte*, *bolskifte* and *tegskifte* in Scandinavia, but are also documented elsewhere, especially in midland England (Hall, 1995, 115–24). The arable fields usually filled a high proportion of the territory belonging to each village, leaving only a limited area of meadow, pasture and wood, which made the access to common grazing on the fields themselves especially important, both for the feeding of the livestock and the dunging of the land. All such open fields operated within individual blocks of territory – sometimes equivalent to ecclesiastical parishes, sometimes smaller units (often referred to as 'townships') – which were often no more than 10 square km in extent, which were mapped in modern times, but which had boundaries that had been set out in the early middle ages. They are described by many of the contributors. It was within these territories that distinct peasant communities and their landholdings developed, many becoming villages farming these kinds of extensive open field.

Having agreed on the characteristics of the 'ideal type', historians studying the fields in any particular locality have generally been anxious to demonstrate that they had distinctive and special features. Even those working on the 'ideal', as outlined above, point to the many departures from the conventional topography and management of the system. In areas of very light soil, for example, where nutrients were rapidly leached out of the land, the pastures might be more extensive, and here complicated arrangements often existed by which common flocks were grazed on the pastures by day, and brought to temporary enclosures on the arable by night, where they dunged the land intensively. In some places the demesne land was dispersed in parcels among the lands of the peasants, but in others it was held in consolidated blocks which might or might not be subject to the rotations observed by the rest of the cultivators. The rotations could themselves be modified in many ways, most commonly by fencing off part of the fallow field in order to plant an extra crop. The regularity with which the strips were allocated to different cultivators could be strictly disciplined, following a regular sequence throughout the fields, but might only in part observe a regular sequence, or display no discernible pattern of organization at all.

Beyond the areas in which this 'ideal type' could be found, open fields took an even wider range of forms and often evolved, over time, from one 'form' to another. In some districts and some periods, they covered extensive areas of ground but rather than having two or three great cropping sectors, communal rotations were organized in discontinuous blocks, were more complicated or – in some cases – were limited in character, so that in extreme cases (as in parts of eastern England) there were no agreed cycles of cropping at all, although fields were grazed in common after the harvest. The much smaller areas of open field found in parts of north and west Europe during certain parts of their history often consisted of islands of cultivation surrounded by forests or were associated with large areas of grass leys which were only intermittently cultivated according to the system conventionally called *infield-outfield*. In the more densely-settled and intensively-farmed regions with lighter soils, however, small open fields existed within a rather different matrix – networks of small hedged enclosures (*bocage*); while in areas with more heavy loamy soils, such as parts of Flanders, Wallonia and Northern France, a *patchwork* open fields (as Alain Derville called them in 1988) could be found, in which numerous small open fields were separated from each other by roads or strips of 'bocage'. Properties were spread over numerous micro open-field 'patches' within which complex collective crop rotations were followed. In all areas of multiple and complex field systems, or mixed ones, rotations often varied greatly from the conventional two-or three-course regimes and could involve multiple courses of cropping, with less frequent fallowing.

To some extent, but only to some extent, these deviations from the 'ideal' were associated with more dispersed patterns of settlement. Scattered farms or small hamlets, often clustered around areas of common pasture, were found in addition to, or instead of, the larger nucleated villages from which – in the English Midlands, northern France or much of west Germany – the large continuous areas of open arable, grouped into two or three great fields, were usually cultivated. In some cases, the lands of separate small settlements might be cultivated separately, as miniature open-field systems, but in others fields were shared between neighbouring settlement clusters and in some the relationship between farms and fields was highly irregular, even chaotic, in character. Even in the well-developed and 'extensive', ideal systems, while fields belonged to a particular village territory, adjacent villages might have coordinated their arrangements, for example by intercommoning, so that areas of pasture were shared by the flocks and herds from a number of settlements.

In short, close examination of open fields reveals a plethora of different regional types, some of which may represent earlier evolutionary stages of those forms found in other districts. In this book, for example, five principal types are identified for Flanders (albeit not all were in contemporaneous existence), and five have been defined for Sweden also. The authoritative book on English field systems (Baker and Butlin, 1973), described many forms in each county, which must amount to dozens of variants over the whole country. If other European countries and regions are taken into account, many of them described in this book, an even greater number of types and sub-types could be identified. Such a mass of local and regional variation threatens to overwhelm the researcher, and it is therefore perhaps useful to distinguish between two broad types of field system which stand out from the individual studies presented here. On the one hand we have big open fields, highly communal in organization and often highly regular in their layout (as in northern France and central England). On the other there were small open fields, often laid out and managed in less regular ways and, in some cases, lying interspersed with fields occupied and farmed on an individual basis (as in Flanders and much of Scandinavia in the early modern period).

It should be emphasized that in some regions of Europe open fields never existed, either because all of the cultivated land was enclosed, or because the economy was entirely pastoral. But in most of north-western Europe there was, during a large period of history, some communal dimension in agricultural management, involving subdivided fields under some form of agreed rotation, and a mixture of land uses. Even the *bocage* landscapes of western France (Brittany, Normandy), fully enclosed by late medieval and modern times, are thought by some to have developed from countryside once dominated by open fields. (Lavigne, 2003; Leturcq, 2003, Chouquer, 1997, Watteaux, 2005). The contributions on Bolivia and Japan in this book remind us, moreover, that such arrangements were not peculiarly European, and that forms of land holding and farming broadly analogous to European open fields can be found on other continents, in very different ecological circumstances.

III. Chronology

The dating of the adoption of open fields in Europe poses many difficulties in part because of the lack of precise evidence, but also in part because different elements of open-field arrangements seem to have emerged over a long period of time. Many historians have argued that not all the variant types which we have just briefly noted, and which are explored in more detail in the various contributions to this volume, necessarily came into existence at the same time. Some, indeed, have suggested that the simple, smaller and often less systematically regulated systems may have developed into the ideal type of two-or three-field, village based systems (Thirsk, 1964). However, it is possible that in some circumstances evolution may have occurred in the opposite direction, and there are clear dangers in embracing a simplistic teleological model, and assuming a necessary and inevitable development from 'primitive' open-field systems towards more complex, 'champion' arrangements.

There is still much dispute about the origins and development of open-field systems. Most researchers accept that they were not static in character but developed over time, as wider structures of social and economic organization changed (see e.g. Warde, 2002 for a discussion of south-west Germany). Most landscape historians believe that open fields as defined above did not have their origins in the prehistoric or Roman periods, even if systems of land division originating before the Middle Ages may sometimes have formed the framework in which they developed, or were created – as some have argued in England (Rippon et al, 2015) and in France, on the basis of area excavations associated with large scale public works. Some have even recently reasserted the prehistoric origins of systems of intermingled strips in Britain, and in Scandinavia, but the evidence so far adduced has been rather thin (Oosthuizen, 2013). In England, areas of intermixed arable first appear to be mentioned in a law of Ine, a seventh-century king of Wessex, and there are clear references to micro-open fields in Carolingian documents for estates in Flanders from the ninth century. References to strips and intermingled properties also appear in charters from the chalklands of southern and midland England from the period 900–1020, especially in their boundary clauses, suggesting that – in these districts – extensive areas of open arable, embracing significant tracts of land, must have existed by this time (Rackham, 1986, 172–76).

In short, there seems to be a consensus that important dimensions of the open fields were in existence by 900, but they may have originated at an earlier date. Amongst some archaeologists it is fashionable to regard the 'long eighth century' as an important period of innovation, which might have included the early stages of open-field farming (Rippon, 2007). In other parts of Europe researchers have focussed on a later period, and especially on the twelfth and thirteenth centuries, which is favoured in parts of France and Germany. Even in many regions of Belgium (as shown in the two contributions dealing with this country presented in this volume), where there are hints of origins before 1000, it can be argued that fully fledged open fields, of the kind we encounter as documents become abundant and detailed later in the Middle Ages, did not emerge until the twelfth or thirteenth centuries. In other regions, however, much later origins for particular kinds of open field – in the late Middle Ages or even the start of the early modern period – have been proposed (as for some areas of poor sandy soil in the Netherlands: see Spek, 2004). In parts of Scandinavia an even later chronology has been proposed, focussing on the sixteenth and seventeenth centuries, or at least it is argued that a fully-fledged open field system with all of its component parts cannot be observed until that period. It might be objected that to some extent this impression is an illusion, created by the paucity of earlier documentary evidence, for this becomes only available in parts of Scandinavia, in quality and abundance, towards the end of the Middle Ages, or even in the early modern period. This said, open fields appear to have been laid out afresh in Poland, and elsewhere in eastern Europe, as late as the sixteenth century (Renes, 2010: 59–60). Evidently, closely comparable forms of field systems developed, in different parts of Europe, at very different times; and variety, complexity, and constant development seem to characterize the development of Europe's open fields.

In part, disagreements over chronology may thus have arisen because different things are being discussed in different contexts. Some forms of open field may have been widespread in many regions by the ninth century, but the more complex systems documented from the thirteenth and fourteenth century – and especially the extensive and highly regular two- and three-field systems, often farmed from large nucleated settlements – may only have developed from the eleventh century. This said, as already noted, there are clear dangers in assuming a universal movement from 'primitive' to 'complex' systems.

In examining when different kinds of field system emerged in particular areas, we also need to take into account changes in the character of rural settlements. In most although by no means all parts of northern Europe there was clearly a strong coincidence between big open fields and large nucleated villages. Villages were of course evolving over centuries, so that in many cases their chronology is uncertain. The growth of Kootwijk in the Netherlands has been dated, from six or so households in the eighth century to twenty households in the tenth centuries, after which it was abandoned. Its associated fields are thought to have been managed on a two-course rotation, and may have been open fields (Hamerow, 2002: 70–75, 137). The emerging orthodoxy in England and France is that villages grew gradually around well favoured nuclei from the seventh or eighth century, and were still gaining new farms in the twelfth and thirteenth centuries, necessitating progressive re-ordering of their associated fields. In Flanders it has similarly been argued that the micro-open fields called *dorpskouters* were developing in the eleventh century, a period of population increase and settlement nucleation, involving the emergence of villages focused on small greens usually termed *dries*-hamlets (for Belgium, Dussart and Claude, 1976 is still instructive).

In Scandinavia most rural settlement in the Middle Ages took the form of quite small hamlets, with three or four, and more rarely up to ten farms, which were associated with small open fields. Here, too, the late and limited appearance of large open fields, organized and laid out in ways similar to those in northern France or western Germany, was closely associated with the development of substantial villages. Nevertheless, there was no necessary or inevitable connection between large villages and large and complex open fields. A number of quite large villages have been excavated in Denmark, the Netherlands and western Germany dating from the fifth, sixth and seventh centuries, with mostly unknown field arrangements (Hamerow, 2002: 55–70). And from a later period not all extensive and continuous open fields are associated with nucleated villages: in parts of eastern England they appear to have belonged to communities living in dispersed settlements. This said, it is hard to escape the impression that the emergence of larger villages was to some extent associated with the development of larger and more complex field systems, and indeed, it is difficult to see how the two could not have been at least loosely connected.

IV. Who made the open fields?

This may really be two if not more questions, for it is necessary to distinguish the agents responsible for the emergence of the basic elements shared by all forms of open-fields, including the simplest and smallest examples, from the individuals or institutions that directed their development or reorganization into more complex systems. The origins of the basic building block of the open fields, the strips or other small parcels, should probably be sought in the culture of peasant farmers. Two main explanations for their origins have been proposed. The first is that peasant holdings were subject to partible inheritance, and this led to smaller and smaller subdivisions, progressively converting networks of enclosed fields into a multitude of small strips: once subdivision had reached a critical stage, hedges or walls around plots of land would have been neglected, or removed. The second relates to organized land reclamation or assarting. When new land was cleared by groups of peasants they shared it by dividing it into narrow parcels, in part because in many cases reclamation was a communal enterprise and the areas in question had formerly been exploited as communal grazing land. Neither explanation is entirely convincing. The division of newly assarted land would not apply where much of the land that made up the open fields, in northern France for example, had been cultivated continuously for centuries and not much land was cleared at a late date: in some areas, as we have seen, open fields may have developed from very ancient systems of land division. Moreover, the standard *mansus* or virgate holdings which appear in the documents between the ninth and the eleventh centuries in many regions were not frequently divided on inheritance, as both lords and tenants had an interest in the preservation of their integrity. Additional incentives to the division of holdings into strips might have been created when – at a rather later period – portions of property were bought and sold on the market.

All this assumes, of course, that intermingled strips arose gradually and organically, and that the very regular open-field systems which came to exist in many parts of Europe represent re-modellings of more chaotic and irregular arrangements of land farmed as intermingled strips. But it is also possible that open field structures may have been created directly from systems of enclosed fields, cultivated on an individual basis, by powerful lords or village communities who saw some clear economic advantages in a sudden conversion to communal farming. We also need to be aware that demographic expansion would not inevitably, in all situations, have led to the emergence of intermingled properties.

As Eric Kerridge noted, this would only occur if there was an increase, not just in the number of people, but in the number of separate *farms*. This, he further argued, would not necessarily happen even under conditions of partible inheritance, for land might be held jointly, by members of a family – as seems to have occurred into the thirteenth century in parts of south-east England. The development of open fields was thus, he suggested, intimately associated with the allocation or division of land in the context of individual tenures – with societies in which each piece of arable land and its produce was associated with a single cultivator and his immediate family. The first open fields might thus have originated not just through partible inheritance, but through the dissolution of joint family properties, or from the allocation of land to groups of slaves or colonists: the necessary preconditions were individualized tenure and a desire for fairness (Kerridge, 1993, 46–49). Only a particular interest in the latter could explain why, for example, at the death of a landholder, a patrimony was allocated by dividing up each piece of his land between heirs, rather than by giving to each a single separate block of his land. It is interesting that very similar developments – the replacement of jointly-held family farms by individual tenures – appear to have encouraged the emergence of intermingled parcels in late sixteenth and seventeenth-century Japan, as the important contribution in this volume demonstrates.

It is also worth noting that open fields only emerged in the context of permanent cultivation, in plots of land which were cropped on a regular basis, if in some cases on a long rotation. They do not seem to have arisen where cultivation was effectively impermanent – as, in particular, in areas of Scandinavia where swidden systems operated. It is probable that a shift to permanent cultivation led, as we have seen, to the emergence of the micro-open fields called *kouters* in the sandy part of Flanders, *akkers* of Limbourg, the *essen* or *enken* in the Netherlands and in many parts of Germany, Scandinavia and north-west France (the *gaigneries* or *gagneries* in the Loire area, *îles* in the area of the Brière in Bretagne, the *champagnes* north of Nantes; Maheux, 2004). All appear to have originally been 'infields' surrounded by extensive areas of less permanent cultivation. The communal organization of such fields was presumably stimulated by the need to organize systematic manuring, as illustrated in early texts from Flanders and Drente, and thus regular systems of fallowing. Manuring may also have played a key role in the development of collective management in Drenthe in the Netherlands and in Holstein, and parts of Lower Saxony and Westphalia, in Germany, where manuring with *plaggen* (= sods) mixed with stable dung became the rule on the infields in the early modern period.

Those who were responsible for the emergence of more extensive, regular and communal systems cannot be easily identified. Many historians have suggested that these must have been brought into existence under the direction of the social elite – by kings, aristocrats, ecclesiastical magnates or local lords. The most coherent case for their importance in the genesis of open fields was made by Verhulst when he argued that Flemish (seigneurial) or (*hof-*)*kouters* (Latin *culturae*) were cultivated as micro-open fields with a three - course rotation on estates as early as the Carolingian period. This farming method was then adopted more generally with the appearance of the *dorpskouter* (micro-open fields belonging to villages) from the eleventh century (see the contribution on Flanders in this book). In England it has been suggested by many that the larger and more regular open fields could not have evolved gradually, but had to be imposed by someone with authority. Across much of western Europe, local lords were gaining control in the period between the ninth to eleventh centuries of relatively small estates – a 'cellular structure of local power', as Wickham calls it – exactly at the period when extensive open fields are thought by many to have begun to develop within these units of land holding (Wickham, 2016: 104–05). The instinct among some historians is to believe that lords favoured order and discipline among their subordinates, and that open fields provided a form of social control. They also wished to cultivate their demesnes efficiently, and following a regular rotation should help to secure that aim. They also expected that if the peasants were observing a coherent agricultural routine, they would be able to afford the resources (such as ploughs, carts and draft animals) to work on the lord's demesne, and would produce a surplus of which the lord could demand a share as a rent in kind, or levy a cash rent which the peasants could pay by selling surplus produce.

At a more basic level, many historians have doubted whether peasant communities, acting on their own, would have been able to plan and enforce the reallocation of properties required for the emergence of these highly communal systems. In England, many historians and archaeologists have drawn attention to the apparently regular layout of many villages, suggesting that they were re-planned at a stroke by great lords when farming was re-organized on communal lines and open fields created. In parts of England, France and Germany, moreover, the 'long furlongs' – bundle of strips extending uninterrupted for hundreds of meters – have likewise been seen as evidence of lordly planning on a grand scale (Hall, 1995), although in some wetland areas they may reflect more organic patterns of development, dictated by the needs of drainage (Soens et al., 2015). But above all it is the very regular arrangements of holdings found in many extensive and complex open fields, with sequences of properties appearing recurrently throughout the furlongs, which seem to signal that they must have been laid out according to some carefully planned and predetermined scheme.

There are, however, a number of obstacles to accepting uncritically this simple 'top-down' view. They include, in particular, the lack of coincidence between seigneurial power and the adoption of extensive, communal open fields. In many parts of Europe, villages of freemen practised open-field farming. Many open-field villages lacked any seigneurial presence, with no demesne or resident noble family to take the initiative. Where the estates of English lords were scattered over a number of landscape regions we find that each portion simply shared the characteristic field types found there, from fully fledged open fields, to various types of *bocge* and patchwork open fields – clearly suggesting that lords did not impose their will uniformly, but accepted the prevailing culture and wisdom about the best ways to farm in different environments. In England, Harold Fox demonstrated that the descriptions we have of the laying out of fields (admittedly few in number) appear to show groups of peasants making the key decisions (Fox, 1981).

If, as seems likely, local communities played a key role in the creation, re-modelling and development of complex open fields, this was usually carried out without leaving any trace in written records. The same is true of most aspects of the day-to-day running of these systems, once they had come into existence. Even when in England numerous by-laws appear in manor court records, the village meetings which probably framed this legislation presumably also made many short-term oral decisions. In most parts of Europe the existence of these important groups of wise elders has to be presumed from the survival and apparently smooth functioning of many thousands of open fields. We are better informed about the Bolivian and Japanese fields discussed in this book, where communities were heavily involved in maintaining the routines of cultivation and pasturage, and in resisting wrong-headed (as they thought) attempts at reform from outsiders. Perhaps in a European context we should not necessarily view the origins of field systems in terms of a simple dichotomy between 'lords' and 'peasants'. Lords and tenants were often in a state of hostility, but they could co-operate, and it may be that they reached agreement on some of the measures needed to begin, change or simply manage a field system, as their interests might coincide on such practical issues. Our consensus is to accept that in certain regions and certain periods lords influenced open-field farming, but that peasant communities played the more important role in its origins and development.

Within the peasant community, who would have taken the lead in innovations in technical matters? The wealthier peasants might be expected to be the most active as they held most of the land and had authority over their neighbours. They certainly played a key role in enforcing the rules and regulations which ensured the smooth running of the system, once it had come into existence. Property relations were also involved, and one might suppose that the freeholders, whose land was being brought under community control, would have been an important factor in any reordering of boundaries and allocation of land. The Bolivian example discussed in this volume offers a salutary comparison, in which successive land reforms sometimes promoted individual rights, and sometimes supported communal property, but the agricultural system survived regardless. Could it be that in societies where inequality was huge, communal practices were abolished earlier or could never develop, in contrast to areas where the differences between large and small peasants where smaller and where rules of solidarity made more sense?

V. The rationale behind open fields: why were they created and maintained?

Having examined some of the mechanisms that led to the formation and continuation of open fields, the crucial 'why' question now needs to be addressed. All of the contributors to this volume agree that the various types of open field were, to some extent at least, a response to environmental circumstances. In Scandinavia the ability to grow crops, or particular types of crop, was limited by the shortness of the growing season and the length of time that land lay under snow. In more climatically favoured regions open fields were most likely to develop in river valleys or plateaus or on undulating gentle slopes, rather than on mountain sides and in broken, hilly country. When the polders in Flanders were made out of reclaimed wetlands they developed as an essentially 'open' landscape, but not as an open *field* landscape. The soils on which open fields were established tended to be of a quality that was suited to corn growing, and it has been said that open fields occupied the best quality land: when plots of new land in the twelfth and thirteenth centuries were made by clearing woodland, wastes or moors, they often resulted in areas of poor land which were tilled by individuals, or in some subdivided field not resembling open fields. This said, in continental Europe open fields are found on all types of soil, and in Sweden and elsewhere settlements with open fields could be found next to those without them, even though both occupied the same types of soil. Indeed, as we emphasized earlier, open-field agriculture was much concerned, in many areas, with changing the character and quality of the soil. Moreover, in most regions soils and topography display a significant if sometimes subtle degree of variation, so that open fields might have very selective distributions, compared with other kinds of land use. In (early) medieval Flanders, for example, most micro-open fields or kouters developed in patches of rather light, sandy soil with a slightly convex, lens shaped topography, providing good natural drainage.

It has been argued that regions with extensive open fields were often characterized by particular types of clay soils which were difficult to cultivate in a wet spring. This encouraged peasants to live in close proximity in large hamlets or villages in order to assemble plough teams efficiently on the days when ploughing was possible, with their lands scattered evenly all around to allow each a reasonable chance of being ploughed and harrowed in time for seeding. In areas of very light soil, in contrast, open fields may have arisen through the need to organize communal flocks, and communal fold to supply the arable with regular inputs of dung. All contributors thus agree that physical environments were a factor in the shaping of different varieties of open field, but they disagree on the amount of importance that should be attached to soils and topography. Even in the extreme circumstances of the Bolivian Andes, the most challenging environment that appears in this book, we are warned against presuming that ecology explains the history of agriculture, which owed more to property rights and the systems of land management.

Open fields in their simplest forms existed in many contexts – almost everywhere in Europe, in fact, that arable farming was practised – but the development of large and complex systems was, according to many commentators of our book, closely associated with the growth of population and the extension of the cultivated area. As population grew the number of farms increased, whether by adding new land or by the subdivision of existing holdings. In both the Netherlands and Scandinavia examples are known of a single farm growing by subdivision among heirs until it became a hamlet large enough to cultivate an extensive area of arable. Population growth was not just a factor because it increased the number of mouths to be fed, but also because labour became more plentiful, enabling an increased number of hectares to be tilled, and for existing land to be tended more intensively by means of manuring and weeding. Throughout Europe the twelfth and thirteenth centuries are acknowledged to have been a period of demographic expansion, and this is the same period when open fields were being established in new areas, and developing into their fully mature forms where already in existence. In many areas, moreover, demographic expansion was associated with growing levels of inequality within peasant society, so that larger numbers were dependent on the social benefits afforded by open-field farming for their survival.

As the arable area expanded at the expense of grassland and woodland, the reduction in the amount of grazing needed to be remedied in order for an adequate number of livestock to be maintained. One solution was to use part of the cultivated land as pasture when it was not bearing crops, that is when it lay fallow. To achieve this aim the cultivators had to accept a degree of community control over their use of land – to agree, in particular, that a half or a third of their holding would be fallowed each year. This *vaine pature*, to use the French term, meant that their land had to be distributed equally over the various fields, and the planted field would have to be fenced each year to keep the animals out of the growing corn. These decisions were made in the context of developing ploughing technology, as at a date given by some as after 800, heavy ploughs were being used which made extensive cultivation more efficient, yet at the same time placed the burden on the cultivators of assembling sufficient oxen, sometimes as many as eight, which made it even more necessary to allow enough grazing for so many of these large beasts.

Yet while this model – of population growth leading to increasingly regular and extensive field systems – is often repeated, even in the English Midlands, where the landscape was dominated by extensive and complex open fields, the proportion of land devoted to arable varied a great deal: while it occupied more than 80 per cent of the available land in some areas, elsewhere it could account for less than 50 per cent. In England as a whole there was a poor correlation between field types and the extent of cultivation, for some of largest areas under the plough were to be found in eastern counties, beyond the region with large open fields. And as an explanation for the genesis of open field farming *per se* the argument that regular fallows and strict rotations were forced on cultivators by the expansion of the arable at the expense of pasture becomes even less tenable. In parts of Sweden open fields might occupy less than 10 per cent of the land area: in many places not only were villages and their fields surrounded by large expanses of grazing land, but some communities were also able to take the flocks and herds to remote upland or forest pastures, where the herdsmen lived in temporary accommodation. The same situation prevailed in the Bolivian Andes.

More importantly, in general terms there is no very neat coincidence between areas of high population density in medieval Europe, and types of field system. One of the most densely populated districts, the more sandy part of Flanders, was in the twelfth and thirteenth centuries also home to the 'kouter in bocage' field type, with only small areas of open field (the *kouters*) and much land held in separate enclosed parcels, in part perhaps because a shortage of fuel placed a particular value on hedges. Conversely, parts of Europe with very low population densities – such as the Breckland of eastern England – were characterized by extensive open fields, while in Bolivia, as explained in one of the contributions to this book, open fields of a kind developed in a landscape which was and is, in European terms, very sparsely populated.

While population pressure and the expansion of arable land may in some cases have been significant factors in the emergence of the more extensive and regular forms of open field, other economic, environmental and agrarian influences must therefore have played an important role. Perhaps the most significant population increase to impact on the development of open fields was not the growth in the number of local cultivators, but in that of town-dwellers, again a pronounced trend in the twelfth and thirteenth centuries. In other words, it was the rising demand for food from those who lived away from the land which stimulated rural output. One contributor to this volume has pointed to the proximity of open-field regions to large towns, for example in the Netherlands and in Flanders. This coincidence cannot be observed in all regions, however, as in the Midlands of England with their complex and extensive open fields the proportion of town dwellers in the fourteenth century amounted to about a fifth of the whole population, which was near to the European average, and only part of the region supplied the London market. By contrast, the districts near to London which played a major role in supplying the demand from the city's grain merchants, based their production on irregular field systems with much enclosed land. Some contributors to this book certainly make a convincing case for regarding urban expansion as playing an important role in stimulating the demand for grain, and thus the development of field systems, but the connections may often have been complex. Many of the most urbanized regions – such as the Low Countries – were importing grain from producers located at a distance, in the Baltic areas or the plains of Northern France (Tits-Dieuaide, 1975). By the seventeenth century all of the cereals consumed in Amsterdam, for example, were imported from Baltic areas (van Tielhof, 2002). Open fields may have been encouraged by this long distance trade in these regions, but they did not develop in the immediate hinterlands of the towns in question (see e.g. Sivéry, 1990:44 ff.). Near the urban centres, holdings were often small, precluding specialization in grain production. Surplus labour in small family holdings was directed towards protoindustrial activities, such as textile making, while stall-fed cattle husbandry, typical of regions in the neighbourhood of large towns, allowed intensive manuring. Open fields played only a small role in such commercially oriented peasant economies.

It has been suggested that open fields advanced in the twelfth and thirteenth centuries to allow cultivators to produce a surplus and benefit from rising prices. Hired labour, another aspect of the impact of market forces in the countryside, is thought to have influenced the modifications of open fields in Sweden, where it is suggested that the number of scattered strips was reduced in order to save on labour and therefore wages. It has also been said, in a contrary argument, that strips dispersed over the fields allowed labour to be used efficiently. The best indication that peasants working open fields were profiting from the market comes from the increased use of their strips in the high and late Middle Ages to plant fodder crops. This indicates that only part of peasant holdings was devoted to feeding families, and the fodder represented a surplus that could be translated into cash from the sale of animals and animal products.

Further economic reasons for adopting various kinds of open field, and especially the more complex forms, relate to the problems caused by a shortage of capital. The best indication of this comes from the smallholders in parts of Flanders, who cultivated their plots with spades rather than ploughs. Their holdings were small and they had excess time on their hands: it therefore made sense to cultivate by hand and avoid the expense of hiring ploughs and ploughmen. Those peasants who did possess holdings large enough to depend on plough cultivation (above 2ha) might still lack the necessary implement, which required expensive iron parts, and they were unlikely to own a complete team of horses or oxen. Sharing or hiring the equipment and animals – often via an exchange of labour with the larger farms (Lambrecht and Winter, 2013) – was more easily arranged in a nucleated village with inhabitants all committed to large expanses of unhedged intermixed strips. Fencing and hedging reveals clearly the relationship between the open fields and the management of scarce resources, as erecting a collectively agreed fence around the sown field required a relatively small expenditure of wood and equipment by each of the cultivators, compared with the cost of hedging and fencing numerous small enclosures.

The great dilemma facing those researching open fields is our lack of knowledge of the level of productivity achieved in them: this, indeed might resolve the problem of the rationale behind the adoption of the various 'systems'. If we could show that the yields in open fields were greater than those in alternative systems, then the question would be settled. In England, where there are thousands of records of the productivity achieved on lords' demesnes, it has been shown that while some quite high yields be found in areas of extensive and regular open field, the highest figures are consistently located in the far east of England, in north-east Norfolk, east Kent and south-east Sussex, where fields were not enclosed, but fall into the category of 'patchwork' systems (Campbell, 2000). Likewise examples of very high yields from Southern Flanders, Artois or the vicinity of Lille in north France come from a region that lay outside the area dominated by large open fields, where various forms of 'patchwork' and 'micro-open fields' could be found (Derville, 1999; Delmaire and Delleaux, 2015: 155–56, 163–65; van Bavel and Thoen, eds, 1999). The evidence, from England especially, is heavily biased towards lords' demesnes, and it might be argued that if we had more evidence of what was achieved on peasant holdings our impressions might be different. On the other hand, eighteenth-century figures for labour productivity provide little support for the argument that small peasant holdings were intrinsically more efficient (Segers and Karel, 2015).

We might say that 'classic' open fields gave the opportunity to extend cultivation over a large area, keep up a reasonably high cropping rate (often two years out of three), ensure through fallow grazing that a balance was maintained between arable and pasture, and allow some cultivation of fodder crops. The cultivators gained an income from the profits of animal husbandry as well as cereals, and while the land was not abundantly manured, it received at least enough to prevent yields from declining drastically. The impression is sometimes given that open fields offered, not abundant harvests and great prosperity, but a safe, steady, mediocre routine in which the land was not exhausted and famine was avoided. This generalization is perhaps unfair because it does not take into account the flexibility of open fields and their ability to absorb technical changes. The various contributions to this book have shown that rotations have often been adapted, usually becoming more intensive over time, progressing from two-course to three-course to four, though in one case discussed there was a reversion from three to two. The most frequently encountered modification, in Flanders, in midland England and in Sweden was to fence in part of the fallow field and plant an extra crop, often peas. In general, the advance of leguminous crops such as peas and beans points to a concern to increase animal numbers. Room was found in the more flexible, irregular and patchwork fields for industrial crops such as flax and hemp. At the end of the Middle Ages, when the demand for animal products increased and grain prices fell, parts of the open fields were often converted to pasture but the cultivated sections continued to function much as before. No doubt such changes had to be approved by a village meeting, and examples have been given of communities resisting imposed changes, but they were clearly capable of being persuaded.

So far we have concentrated mainly on environmental, agrarian and economic factors. But perhaps the chief merit of open fields lay not in their efficiency and productivity, but in the way that they delivered a fair and relatively equitable division of resources. The high value placed on such matters may, as we have suggested, explain the original division of properties into strips. It also helps to explain many aspects of the more complex and developed systems into which these came to be incorporated. Peasants had ideals of morality and equality, which were expressed in the standard size of holdings which prevailed in many open-field villages. Some holdings were larger than others, but access to common assets was divided according to agreed rules, so that the number of animals that an individual could keep, or their contribution to the village's tax payment were decided fairly. The operation of the fields depended on co-operation and mutual respect, but in case of failure to comply with the rules, the majority could use sanctions such as financial penalties. In the same spirit the scattered strips ensured that everyone had approximately the same balance of good, bad and indifferent land. That combination of varied land was one way in which the system protected its members from the worst effects of extreme weather conditions and thereby ensured the commitment of all participants to the common good.

In nineteenth- and twentieth-century Japan the scattering of parcels of land over a large area became one of the most contentious problems in agricultural reform. Ignoring all of the arguments about efficiency, the peasants clung to the custom of dispersal with a tenacity which suggests that it was a matter of ideology rather than an issue in practical farming. In Japan there was no sharp distinction between peasants and smallholders, and everyone with land in the fields was concerned with the maintenance of the irrigation and drainage ditches. In Europe the division between those with sufficient land to provide most of their subsistence, and the smallholders who depended on employment, often by their neighbours, potentially presented most villages with a contentious social issue. Cottagers were sometimes prevented from grazing their livestock on the commons, and could suffer from other forms of discrimination, but if they felt a strong sense of alienation it was rarely expressed. In Flanders the polders settlements, where open fields were absent, had the most unequal distribution of wealth, and the most acute problems of poverty. Was the comparative social harmony associated with open fields one reason for the adoption and retention of that method of managing agriculture? It is certainly true that we should not consider the operation of field systems as something separate from the wider life of the village, or as divorced from other communal activities. In general, a well-developed open field system was associated with the effective communal management of wastes and other resources (Dahlman, 1980).

We should also note that in some cases open fields of different kinds may simply have been adopted by emulation, or spread by conquest. Researchers have suggested that in early medieval Flanders the practices on the great estates in the ninth century appear to have spread among peasant cultivators two centuries later. One means by which the technique was transmitted was through peasants doing labour services on the demesne, sometimes (in the lot-corvee custom) being given responsibility for a strip of demesne which they had to cultivate according to the prevailing methods used by the lord. They might then adopt the same methods on their own holdings. In Finland it is said that open fields spread from Sweden, as if by a process of imitation. More direct diffusion may have taken open field farming to Ireland when that island was invaded by Anglo-Normans in the twelfth century, and German settlers imposed open fields on the new villages that they founded east of the Elbe when parts of eastern Europe were subject to a large scale migration (Nitz, 1988). In these latter cases the new fields may have had both a symbolic significance, expressing the dominance of one ethnic group by another; and a practical function, in facilitating the extraction of labour and money rents from the subjugated.

Perhaps what the contributions to this volume show, above all, is that there was not one simple 'rationale' behind open-field agriculture. Open fields of different kinds developed, or were adopted, in different areas and in different periods for a range of reasons, and in response to a range of influences: environmental, economic, agricultural/technological and social. They also evolved over time in response to a similar range of factors. At the most basic level, however, we can say that open fields were associated with societies of small farmers, in which individual families were involved in the permanent cultivation of particular holdings, and in which mixed farming, with an arable bias, was practised. Open fields emerged in 'pre-capitalistic' societies that regarded survival rather than profit as a priority. Nevertheless the market might still have been an important factor. In these societies it was advantageous to undertake many tasks on a communal basis, and in addition equipment was often shared. The initial emergence of intermixed strips embodies concepts of fairness, while the layout and management of the more complex forms of open field similarly expresses a desire for a degree of community cohesion. Bloch's 'cohésion sociale' (Bloch 1931:46) was probably more important than any simple notion of 'risk aversion' of the kind suggested by McCloskey (1991). Open fields not only ensured a reasonable living for local people, but also articulated their economic and fiscal relationships with social superiors and the state, providing a basis for surplus extraction which allocated burdens according to their capacity to pay or perform tasks. Beyond this, their endless variety of forms reflects the diverse environments, and societies, which were to be found across medieval Europe.

VI. Why did open fields end?

Our summary, offered above, of the kinds of social and economic contexts in which open fields flourished also supplies much of the answer to this question, for as these circumstances changed significantly, open fields began to lose their rationale. Already, by the high Middle Ages, some of the putative advantages accruing from open-field farming were being eroded. In some areas, such as midland England, many peasants possessed their own ploughs and teams, and did not need to cooperate with neighbours in joint ploughing. In many regions of Europe the number of farms was gradually reduced through the sixteenth and seventeenth centuries and, as the larger holdings became larger, they were increasingly dependent on paid workers, many of whom also supported themselves on the produce of small-holdings and income from proto-industrial activities. As a result, communal agriculture lost much of its rationale, and the disadvantages of the system – in terms of trespass, straying livestock, time spent moving from plot to plot – began to outweigh any perceived advantages. Enclosure not only allowed farmers freedom to innovate or specialize but also – as was noted in the contribution about Flanders – was often associated with the planting of hedges, and thus with an enhanced supply of fuel. In post-medieval England, advocates of enclosure frequently praised the way that enclosure provided villagers with increased quantities of firewood.

In the more complex agricultural economy which developed in many parts of Europe from the late middle ages, not only did farm size often increase but some districts began to specialize in particular forms of farming to a greater extent than before. In particular, there was an expansion of livestock grazing in many districts, sometimes to the virtual exclusion of arable husbandry. This rendered open fields redundant and, indeed, effectively necessitated their enclosure. This phase of increased specialization was soon followed, from the seventeenth century, by the development of new kinds of crop rotation, involving the use of new fodder crops. Originating in Flanders, these new practices were rapidly adopted in England and, more gradually, elsewhere in western Europe. Not only did the resultant increases in winter feed reduce the need for regular fallowing; the times at which the new crops were planted were incompatible with existing three-course rotations, and while field courses could be adapted to incorporate them, it was easier to achieve this within enclosed fields.

Social and ideological changes also undermined the open fields. Enlightenment ideas about absolute ownership of property, and the increasing unease over situations in which individuals' ability to maximize profit from their own holdings was constrained by the rights and decisions of others, challenged the whole ethos of communal agriculture. The emergence, across much of western Europe, of more socially polarized rural communities likewise eroded any enthusiasm for the communal values with which open fields were associated, while new forms of taxation rendered redundant the fiscal advantages of the regular systems of landholding found in many open-field systems, and new forms of surveying undermined the tradition of allotting or dividing landed property in the form of strips. To some extent, the mounting opposition to open fields, and to their associated areas of common grazing, became a fad – enclosure was everywhere advocated by the social elite regardless of any real advantages it might bring. State-sponsored reforms in modern times, in Japan as well as in Europe, chose to identify the scattering of strips as a fundamental weakness, and so we find legislation in eighteenth-century Scandinavia and more recently in Japan for the consolidation of holdings. In England the enclosure movement, most visible in a succession of parliamentary acts, enforced both consolidation and the fencing of fields in the eighteenth and early nineteenth centuries.

Nevertheless, it is worth emphasizing just how late the open fields, and especially the large and complex open fields associated with big villages, often survived. Even in England at the start of the eighteenth century some agricultural writers, such as Timothy Nourse, could argue their superiority over enclosures (Nourse, 1700). In many places within Europe open fields continued to be cultivated well into the twentieth century – as at Bygrave, Clothall and Wallington, only 60 km from London; while elsewhere in England, as at Belton in Lincolnshire, large areas of open field survive to this day (Rowe and Williamson, 2013, 50–53; Fulton, 2010). The rationale of open-field agriculture may be a matter for debate among historians, but for farmers in many contexts – even many of the contemporaries of Arthur Young – it was a system which evidently made perfectly good sense.

Bibliography

Note: The contributions to this book referred to in this concluding chapter are not mentioned in this bibliography.

Baker, A. R. H. and Butlin, R. A., eds (1973), *Studies of field systems in the British Isles*, Cambridge.

Bavel, B.van and Thoen, E., eds (1999), *Land productivity and agro-systems in the North Sea area: Middle Ages-20th century, Elements for comparison*, CORN Publication Series 2, Turnhout.

Bloch, M. (1931), *Les caractères originaux de l'histoire rurale française*, tome 1, Oslo.

Campbell, B. M. S. (2000), *English seigniorial agriculture, 1250–1450*, Cambridge.

Chouquer, G. (1997), *Les formes du paysage. Tome 1: Etudes sur le parcellaire. Tome 2: Archéologie du parcellaire*, Paris.

Dahlman, Carl J. (1980), *The open field system and beyond: a property rights analysis of an economic institution*, Cambridge.

Delmaire, B. and Delleaux, F. (2015), 'Northern France, 1000–1750' in: E. Thoen and T. Soens ed., *Struggling with the environment: land use and productivity. Rural economy and society in North-Western Europe, 500–2000,* Turnhout, pp. 149–82.

Derville, A. (1999), *L'agriculture du Nord au Moyen-Âge: Artois, Cambrésis, Flandre wallonne*, Villeneuve d'Ascq.

Dussart, F. and Claude, J. (1976), *Les villages de "dries" en Basse et Moyenne-Belgique*, Centre belge d'histoire rurale n° 51, Liège.

Fox, H. S. A. (1981), 'Approaches to the adoption of the Midland system' in: T.Rowley ed., *The origins of open-field agriculture*, London, pp. 64–111.

Fulton, T. (2010), 'The fields of Belton in Axholme', *Landscape History*, 31, pp. 71–73.

Hall, D. (1995), *The open fields of Northamptonshire*, Northampton.

Hall, D. (2014), *The open fields of England*, Oxford.

Hamerow, H. (2002), *Early medieval settlements. The archaeology of rural communities in north-west Europe 400–900*, Oxford.

Kerridge, E. (1992), *The common fields of England*, Manchester.

Lambrecht, T. and Winter, A. (2013), 'Migration, poor relief and local autonomy: settlement policies in England and the Southern Low Countries in the eighteenth century', *Past and Present*, 218, pp. 91–126.

Lavigne, C. (2003), 'De nouveaux objets d' histoire agraire pour en finir avec le bocage et l'openfield', *Etudes rurales*, 167–68, pp. 133–86.

Leturcq, S. (2003), 'Pour une meilleure compréhension des openfields médiévaux et modernes: une approche géographique des territoires agraires' in *Actualités de la recherche en Histoire et Archéologie agraires. Actes du colloque international AGER V, septembre 2000*, Besancon, pp. 23–29.

Maheux, H. (2004), 'Champs ouverts, habitutes communautaires et villages en alignements dans le nord de la loire-Atlantique: des micro-sociétés fossilisées dans l'Ouest bocager.' *In Situ*, 5 http://insitu.revues.org/2350 (consulted 2016)

McCloskey, D. N. (1991), 'The prudent peasant: new findings on open fields', *Economic History Review*, 51, pp. 343–55.

Nitz, H-J. (1988), 'Introduction from above; intentional spread of common-field systems by feudal authorities through colonization and reorganization', *Geografiska Annaler*, series B, 70, pp. 149–59.

Nourse, T. (1700), *Campania felix*, London.

Oosthuizen, S. (2013), *Tradition and transformation in Anglo-Saxon England: archaeology, common rights and landscape*, London.

Rackham, O. (1986), *The history of the countryside*, London.

Renes, H. (2010), 'Grainlands. The landscape of open fields in a European perspective', *Landscape History*, 31, pp. 37–70.

Rippon, S. (2007), 'Emerging regional variation in historic landscape character: the possible significance of the "long eighth century"' in: M. Gardiner and S. Rippon ed., *Landscape history after Hoskins, Vol. 2: Medieval landscapes*, Macclesfield, pp. 105–21.

Rippon, S., Smart, C. and Pears, B. (2015), *The fields of Britannia*, Oxford.

Rowe, A. and Williamson, T. (2013), *Hertfordshire: a landscape history*, Hatfield.

Segers Y. en Karel E. (2015), 'The Low Countries 1750–2000', in: E. Thoen, T. Soens ed., *Struggling with the environment: land use and productivity. Rural economy and society in North-western Europe 500–2000*, Turnhout, pp. 261–306.

Sivéry, G. (1990), *Terroirs et communautés rurales dans l'Europe occidentale au moyen âge*, Lille.

Soens, T., Tys, D. and Thoen, E. (2014), 'Landscape transformation and social change in the North Sea polders, the example of Flanders (1000–1800 AD)', *Siedlungsforschung: Archaelogie, Geschichte, Geographie*, pp. 133–60.

Spek, Th. (2004), *Het Drentse esdorpenlandschap. Een historisch-geografische studie*, Utrecht.

Thoen, E. and Soens T. (2015), 'The Low Countries, 1000-1750', in: E.Thoen and T. Soens, eds, *Struggling with the environment: land use and productivity. Rural economy and society in North-western Europe, 500–2000*, Turnhout, pp. 221–58.

Tielhof, M. van (2002), *The 'mother of all trades'. The Baltic grain trade in Amsterdam from the late 16th to the early 19th century*, Leiden, Boston, Köln.

Turner, S. (2011), 'Paysages et relations: archéologie, géographie, archéogéographie', *Etudes rurales*, 188, pp. 143–54.

Rowe, A. and Williamson, T. (2013), *Hertfordshire: a landscape history*, Hatfield.

Thirsk, J. (1964), 'The common fields', *Past and Present*, 29, pp. 3–29.

Tits-Dieuaide, M. J. (1975), *La formation des prix céréaliers en Brabant et en Flandre au XVe siècle,* Brussels.

Watteaux, M. (2005), 'Sous le bocage, le parcellaire[…]', *Etudes Rurales*, 175–76, pp. 53–80.

Warde, P. (2002), 'Common rights and common land in south west Germany, 1500–1800' in M. De Moor, L. Shaw-Taylor, P. Warde ed., *The management of common land in north west Europe, c. 1500–1850*, CORN Publication Series, Turnhout, pp. 195–224.

Wickham, C. (2016), *Medieval Europe*, New Haven and London.

Young, A. (1932), *Tours in England and Wales*, London.